Imidazole and Benzimidazole Synthesis

BEST SYNTHETIC METHODS

Series Editors

A. R. Katritzky
University of Florida
Gainesville, Florida
USA

O. Meth-Cohn
University of Sunderland
Sunderland
UK

C. W. Rees
Imperial College of Science
and Technology
London, UK

R. F. Heck, *Palladium Reagents in Organic Synthesis*, 1985
A. H. Haines, *Methods for the Oxidation of Organic Compounds: Alkanes, Alkenes, Alkynes, and Arenes*, 1985
P. N. Rylander, *Hydrogenation Methods*, 1985
E. W. Colvin, *Silicon Reagents in Organic Synthesis*, 1988
A. Pelter, K. Smith and H. C. Brown, *Borane Reagents*, 1988
B. Wakefield, *Organolithium Methods*, 1988
A. H. Haines, *Methods for the Oxidation of Organic Compounds, Alcohols, Alcohol Derivatives, Alkyl Halides, Nitroalkanes, Alkyl Azides, Carbonyl Compounds, Hydroxyarenes and Aminoarenes*, 1988
H. G. Davies, R. H. Green, D. R. Kelly and S. M. Roberts, *Biotransformations in Preparative Organic Chemistry: The Use of Isolated Enzymes and Whole Cell Systems*, 1989
I. Ninomiya and T. Naito, *Photochemical Synthesis*, 1989
T. Shono, *Electroorganic Synthesis*, 1991
W. B. Motherwell and D. Crich, *Free Radical Chain Reactions in Organic Synthesis*, 1991
N. Petragnani, *Tellurium in Organic Synthesis*, 1994
T. Imamoto, *Lanthanides in Organic Synthesis*, 1994
A. J. Pearson, *Iron Compounds in Organic Synthesis*, 1994
P. Metzner and A. Thuillier, *Sulfur Reagents in Organic Synthesis*, 1994
B. Wakefield, *Organomagnesium Methods in Organic Synthesis*, 1995
A. Varvoglis, *Hypervalent Iodine in Organic Synthesis*, 1996

SUB-SERIES:
KEY SYSTEMS AND FUNCTIONAL GROUPS

Series Editor

O. Meth-Cohn

R. J. Sundberg, *Indoles*, 1996
M. R. Grimmett, *Imidazole and Benzimidazole Synthesis*, 1997

Imidazole and Benzimidazole Synthesis

M. Ross Grimmett
Department of Chemistry
University of Otago
Dunedin, New Zealand

Academic Press
Harcourt Brace & Company, Publishers

**London · San Diego · New York · Boston · Sydney
Tokyo · Toronto**

This book is printed on acid-free paper.

Copyright © 1997 by ACADEMIC PRESS

All Rights Reserved.
No part of this publication may be reproduced or transmitted in any form or by any means electronic or mechanical, including photocopying, recording, or any information storage and retrieval system, without permission in writing from the publisher

Academic Press, Inc.
525 B Street, Suite 1900, San Diego, California 92101-4495, USA
http://www.apnet.com

Academic Press Limited
24-28 Oval Road, London NW1 7DX, UK
http://www.hbuk.co.uk/ap/

ISBN 0-12-303190-7
Library of Congress Cataloguing in Publication Data

This book is a guide providing general information concerning its subject matter; it is not a procedural manual. Synthesis is a rapidly changing field. The reader should consult current procedural manuals for state-of-the-art instructions and applicable government safety regulations. The publisher and the author do not accept responsibility for any misuse of the book, including its use as a procedural manual or as a source of specific instructions

A catalogue record for this book is available from the British Library

Typeset by Laser Words, Madras, India
Printed in Great Britain by
Hartnolls Limited, Bodmin, Cornwall

97 98 99 00 01 02 EB 9 8 7 6 5 4 3 2 1

Contents

Foreword vii

Detailed Contents ix

Abbreviations and Acronyms xxi

Chapter 1. Introduction 1

Chapter 2. Ring Synthesis by Formation of One Bond 3

Chapter 3. Ring Syntheses Involving Formation of Two Bonds: [4 + 1] Fragments 63

Chapter 4. Ring Syntheses Involving Formation of Two Bonds: [3 + 2] Fragments 103

Chapter 5. Ring Syntheses which Involve Formation of Three or Four Bonds 151

Chapter 6. Syntheses From Other Heterocycles 167

Chapter 7. Aromatic Substitution Approaches to Synthesis 193

Chapter 8. Synthesis of Specifically Substituted Imidazoles and Benzimidazoles 227

Index of Compounds and Methods 249

Dedication

To my wife, Anne, whose support and understanding has made it possible for me to take on a succession of such projects, and to those stalwarts of Heterocyclic Chemistry, Ken Schofield and Alan Katritzky, who have been my catalysts over many years.

Foreword

This is the second book in the sub-series of *Best Synthetic Methods — Key Systems and Functional Groups* and is particularly close to my heart. As a young research chemist I recall we formulated the 'Benzimidazole Rule'. This stated that given the right number of carbons and nitrogens, any starting material would ultimately end up as a benzimidazole! A good example is shown below:

The imidazole and benzimidazole ring systems are, of course, not just interesting and a source of endless research pleasure but are key systems both in nature (such as the amino acid histidine, vitamin B_{12}, a component of DNA base structure and purines, histamine, biotin, etc.) and thus obviously in pharmaceutical, veterinary and agrochemical products such as cimetidine (tagamet), azomycin, metronidazole, misonidazole, chlotrimazole, thiabendazole, benomyl to name but a few. To underline this ubiquity of their medicinal applications, over one third of the pages of the excellent compilation of the *Drug Compendium* in Volume 6 of *Comprehensive Medicinal Chemistry* (Pergamon Press, 1990) contain imidazole or benzimidazole units. Other important uses of these systems include high temperature polymer products and dyestuffs.

Ross Grimmett is an established master in the field, having compiled several authoritative reviews of these systems. He has now lent his effort to doing synthetic chemists a great favor with this wide-ranging 'cookery book' where every type of synthetic problem, both of ring synthesis and of incorporation of key functionality have been addressed and lavishly exemplified with actual preparations. I hope you will enjoy it as much as I have.

Otto Meth-Cohn
September 1997

Detailed Contents

1 Introduction	1
References	2
2 Ring Synthesis by Formation of One Bond	3
2.1 Formation of the 1,2 (or 2,3) bond	3
2.1.1 Imidazoles	3
5-Chloro-1-methyl-2-phenylimidazole	4
1-Benzyl-5-Chloroimidazole	5
4-Ethoxy-2-phenylimidazole	5
N-*Methyl*-N-*[2-[[(methylamino)phenyl-methylene]amino]-1-cyclohexen-1-yl]-benzamide*	8
1-Methyl-2-phenyl-4,5-tetramethyleneimidazole	9
3-Benzoylaminobutanone	9
General method for preparing imidazoles or 1-imidazolamines	10
4,5-Dicyano-2-phenylimidazole	12
4,5-Dicyanoimidazole	13
Tetrahydrobenzimidazole	15
2,4(5)-Dialkyl-5(4)-arylthioimidazoles	15
1-t-Butyl-4-isopropyl-5-trifluoromethyl-imidazole	17
5-Methoxy-4-methylimidazole	17
4-Benzylthio-1-methyl-5-[(2-tetrahydropyranyl)oxy]methylimidazole	18
2.1.2 Benzimidazoles	19
2-Methylbenzimidazole	20
1,2-Dimethylbenzimidazole 3-oxide	21
N-*Ethoxycarbonyl*-N-*alkyl*-o-*nitroanilines*	22
N-*Ethoxycarbonyl*-N-*alkyl*-o-*phenylenediamine*	22
1-Alkyl-1,3-dihydro-2H-benzimidazolin-2-ones	23
N-(2-Aminophenyl)-*1*H-*pyrrol-1-amine*	24

N-*(2-Acetylaminophenyl)-1H-pyrrol-1-amine*	24
1-(1-Pyrryl)-2-methylbenzimidazole	24
1-Acetyl-2-methylbenzimidazole	25
2-Aminobenzimidazole	28
2-Phenylaminobenzimidazole	28
N-*Cyanomethyl*-o-*nitroaniline*	32
2-Cyanobenzimidazole N-*oxide*	32
Hydrolysis of the 2-cyano or 2-carbethoxy group	32
Oxidative cyclization of acylated N,N-*diakyl-*o-*aminoanilines*	36
References	36
2.2 Formation of the 1,5 (or 3,4) bond	40
2.2.1 Imidazoles	41
General method	41
Preparation from α-*aminoacetals*	42
N-*(2,2-Dimethoxyethyl)dichloro-acetamidine*	44
Imidazole-2-carbaldehyde	44
General method for cyclization of an enaminone	45
*Ethyl (Z)-*N-*(2-amino-1,2-dicyanovinyl) formimidate*	49
*(Z)-*N-*(2-Amino-1,2-dicyanovinyl) formamidine*	50
5-Amino-4-(cyanoformimidoyl)imidazole	50
5-Amino-1-aryl-4-(cyanoformimidoyl) imidazoles	51
5-Amino-1-aryl-4-cyanoimidazoles	51
4,5-Dicyano-1-methylimidazole	51
(Z)-N^3-*(2-Amino-1,2-dicyanovinyl)-formamidrazone*	52
1,5-Diamino-4-cyanoimidazole	52
2.2.2 Benzimidazoles	54
Oxidation of N-*phenylphenylacetamidine; 2-benzylbenzimidazole*	55
References	55
2.3 Formation of the 4,5 bond	57
N-*Ethoxycarbonylmethyl*-N'-*cyano*-N-*phenylformamidine*	59

DETAILED CONTENTS xi

        *Ethyl 5-amino-2-phenylimidazole-
          4-carboxylate* 59
        N,N-*Dimethyl-(1-t-butyl-3,3-bis[di-
          methylamino]-2-aza-3-propyeniden)-
          ammonium perchlorate* 60
        *1-Methyl-2,4-bis(dimethylamino)-
          imidazole* 61
  References 61

**3 Ring Syntheses Involving Formation of Two Bonds: [4 + 1]
Fragments** **63**
  3.1 Formation of 1,2 and 2,3 bonds 63
    3.1.1 Imidazoles 63
        1-Methyl-2-imidazoline-4-carboxylic acid ... 64
        *Methyl 1-methyl-2-imidazoline-
          4-carboxylate* 64
        Methyl 1-methylimidazole-4-carboxylate 65
        2-t-Butylamino-4,5-dicyanoimidazole 67
        4,5-Dicyano-2-sulfonylaminoimidazoles 67
        4-Cyanoimidazole-5-carboxamide 68
        General method 70
    3.1.2 Benzimidazoles 71
        2-Trifluoromethylbenzimidazole 72
        5-Methoxy-1-methylbenzimidazole 74
        2-t-Butylbenzimidazole 75
        2-(2'-Pyridyl)benzimidazole 75
        *2-Methyl-5,7-dinitro-1-phenyl-
          benzimidazole* 75
        General procedure 75
        2-Phenylbenzimidazole 76
        2-Ethylbenzimidazole 76
        4-Hydroxybenzimidazole 77
        Ethyl 2,4-dinitrophenylacetimidate 78
        2-(2',4'-Dinitrobenzyl)benzimidazole 79
        *2-[(2'-Carbomethoxyphenoxy)methyl]-
          benzimdazole* 79
        2-Trichloromethylbenzimidazoles 79
        *5(6)-Carbethoxy-2-(4'-hydroxyphenyl)-
          benzimidazole* 80
        1-Methyl-4-nitrobenzimidazole 80
        Benzimidazolone — method A 81
        Benzimidazolone — method B 81

1-Trifluoroacetyl-5-fluorobenzimidazolin-2-thione	81
2-Methyl-4-nitrobenzimidazole	82
General procedure	83
General method	83
Dimethyl N-aryldithiocarbonimidates	84
2-Phenylaminobenzimidazole	84
2-Methylsulfonylaminobenzimidazole	85
2-Amino-1-benzylbenzimidazole	85
2-Aminobenzimidazole	86
2-Amino-1-(2'-pyridyl)benzimidazole	86
General method	86
1,2-Diamino-5-trifluoromethylbenzimidazole	87
General procedure for preparation of benzimidazoles	88
1-Hydroxybenzimidazole 3-oxides	89
References	89
3.2 Formation of 1,2 and 1,5 bonds	94
Ethyl 5-amino-1-benzylimidazole-4-carboxylate	95
General method for synthesis of ω-acylamides	96
N-Methylbenzamide acetone	97
General method for synthesis of 2,5-disubstituted 4-(2'-thienyl)-imidazoles	97
1-Isocyano-2-phenyl-1-tosylethene	99
1-Cyclohexyl-5-phenylimidazole	99
Methyl (E)- and (Z)-3-Bromo-2-isocyanocinnamate	99
Methyl 1-phenethyl-5-phenylimidazole-4-carboxylate	100
General procedure for synthesis of imidazoles	101
References	101

4 Ring Syntheses Involving Formation of Two Bonds: [3 + 2] Fragments **103**

4.1 Formation of the 1,2 and 3,4 (or 1,5 and 2,3) bonds 103

General method 104

Acetamidoacetone	105
Aminoacetone hydrochloride	105
α-Benzoylaniline methyl ester hydrochloride	105
Typical hydrolysis procedure to α-aminoketone hydrochloride	106
Alkylaminoethanal dimethylacetals	106
Methyl 2-mercapto-4-phenylimidazole-5-carboxylate	106
5-t-Butyl-1-methoxyethylimidazolin-2-thione	108
Methyl imidazole-4-carboxylate	109
1-Ethyl-5-methyl-4-phenylimidazolin-2-one	110
2-Aminoimidazole	110
Ethyl 2-amino-1-methylimidazole-5-carboxylate hydrochloride	110
2-Methylimidazole-d_3	111
2-(p-Fluorophenyl)-4-methylimidazole	112
General procedure for 1,4-disubstituted imidazoles using formamide	112
General procedure for 1-alkyl- and 1-aryl-2-methylthioimidazoles	113
General method for 1-substituted 5-amino-imidazole-4-nitriles	113
1-Isopropyl-2,5-dimethylimidazole	114
1-Hydroxy-2,4,5-trimethylimidazole 3-oxide	115
1-Benzyl-4,5-dimethylimidazole 3-oxide	115
General method for synthesis of 4-acylimidazoles	116
4-Acetyl-5-methyl-2-vinylimidazole	116
References	117
4.2 Formation of 1,2 and 4,5 (or 2,3 and 4,5) bonds	119
5-(p-Nitrophenyl)-1-phenylimidazole	120
1-t-Butyl-5-methylimidazole	120
α-Tosylbenzyl isocyanide	121
1-t-Butyl-5-methyl-4-phenylimidazole	121
1,5-Diphenyl-4-tosylimidazole	122
1-(4'-Pyridinyl)-4-tosylimidazole	122
1-Methyl-4-tosylimidazole-5-thiol	123

	General preparation of a 4-tosyloxazoline	125
	Conversion of a 4-tosyloxazoline into a 1,4-disubstituted imidazole	126
	Ethyl 1-phenyl-5-phenylaminoimidazole-4-carboxylate	127
	1-Methyl-2,5-diphenyl-4-styrylimidazole	128
	5-Benzyl-4-butylimidazole (method A)	129
	General procedure (method B)	129
	N,α-Diphenylnitrone	130
	4,5-Diaryl-1-methylimidazole	130
	Methyl 5-diethoxymethylimidazole-4-carboxylate	132
References		133
4.3 Formation of 1,5 and 3,4 bonds		134
	Imidazole-4-methanol hydrochloride	136
	4-Methyl-2,5-diphenylimidazole	137
	N-Chloroamidines (general procedure)	138
	1,2,5-Trisubstituted imidazoles (general procedure)	138
	Aromatization of 5,5-disubstituted 2-imidazolines	139
	3-Bromo-4-ethoxy-3-buten-2-one	141
	4-Acetyl-2-methylimidazole	141
	3,4-Dichloro-3-buten-2-one	141
	3-Chloro-4,4-dimethoxy-2-butanone	142
	4-Acetyl-2-phenylimidazole	142
	α-Amino-α-carbethoxy-N-methylnitrones (general method)	143
	2-Carbethoxy-4-carbomethoxyimidazole	144
	5-Methyl-2-methylamino-4-phenylimidazole	145
	General procedure	145
	2-Anilino-1-phenylimidazolin-4-one	146
	1-Cycloheptyl-4,5-diphenylimidazole	148
References		149

5 Ring Syntheses Which Involve Formation of Three or Four Bonds ... **151**

5.1 Formation of 1,2, 3,4 and 1,5 bonds or 1,2, 2,3, 3,4 and 1,5 bonds ... 151

	2-(4'-Bromophenyl)-4,5-bis(4''-methoxyphenyl)-imidazole	154

DETAILED CONTENTS xv

    Ethyl 2,4-diphenylimidazole-
     5-carboxylate 155
    1-Hydroxy-2,4,5-trimethylimidazole
     3-oxide 155
    4-Isopropyl-2-methylimidazole 156
    4,5-Bis-(p-fluorophenyl)imidazole 159
    4-Phenylimidazole 159
    (R)-1(α-Carboxy-γ-methylthio)propyl-
     4,5-dimethylimidazole 3-oxide 160
    erythro-5-Amino-1-(2-hydroxy-3-nonyl)-
     imidazole-4-carbonitrile 161
    erythro-1-(2-Hydroxy-3-nonyl)imidazole-
     4-carbonitrile 161
    General procedure for 2,4,5-triaryl-
     imidazoles 162
    2-(4'-Pyridyl)-4-(4'-fluorophenyl)-
     5-phenylimidazole 163
  References 163

6 Syntheses From Other Heterocycles **167**
 6.1 Imidazoles 167
  6.1.1 From three-membered rings 167
  6.1.2 From five-membered rings 168
   6.1.2.1 Pyrazoles 168
   6.1.2.2 Dehydrogenation of imidazolines 168
    Methyl 1-methylimidazole-4-carboxylate 169
    2-Methylimidazole 169
    2,4-Diphenylimidazole 169
   6.1.2.3 Triazoles and tetrazoles 170
    2-Tri-n-butylstannyltetrazole 171
    General procedure: ring-opening
     of epoxides 171
    1-Vinyltetrazole 171
    Photolysis of 1-vinyltetrazole 171
   6.1.2.4 Benzimidazoles 172
    Imidazole-4,5-dicarboxylic acid 172
   6.1.2.5 Isoxazoles 172
    N-(5-Methyl-4-isoxazolyl)acetamide 174
    General procedure for isoxazole–imidazole
     interconversion 174
    N-(4-Isoxazolyl)formamide 175
    Imidazole-4-carbaldehyde 175

	4-(N-Benzylamino)isoxazole	175
	N-Benzyl-N-(4-isoxazolyl)formamide	175
	1-Benzylimidazole-5-carbaldehyde	176
	1-Benzylimidazole-4-carbaldehyde	176
	4-Acetylimidazole	176
	N-(5-Methyl-4-isoxazolyl)acetamide	177
	4-Acetyl-2-methylimidazole	177
	1,4-Diacetyl-2-methylimidazole	177
	5-Acetyl-1,2-dimethylimidazole	177
6.1.2.6	Oxazoles and oxazolines	178
	2,4,5-Triphenylimidazole	179
	3-Benzyl-1,4,5-triphenylimidazolium perchlorate	179
6.1.2.7	Thiazoles	179
6.1.2.8	Oxadiazoles or thiadiazoles	180
6.1.3 Ring contractions		180
6.1.3.1	Pyrimidines	180
	2,5-Dimethyl-4-methylcarbamoyl-1-phenylimidazole	182
	2-Methoxy-5-methyl-4-phenyl-carbamoylimidazole	182
6.1.3.2	Pyrazines	182
	2,5-Dialkyl-1-cyanoimidazoles	182
	2,5-Dialkyl-4-chloro-1-cyanoimidazoles	183
6.1.3.3	Other six- and seven-membered rings	183
References		183
6.2 Benzimidazoles		186
6.2.1 From imidazoles		186
6.2.2 From 1-aryltetrazoles		186
6.2.3 From 1-(2-nitroaryl)-1,2,3-triazolines		186
	General procedure for synthesis of 1-alkyl-2-aminobenzimidazoles	186
6.2.4 From aryl-1,2,4-oxadiazol-5-ones		187
6.2.5 From benzo five-membered heterocycles		187
6.2.5.1	Indazoles	187
6.2.5.2	Benzoxazoles	187
6.2.5.3	Benzofuroxans (benzofurazan 1-oxides, 2,1,3-benzoxadiazole 1-oxides)	188
	1-Hydroxybenzimidazole 3-oxides	189
	2-Aminocarbonyl-1-hydroxybenzimidazole 3-oxide	189
6.2.6 From benzo six-membered heterocycles		190

	6.2.6.1 Quinoxalines	190
	6.2.6.2 Miscellaneous ring contractions	190
References		190

7 Aromatic Substitution Approaches to Synthesis 193
References 194
7.1 Synthesis of N-substituted imidazoles by electrophilic
    substitution 195
    7.1.1 N-Alkylimidazoles 195
                 Deprotonation 196
                 1-Methylimidazole 196
                 1-n-Propylimidazole 196
                 1-Benzyl-2,4,5-tribromoimidazole 196
                 1-Benzylimidazole (using phase transfer
                    catalysis) 197
                 1-[(Dimethylamino)methyl]imidazole 200
                 1-(N,N-Dimethylsulfamoyl)imidazole 200
                 1-Ethoxymethyl-2-phenylimidazole 200
                 4-Iodo-1-tritylimidazole 200
                 General procedure: 1-SEM-protected
                    imidazoles 201
                 Quaternization (Begtrup method) 202
                 1,3-Dimethylimidazolium iodide 202
                 1-Ethyl-3-methylimidazolium tetraphenyl-
                    borate 202
    7.1.2 N-Alkylbenzimidazoles 202
                 1-Hydroxymethylbenzimidazole 203
    7.1.3 N-Arylimidazoles 203
                 1-Phenylimidazole 204
                 Alternative modified Ullmann procedure 204
                 General procedure for 1-arylimidazoles (and
                    1-arylbenzimidazoles) 204
                 Phase transfer method 204
    7.1.4 N-Acylimidazoles and N-acylbenzimidazoles 205
                 1-(p-Toluoyl)imidazole 205
                 1-Benzoyl-4-phenylimidazole 206
                 1-Ethyl-5-phenylimidazole 206
                 1-(Diethoxymethyl)imidazole 206
                 1-Benzoylimidazole (Begtrup method) 206
    7.1.5 Other N-substituted imidazoles 207
                 1-Trimethylsilylimidazoles 208
                 N,N-Dimethylimidazole-1-sulfonamide 208

1-Benzenesulfonylimidazole	208
1-Methylsulfonylimidazole (Begtrup method)	208
2,4,5-Tribromo-1-vinylimidazole	208
Devinylation method: 4-bromoimidazole-5-carbonitrile	209
1,4-Dinitroimidazole	209
References	209
7.2 Synthesis of *C*-substituted imidazoles and benzimidazoles by electrophilic substitution	211
7.2.1 Conventional electrophilic substitutions	212
7.2.1.1 Imidazoles	212
2,4,5-Tribromo-1-methylimidazole	212
4-Bromoimidazole	212
2-Ethyl-4-bromoimidazole	213
4-Chloro-5-iodoimidazole	213
4-Chloroimidazole	213
4-Nitroimidazole	214
1-Benzyl-2-hydroxymethylimidazole	214
2-Benzoyl-1-methylimidazole	215
7.2.1.2 Benzimidazoles	215
7.2.2 Electrophilic substitutions which involve metallic derivatives	215
7.2.2.1 Imidazoles	216
4,5-Dibromo-1-methylimidazole-2-carboxylic acid	217
4-Bromo-1,5-dimethylimidazole	218
N,N-Dimethyl-2,5-dimethylimidazole-1-sulfonamide	218
2-Nitroimidazole	218
General method: 4(5)-substituted imidazoles	219
General procedure: 1-protected 4-imidazolyl-carbinols	220
7.2.2.2 Benzimidazoles	220
General method: 2-substituted benzimidazoles	220
References	221
7.3 Other substitution methods	222
7.3.1 Nucleophilic substitution approaches	222
4-Nitro-1-phenylimidazole	223

      2-Amino-1-benzylimidazole 223
      Ethyl 2,4-difluoroimidazole 5-carboxylate ... 223
  7.3.2 Radical substitution methods 224
      Trifluoromethylation of 4-methylimidazole ... 224
References 225

8 Synthesis of Specifically Substituted Imidazoles and Benzimidazoles 227

 8.1 Halogen groups 227
  8.1.1 Fluoro derivatives 227
      2-Fluoro-1-methylimidazole 228
      1-Methyl-2-trimethylstannylimidazole ... 228
  8.1.2 Chloro derivatives 228
      4(5)-Chloroimidazole 229
  8.1.3 Bromo derivatives 230
      4-Bromo-2-ethylimidazole 230
      1-Benzyl-4-bromoimidazole 230
  8.1.4 Iodo derivatives 231
      2-Iodoimidazole 232
      4,5-Diiodoimidazole 232
      2,4,5-Triiodoimidazole 233
References 233
 8.2 Nitrogen groups 234
  8.2.1 Nitro derivatives 234
  8.2.2 Amino derivatives 234
References 236
 8.3 Acyl groups 236
  8.3.1 Carboxylic acid derivatives 236
      Diimidazo[3,4-a;3',4'-d]piperazin-
      2,4-dione 237
      4-(2',2'-Dimethyl)imidazolecarbo-
      hydrazide 238
      1-Methylimidazole-4,5-dicarboxylic
      acid 238
      Alternative procedure using diglyme 238
      1-Methylimidazole-5-carboxylic acid 239
      1-Methylimidazole-4-carboxylic acid 239
      N-(2,2-Dimethoxyethyl)trichloro-
      acetamidine 240
      Ethyl imidazole-2-carboxylate 240
      Methyl 1-methylimidazole-
      5-carboxylate 242

8.3.2 Aldehyde and ketone derivatives	242
1-Benzylimidazole-5-carbaldehyde	242
Imidazole-2-carbaldehyde	243
2-n-Heptanoyl-1-methylimidazole	244
References	245
8.4 Thiol and other sulfur functions	245
1-Methylbenzimidazole-2-thiol	247
1-Methylbenzimidazole-2-sulfonic acid	247
5-Methylsulfinyl-4-nitroimidazole	247
References	248
Index	**249**

Abbreviations and Acronyms

DAMN	diaminomaleodinitrile
DBU	1,8-diazabicyclo[5.4.0]undec-7-ene
DCCI	N,N-dicyclohexylcarbodiimide
DDQ	dichlorodicyanoquinone
DISN	diaminosuccinonitrile
DMAP	4-(N,N-dimethylamino)pyridine
DME	ethylene glycol dimethyl ether (1,2-dimethoxyethane)
DMF	dimethylformamide
DMK	dimethyl ketone (acetone)
DMSO	dimethyl sulfoxide
HMPT	hexamethylphosphoramide
LDA	lithium diisopropyamide
NBS	N-bromosuccinimide
NCS	N-chlorosuccinimide
NIS	N-iodosuccinimide
NMR	nuclear magnetic resonance
PPA	polyphosphoric acid
R.T.	room temperature
SEM	[2-(trimethylsilyl)ethoxy]methyl
THF	tetrahydrofuran
TLC	thin layer chromatography
TOSMIC	toluene p-sulfonylmethyl isocyanide
UV	ultraviolet

–1–
Introduction

When faced with the challenge of writing a monograph on the synthesis of imidazoles and benzimidazoles, one is immediately aware that although there are general methods available for the bicyclic compounds, the contrary is true for imidazoles themselves. Indeed, it is necessary to consider a number of widely divergent processes each time a synthesis is contemplated. For this reason I have divided the synthetic approaches arbitrarily into methods which make specific bonds in the imidazole product, those which transform another heterocycle, and those which start with the preformed imidazole ring.

With such a myriad of methods it has not been possible to check many of the actual syntheses included, although an endeavour has been made to cross check among different publications, and our experiences in the laboratory at Otago University have been drawn on extensively in assessing methods included. I have provided an index which should help lead research workers to specifically substituted products, so that comparisons can be made of the possible approaches. In addition, care has been taken to ensure that the referencing is sufficiently detailed.

Benzimidazole synthesis has been reviewed in detail in Preston's volumes [7] which contain a wealth of tabular material. Other review material has been less heavily oriented towards synthesis [1–6, 8–15].

When making use of the synthetic details included one needs to be aware of modern safety considerations. In consequence, it may be advisable to seek alternatives to solvents such as benzene, chloroform, carbon tetrachloride and hexane used in the methods described. Some earlier examples quoted could profitably be improved by the use of more modern separation and purification techniques. Yields quoted are those from the original publications; in our experience these have sometimes been shown to be optimistic.

My thanks are due to the University of Otago who granted me a short period of study leave, part of which was devoted to this work, and to the staff of the Chemistry Department at the Open University, Milton Keynes, UK, where I was able to work with few distractions in a friendly environment. I also owe much to Julie Leith, Andrea Krause and Diane Watson who so expertly did most of the word processing and preparation of diagrams at the University of Otago.

REFERENCES

1. K. Hofmann, in *The Chemistry of Heterocyclic Compounds. Imidazole and its Derivatives, Part 1* (ed. A. Weissburger). Interscience, New York, 1953.
2. E. S. Schipper and A. R. Day, in *Heterocyclic Compounds* (ed. R. C. Elderfield). Wiley, New York, 1957, Vol. 5.
3. P. N. Preston, *Chem. Rev.* **74**, 279 (1974).
4. P. N. Preston and G. Tennant, *Chem. Rev.* **72**, 627 (1972).
5. M. R. Grimmett, *Adv. Heterocycl. Chem.* **12**, 103 (1970).
6. M. R. Grimmett, *Adv. Heterocycl. Chem.* **27**, 242 (1980).
7. P. N. Preston, in *Benzimidazoles and Congeneric Tricyclic Compounds* (ed. P. N. Preston). Interscience-Wiley, New York, 1981.
8. M. R. Grimmett, in *Comprehensive Heterocyclic Chemistry* (ed. A. R. Katritzky and C. W. Rees). Pergamon Press, Oxford, 1984, Vol. 4.08 (ed. K. T. Potts), p. 457.
9. M. R. Grimmett, in *Comprehensive Heterocyclic Chemistry*. Elsevier, (ed. A. R. Katritzky and C. W. Rees) Oxford, Vol. 3.02 (ed. I. Shinkai) 1996, p. 77.
10. A. F. Pozharskii, A. D. Garnovskii and A. M. Simonov, *Russ. Chem. Rev. (Engl. Transl.)* **35**, 122 (1966).
11. J. B. Wright, *Chem. Rev.* **48**, 397 (1951).
12. G. V. Nikitina and M. S. Pevzner, *Chem. Heterocycl. Compd. USSR (Engl. Transl.)* **29**, 127 (1993).
13. L. B. Townsend and G. R. Revankar, *Chem. Rev.* **70**, 389 (1970).
14. A. R. Katritzky and J. M. Lagowski, *Chemistry of Heterocyclic N-Oxides*. Academic Press, New York, 1971.
15. V. I. Kelarev and V. N. Koshelev, *Russ. Chem. Rev. (Engl. Transl.)*, **64**, 317 (1995).

–2–
Ring Synthesis by Formation of One Bond

2.1 FORMATION OF THE 1,2 (OR 2,3) BOND

Synthetic procedures which fall under this heading apply to both imidazoles and benzimidazoles, and some are closely related to those methods which involve formation of both the 1,2 and 2,3 bonds. This section should therefore be read closely with Section 3.1.2, particularly with regard to benzimidazole synthesis.

2.1.1 Imidazoles

The earliest method of this type was the old Wallach synthesis, which formed the imidazole ring by cyclization of an N,N'-disubstituted oxamide with phosphorus pentachloride; the usual product was a 1-substituted 5-chloroimidazole (**1**) (Scheme 2.1.1) [1–4]. Although initially thought to be rather limited in its application, the method was later extended to cyclization of higher symmetrical and unsymmetrical oxamides, and now provides access to a variety of 5-chloroimidazoles [5–8].

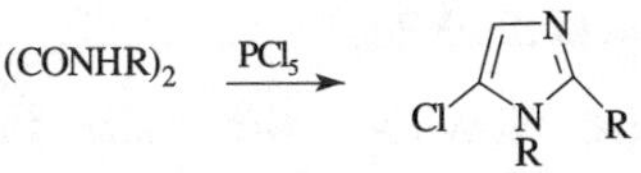

$$(CONHR)_2 \xrightarrow{PCl_5} \text{(1)}$$

Scheme 2.1.1

Symmetrically substituted oxamides can be made in 84–94% yields from diethyl oxalate and primary amines [5], whereas the unsymmetrical N-alkyl-N'-arylmethyloxamides are available in high yields by initial treatment of an ethanolic solution of diethyl oxalate with a benzylamine at 0°C, followed by

TABLE 2.1.1
5-Chloroimidazoles (2) prepared from N,N'-disubstituted oxamides or from N-formylglycine amides

R^1	R^2	Yield %	Ref.
Et	Me	75	5
nPr	Et	70	5
nBu	nPr	72	5
iBu	iPr	92	5
$CH_3(CH_2)_4$	nBu	72	5
CH_2Ph	Ph	65	5
Me	$p\text{-Cl}-C_6H_4$	61	8
Me	$o\text{-Cl}-C_6H_4$	81	8
Me	3-Pyridyl	39	8
$(CH_2)OMe$	Ph	75	8
Me	Ph	75	8
Me	H	30	9
Et	H	50	9
iPr	H	31	9
nBu	H	52	9
iBu	H	30	9
Ph	H	41	9
$PhCH_2$	H	49	9

reaction of the resulting N-substituted oxamic acid ester with a large excess of aliphatic amine [8].

Although yields of imidazoles are not always good, and the 5-chloro products may be contaminated with polychlorinated products, the method can be useful. If symmetrical oxamides are heated at 60–95°C with 2.1 mol eq. of phosphorus pentachloride in phosphoryl chloride (with cooling to moderate the exothermic reaction), followed by further heating at around 100°C for 30 min, high yields of (1) are achievable [5]. Some unsymmetrical oxamides have also been shown to give good yields on occasion [8] (Table 2.1.1).

5-Chloro-1-methyl-2-phenylimidazole [8]

To a stirred slurry of N-methyl-N'-benzyloxamide (192 g, 1.0 mol) in $POCl_3$ (600 ml) PCl_5 (436 g, 2.1 mol) is added over 10 min with cooling. The mixture is then refluxed (2 h) and the solvent is removed at atmospheric pressure. Xylene is added to assist this removal by distillation of the last traces of $POCl_3$. The cooled mixture, which contains around 200 ml of xylene, is extracted successively with water (1 l) and then dilute hydrochloric acid solution. Basification of the combined aqueous solutions precipitates 5-chloro-1-methyl-2-phenylimidazole (143 g), which is recrystallized from boiling heptane as needles, m.p. 106–107°C (75% yield).

2.1. FORMATION OF THE 1,2 (OR 2,3) BOND

Closely related to the Wallach synthesis is the ring closure of an acylated glycine by heating with phosphorus pentachloride. Early examples provided 5-chloro-1-ethyl-2-phenyl- and 5-chloro-1,2-diphenylimidazoles, albeit in rather poor yields [10, 11], but more recently analogous N-formylethanamides have given a number of 1-alkyl- and 1-aryl-5-chloroimidazoles (**2**) in yields between 30 and 52% (Scheme 2.1.2 and Table 2.1.1) [9]. Most recently, the method has been adapted to the preparation of imidazopyridines, by closing a 2-benzylamido side chain on to a pyridine nitrogen [12]. It is now believed that the Wallach reaction involves nitrile ylide species, with the limiting parameter being the electron-demanding nature of the heterocyclic ring which influences the facility of ionization of the halogen atom [12].

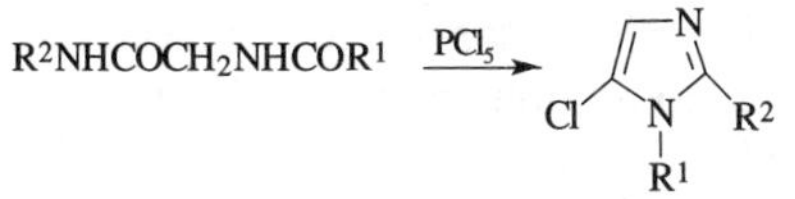

$$R^2NHCOCH_2NHCOR^1 \xrightarrow{PCl_5}$$

(R^2 = alkyl, aryl; R^1 = H, Ph)   (**2**)

Scheme 2.1.2

1-Benzyl-5-chloroimidazole [9]

A mixture of N-formylglycine benzylamide (0.1 mol), phosphorus pentachloride (0.205 mol), and phosphorus oxychloride (40 ml) is cautiously prepared at room temperature. After 3–5 min the mixture warms spontaneously with foam formation as HCl(g) is evolved. Ice water is used to keep the temperature below 60°C. The mixture is then stirred (2 h at 20–25°C and 3 h at 55–60°C). After removal of the phosphoryl chloride under reduced pressure, the cooled residue is treated with crushed ice (50 g), neutralized with aqueous ammonia to pH 8–9, and extracted with chloroform (3 × 60 ml). The extracts are washed with water, dried over sodium sulfate, filtered, the solvent evaporated, and the residue is vacuum distilled to give the product, b_3 146–147°C (49%).

1-Benzyl-5-chloroimidazole cannot be readily made by benzylation of 4(5)-chloroimidazole; the other isomer (1,4) is the major product. Nor is chlorination of 1-benzylimidazole a feasible alternative. (See Chapters 7 and 8.)

4-Ethoxy-2-substituted imidazoles, too, can be made from acylglycinamides in yields of 10–35% [13]. Modification of the reaction conditions by the use of triethyloxonium fluoroborate to induce cyclization has led to the formation of 4-ethoxy-2-phenylimidazole in 80% yield [14].

4-Ethoxy-2-phenylimidazole [14]

A mixture of triethyloxonium tetrafluoroborate (14 g, 74 mmol) and hippuric amide (12 g, 67 mmol) in dichloromethane (100 ml) is stirred for 7 days at room temperature before being poured into a saturated solution of sodium bicarbonate (0.5 g). The crystalline product is filtered and recrystallized from ethanol–isopropyl ether in 82% yield, m.p. 176–177°C. Similarly prepared are: 2-(p-chlorophenyl)-4-ethoxyimidazole (61%), m.p. 209–210°C (from ethanol); 4-ethoxy-2-(o-methoxyphenyl)imidazole (88%), m.p. 101–102°C (from isopropyl ether); 2-butyl-4-ethoxyimidazole (52%), m.p. 65–66°C (from chloroform); and 4-methoxy-2-(2'-thienylmethyl)imidazole (35%), m.p. 124–125°C (from isopropyl ether–ethyl acetate).

One of the more common methods of benzimidazole synthesis is that which cyclizes o-arylenediamines in which one of the amino groups has been acylated or thioacylated. The starting material must be functionalized in such a way that an aromatic product can be formed when cyclization takes place. It is more common with analogous aliphatic diamine derivatives for reduced imidazoles to be formed, but there are a number of synthetically useful processes which cyclize monoacylated alkylenediamines. One example involves monoacylated diaminomaleonitrile (see below), which can be dehydrated to form imidazoles [15].

Such substrates as α-(acylamino)enaminones (**4**) can be made by catalytic reduction of acylated 4-aminoisoxazoles (**3**) [16, 17], or acetamidines made from 4-amino-5(4H)-isoxazolones [18]. Although the starting materials require multistep syntheses, they are quite readily available in high yields, and their reduction and transformation into imidazoles are often quantitative (see Section 6.1.2(e)).

The derived acylaminoenaminones (**4**)–(**6**) can be modified to give rise to a range of substituted imidazole products. If the original isoxazole is alkylated on the 4-amino group prior to acylation it produces acylaminoenaminones (**4**), which are converted by base into 1,2,4-trisubstituted 5-acylimidazoles (**7**) in around 50% yields. If the acylated isoxazole is not N-substituted, treatment of the resulting enaminone (**5**) with a primary amine serves to incorporate a substituted amino group to give enaminone (**6**), which is transformed by ethanolic sodium hydroxide into the isomeric 1,2,5-trisubstituted 4-acylimidazole (**8**) in 75–92% yield. The reaction is less successful for making 4-acylimidazoles with a 5-substituent because of steric hindrance to amine exchange (Scheme 2.1.3) [16, 17]. This approach to the synthesis of 4- and 5-acylimidazoles provides much better control of the regiochemistry than direct alkylation of 4(5)-acylimidazoles. Table 2.1.2 lists some representative examples. (See also Tables 6.1 and 6.2 and Chapter 6), where some synthetic examples are also found.)

4(5)-Acylimidazoles are sometimes available by photochemical rearrangement of the 1-acyl isomers [19], by oxidation of suitable alcohol side chains (see Section 8.5), and by other ring-synthetic methods [20–23], although

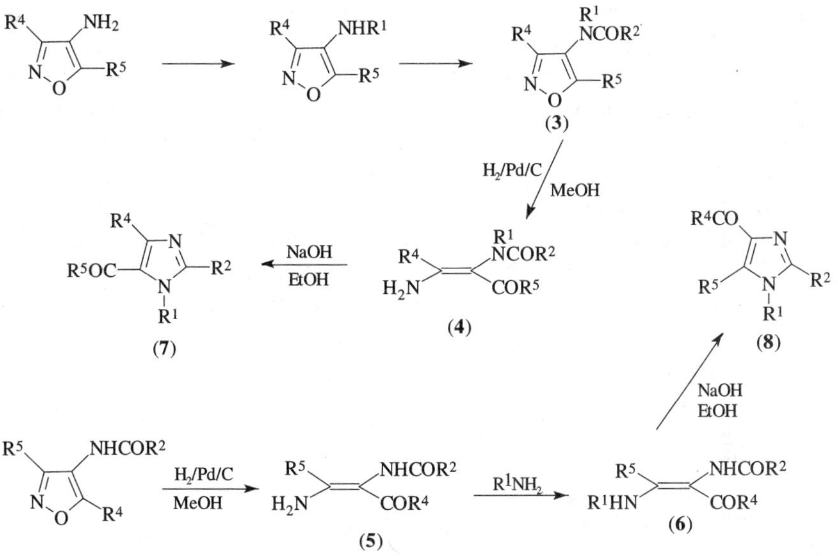

2.1. FORMATION OF THE 1,2 (OR 2,3) BOND

Scheme 2.1.3

TABLE 2.1.2
4-Acetyl-2-substituted imidazoles (**8**) made from acylated 4-aminoisoxazoles (**3**)
(R^1, R^5 = H; R^4 = Me) via the acylaminoenaminones [16]

R^2	Yield %	m.p. (°C)	R^2	Yield %	m.p. (°C)
H	98	169–170	CF_3	83	187–188
Me	78	128–130	CH_2OMe	70	90–92
Et	93	117–119	CH_2NMe_2	52	219–225 (HCl)
t-Bu	98	198–199	Ph	80	155–156

there are no other really general methods [24]. Reactions of 3-bromo-4-ethoxy-3-buten-2-one with amidines are limited by the availability of such butenones and of the amidines [23], but the alternative starting material, 3-chloro-4,4-dimethoxy-2-butanone, allows large-scale synthesis of 4-acetyl-2-methylimidazole [25] (see Section 4.3). Approaches via lithioimidazoles (see Section 7.2.2) allow some versatility, but introduction of an acyl group at C-4 or C-5 is largely limited to instances where the imidazole 2-position is blocked [19, 26–28].

In basic conditions, the adducts which form when secondary enaminones or enaminosulfones react with diethoxycarbonyl- or benzylethoxycarbonyl-1-diazene cyclize to 2-imidazolones [29]. The transient intermediates which form when oximes react with imidoyl chlorides undergo [3,3]-sigmatropic rearrangements to give amidinoamides (**9**). These are capable of smooth acid-catalysed conversion into imidazoles (**10**) at slightly elevated temperatures. The procedure (which might well have been covered in Section 4.1) is equally facile for making 1-methyl- and 1-phenylimidazoles. Oximes which have an α-hydrogen on only one of the substituents give rise to a single product (**9**), but those with hydrogen atoms on both α-substituents give mixtures, as do mixtures of *syn*- and *anti*-oximes. Hydrolytic interception of the amidinoamides (**9**) sometimes leads to α-amidoketones which are also suitable substrates for imidazole synthesis (see Section 3.2). For best results the amidinoamides are made by reaction of a 2:1 mixture of the imidoyl chloride and the oxime in THF with 3 eq. of triethylamine (Scheme 2.1.4) [30].

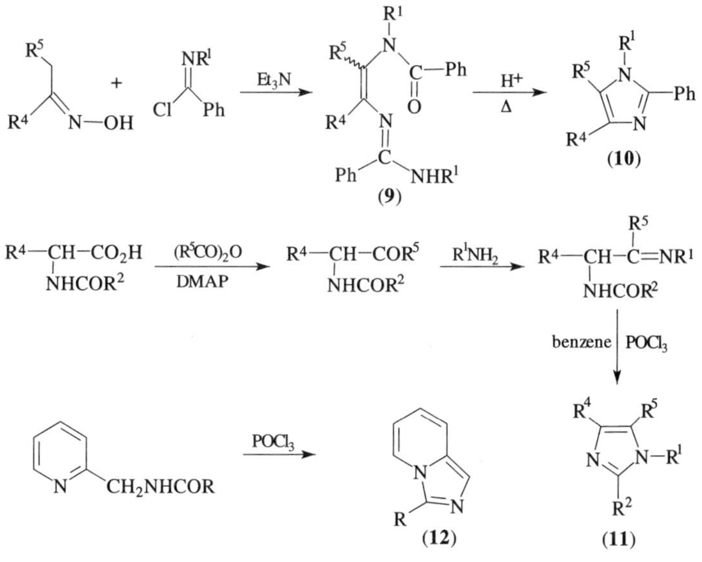

Scheme 2.1.4

N-Methyl-N-[2-[[(methylamino)phenylmethylene]amino]-1-cyclohexen-1-yl]-benzamide (**9**) (R^4, $R^5 = -(CH_2)_4-$; $R^1 = Me$) [30]

To a cooled solution of *N*-methylbenzenecarboximidoyl chloride [30] (3.1 g, 20 mmol) in dry THF (50 ml) at −78°C is added a 3.5 molar excess of triethylamine. The mixture is stirred (0.5 h) before addition of a solution of cyclohexanone oxime (1.13 g, 10 mmol). After refluxing (12 h), water is added and the mixture is extracted with dichloromethane (3 × 50 ml), the combined extracts are washed with saturated sodium chloride solution, and dried (Na$_2$SO$_4$) before removal of the solvent under reduced pressure. Flash

2.1. FORMATION OF THE 1,2 (OR 2,3) BOND

column chromatography using 20% ethyl acetate in hexane gives the product (1.8 g, 52%) as an oil.

1-Methyl-2-phenyl-4,5-tetramethyleneimidazole (**10**) *(R^4, $R^5 = -(CH_2)_4-$; $R^1 = Me$)*

The amidinoamide (1 g, 2.86 mmol) is heated (12 h) in a Dean–Stark apparatus with *p*-toluenesulfonic acid (2.5 molar equivalents) in toluene. The solution is then cooled to room temperature, washed with 1 M sodium hydroxide solution, then concentrated *in vacuo* before chromatographic purification (10% ethyl acetate in hexane) to give white crystals (97%) m.p. 119–120°C. Similarly prepared are other (**10**) (R^1, R^4, R^5, yield given): Me, Ph, Me, 90%; Me, Ph, Ph, 96%; Me, Me, Me, 90%; Me, Et, H, 42%.

When primary amines react with α-acylaminoketones the resulting Schiff bases can be cyclized in the presence of phosphoryl chloride, phosphorus pentachloride, or triphenylphosphine and triethylamine in hexachloroethane to give 1-substituted imidazoles (**11**) (Scheme 2.1.4). The starting α-acylaminocarbonyls are readily prepared from α-amino acids by reduction with sodium amalgam [31, 32] or by the Dakin–West reaction [33, 34], which is most conveniently conducted in the presence of 4-(*N,N*-dimethylamino)pyridine (DMAP) as an acylation catalyst [35–37].

3-Benzoylaminobutanone [34]

N-Benzoylalanine (3.9 g, 20 mmol) and DMAP (0.1 g, 0.82 mmol) are stirred with acetic anhydride (4 ml, 42 mmol) and triethylamine (4 ml, 0.29 mmol). Lively evolution of carbon dioxide and slight warming are observed. After 30 min, glacial acetic acid (30 ml) is added and the solution is allowed to stand (30 min), concentrated *in vacuo*, shaken with 2 M sodium hydroxide solution, and extracted with ether. The ether extracts are shaken with 1 M HCl solution, dried, and evaporated to give 3-benzoylaminobutanone (3.0 g, 78%), m.p. 63°C.

α-Aminoketones can also be prepared from α-chloroketones and potassium phthalimide, and from α-oximinoketones.

The versatility of this reaction is increased when a substituted hydrazine is used instead of the primary amine, providing ready access to 1-aminoimidazoles (**11**) ($R^1 = $ NHR) from the hydrazones [36]. Hydroxylamine converts the α-(acylamino)ketones into oximes. In the presence of phosphoryl chloride or the fluoride ion, imidazole *N*-oxides can be generated. Thus, 2-formyl-4-methyl-1-phenylimidazole 3-oxide has been made in around 50%

TABLE 2.1.3
Imidazoles (11) prepared by cyclization of Schiff bases of α-(acylamino)ketones [36]

R^{1a}	R^2	R^4	R^5	Method[b]	Yield (%)
Ph	Me	Me	Me	A	75
CH_2Ph	Me	Me	Me	A, B	70, 81
CH_2Ph	Ph	Me	Me	B	71
CH_2Ph	Ph	iPr	Me	B	73
CH_2CH_2Cl	Ph	iPr	Me	B	34
CH_2Ph	Ph	CH_2Ph	Me	B	71
$CH_2=CHCH_2$	p-ClC_6H_4	Ph	Me	B	77
NMe_2	Me	H	Me	A	79
NMe_2	Me	Me	Me	B	87
p-TosNH	Me	Me	Me	C	83
p-TosNH	Me	Me	Me	C	54
NMe_2	Ph	Me	Me	B	78
a	Ph	iPr	Me	B	76
b	Ph	iPr	Me	B	61
NMe_2	Me	$MeS(CH_2)_2$	Me	A	75
NMe_2	p-ClC_6H_4	$MeS(CH_2)_2$	Me	A	60
a	Ph	CO_2Me	Me	A	50

[a] a, phthalimido; b, 2,4-$(NO_2)_2C_6H_3NH$.
[b] A, $PPh_3/C_2Cl_6/Et_3N$; B, PCl_5; C, $POCl_3$.

yield by fluoride ion-promoted cyclization of the trimethylsilyl ether of *N*-dichloroacetyl-*N*-phenylaminopropanone oxime [38]. This general reaction type is sufficiently versatile to allow synthesis of the imidazo[1,5-*a*]pyridine (12), where the pyridine ring provides the Schiff base function [39]. Table 2.1.3 lists some examples of the general reaction type.

General method for preparing imidazoles or 1-imidazolamines [36]

The α-acylaminoketimine or α-acylaminohydrazone (10 mmol) and triethylamine (3.67 g, 36 mmol) together with hexachloroethane (3.55 g, 15 mmol) are treated at −25°C in acetonitrile (30 ml) with triphenylphosphine (3.93 g, 15 mmol). After stirring (5 h) at around 23°C, the solid is filtered and washed with a little THF, and the filtrate is evaporated to dryness. The residue is treated with 10% hydrochloric acid (40 ml), diluted with water (250 ml) and exhaustively extracted with ether. The aqueous phase is saturated with sodium chloride, made strongly alkaline with solid KOH, and again exhaustively extracted with ether. The dried ether extracts are concentrated, and the residue is either recrystallized or distilled.

Ring closure of formylglycine amidines, induced by heating alone or in the presence of phosphoryl chloride, bears similarities to the general method

2.1. FORMATION OF THE 1,2 (OR 2,3) BOND

above. The products are 1-alkyl-5-aminoimidazoles, but they are only formed in low yields (5–15%) [40]. In consequence, these compounds are better prepared by reaction of ethyl N-cyanomethylimidates with primary amines or hydrazines (see Section 3.2).

Methods of imidazole synthesis based on the use of diaminomaleonitrile (DAMN; (**13**), $R^1 = H$) have proliferated in recent years. The reagent is now readily available from chemical distributors at a price which makes its use reasonably economical. Some of these methods of synthesis fall under the present heading; others involve 1,5 bond formation (see Section 2.2), or 1,2 and 2,3 bond formation (see Section 3.1.1).

Schiff bases (**14**), which are formed by reaction between DAMN and appropriate carbonyl reagents, are oxidatively cyclized to give a variety of 2-substituted 4,5-dicyanoimidazoles (**15**) (Scheme 2.1.5). Although dichlorodicyanoquinone (DDQ) or diaminosuccinonitrile (DISN) have been used frequently to achieve the oxidative cyclization, long reaction times (17 h to 4 days under reflux) are a disadvantage, and N-chlorosuccinimide (NCS) under basic conditions is more convenient in many cases. The Schiff bases are best formed from aromatic aldehydes, but aliphatic aldehydes and ketones, ketoesters, orthoesters, amides, imidates and cyanogen chloride have all been used [15, 41–49].

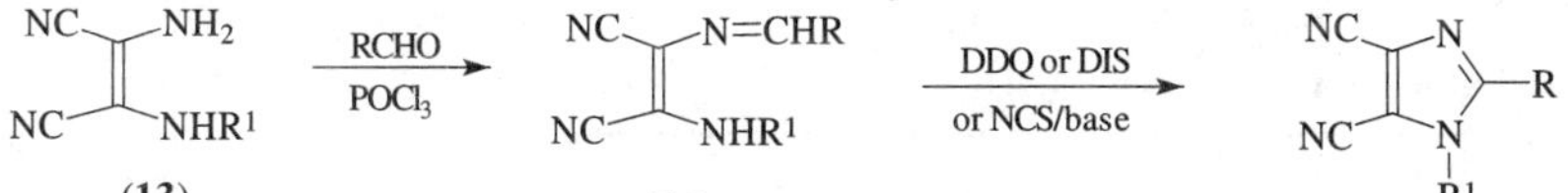

Scheme 2.1.5

If DAMN is mono-N-alkylated before reaction with the carbonyl reagent the method gives 1-alkylimidazoles [15]. For example, 4,5-dicyano-1-(2′,3′,5′-tri-O-benzoyl-β-D-ribofuranosyl)imidazole (57%) has been made by treatment of the ribosylamino-DAMN with triethyl orthoformate at 90°C in anisole under nitrogen for 5 h. The intermediate enamine is not isolated in this case [42]. When DAMN is treated with N,N-dialkylamides in the presence of phosphoryl chloride, the Schiff base (**14**) ($R = NR^2R^3$) cyclizes to give 2-dialkylaminoimidazoles (**15**) ($R = NR^2R^3$), including 4,5-dicyanoimidazole (**15**) ($R = H$) when DMF is used [15]. Imidazoles (**15**) ($R = OR^4$, NR^2R^3) can be made in one step when DAMN reacts with orthoesters or iminoether hydrochlorides. Under mild reaction conditions the intermediate alkoxyimines (**14**) ($R = OR^4$) or amidines (**14**) ($R = NR^2R^3$) can be isolated before oxidation to (**15**) [46, 47, 49]. Table 2.1.4 lists some examples.

TABLE 2.1.4
4,5-Dicyanoimidazoles (15) ($R^1 = H$) by oxidative cyclization of (14)

R	Conditions	Yield %	Ref.
H	DMF, $POCl_3$	90	15, 49
Ph	NCS	66	43
Ph	Heat in BuOH	77[a]	45
p-MeOC$_6$H$_4$	NCS	74	43
tBu	DISN/DDQ	57	15
Me	Heat in BuOH	83[a], 92[b]	45
Et	Heat in BuOH	84[a]	45
iPr	Heat in BuOH	33[a]	45
o-ClC$_6$H$_4$	NCS	64	43
m-ClC$_6$H$_4$	DDQ	40	15
p-ClC$_6$H$_4$	DDQ	33	15
o-FC$_6$H$_4$	DDQ	25	15
m-FC$_6$H$_4$	DDQ	29	15
p-FC$_6$H$_4$	DDQ	24	15
o-BrC$_6$H$_4$	DDQ	34	15
m-BrC$_6$H$_4$	DDQ	50	15
p-BrC$_6$H$_4$	DDQ	50	15
CF_3	DDQ	50	15
Pr	DISN	37	15
Cyclohexyl	DISN	41	15
OMe	DDQ	—[c]	44
OEt	DDQ	54	15
OCH$_2$CH=CH$_2$	DDQ	27	15
NMe$_2$	DDQ	34	15

[a] From ester.
[b] From amide.
[c] Yield not given but appears to be low.

2-Aminoimidazoles are also formed when DAMN is treated with isocyanide dichlorides (e.g. cyanogen chloride) (see Section 3.1.1).

Oxidative cyclization with DDQ of amino(ribosylamino)maleonitrile in methanol gives a 64% yield of 4-cyano-5-methoxy-2-(D-ribotetrahydroxybutyl)-imidazole, whereas an anomeric mixture of 2-(D-erythrofuranosyl)-4-cyano-5-methoxyimidazole is obtained in 77% total yield by heating the open chain species for 3 h at 150°C in a mixture of 2-ethoxyethanol and acetic acid [41].

4,5-Dicyano-2-phenylimidazole [43]

To a solution of *N*-benzylidenediaminomaleonitrile [15] (0.50 g, 2.56 mmol) and nicotinamide (0.37 g, 3.07 mmol) in DMF (5 ml), NCS (0.34 g, 2.56 mmol) is added, and the mixture is stirred at 40°C. Nicotinamide hydrochloride slowly separates from solution. After stirring (3 h), this salt is filtered off

2.1. FORMATION OF THE 1,2 (OR 2,3) BOND

and washed with acetone. The combined filtrate and washings are poured into water (200 ml), and the resulting precipitate is collected and recrystallized from benzene (charcoal treatment may be necessary) to give colourless crystals (0.33 g, 66%), m.p. 220–222°C.

4,5-Dicyanoimidazole [15]

A solution of DAMN (10.8 g, 100 mmol) in DMF (100 ml) is cooled in an ice bath. Phosphoryl chloride (10 ml, 65 mmol) is added dropwise, keeping the temperature below 10°C. After this addition is complete, the solution is heated to 160°C and maintained at that temperature (1 h), then cooled to around 100°C, when most of the solvent is removed on the rotary evaporator. Water (100 ml) is added, and the solution is warmed to 70°C, filtered, and the filtrate is extracted nine times with 200 ml portions of diethyl ether. The combined ether extracts are dried ($MgSO_4$), filtered and evaporated, to give the product (10.7 g, 90%), m.p. 174–175°C. 4,5-Dicyanoimidazole is available in 95% purity from Aldrich.

With β-ketoesters, β-ketoamides and β-diketones, DAMN is converted into enamines, which can be cyclized purely by heating in an alcoholic solvent. The products are 2-substituted-4,5-dicyanoimidazoles (Scheme 2.1.5) [45]. When DAMN reacts with formamidine, the initial condensation product **(16)**/**(17)** can form imidazoles either by loss of ammonia (when 4,5-dicyanoimidazole is formed), or via an isomerization and subsequent cyclization which eliminates HCN to give 4-amino-5-cyanoimidazole **(18)**. With excess formamidine the latter product is converted into adenine. The intermediate amidines **(16)**/**(17)**, as transient intermediates, react with aqueous ammonia to form **(18)**, and with acetic acid to give **(15)** ($R = R^1 = H$) [48, 50]. The transformation of **(17)** to **(18)** is a 1,5 bond formation (Scheme 2.1.6).

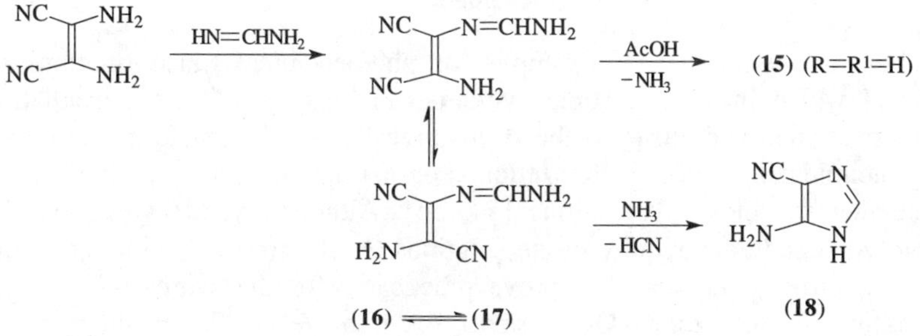

Scheme 2.1.6

Oxidation of the Schiff bases (**19**) made from the monoamide of DAMN (derived from DAMN by hydration of one of the nitrile functions) gives 4-cyanoimidazole-5-carboxamides (**20**) in good yields (~70% or more) (Scheme 2.1.7). Such oxidations can be readily accomplished using hydrogen peroxide with a catalytic amount of sodium molybdate [51].

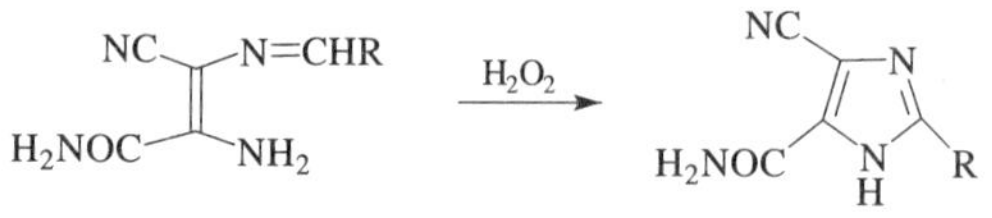

Scheme 2.1.7

The preparation of 2-alkyl- and 2-acyl-1-methylimidazole-4-carboxylates from (Z)-β-dimethylamino-α-isocyanoacrylate (**21**) with an alkyl or acyl halide bears similarities to the foregoing DAMN procedures (Scheme 2.1.8) [52, 53]. The isocyanoacrylates can be made quite readily by reaction of DMF diethylacetal with the carbanion generated when the isocyanide (**22**) is metallated [53].

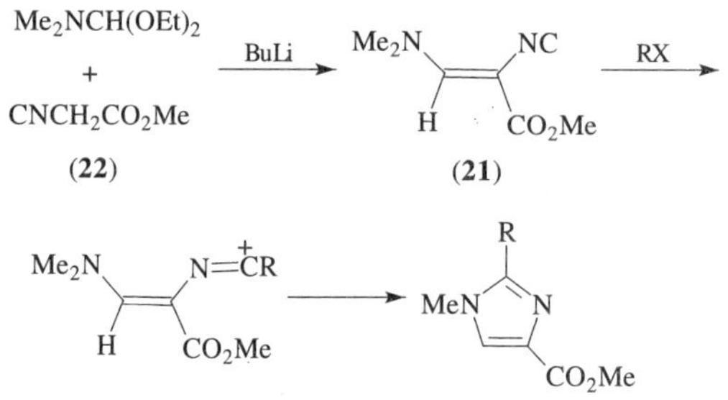

e.g. R (%): Me (80); Et (60); CH_2Ph (50)
COPh (50); COMe (50); CO_2Me (45)

Scheme 2.1.8

There are a number of examples of photochemical transformations of DAMN which may have some synthetic applications. At first, irradiation isomerizes the *cis*-dinitrile to the *trans*-dinitrile, which then forms 5-amino-4-cyanoimidazole (**18**). Both azetidine or azirine intermediates have been postulated for these cyclizations [54, 55]. Although yields are good the photolytic conditions require dilute solutions (~ 10^{-2} mol l^{-1} solutions of the enaminonitriles) and may not prove practicable for the synthesis of larger quantities of imidazoles. On a small scale, however, the method merits consideration particularly since the process works for a wide variety of enaminonitriles (Scheme 2.1.9) [56–58].

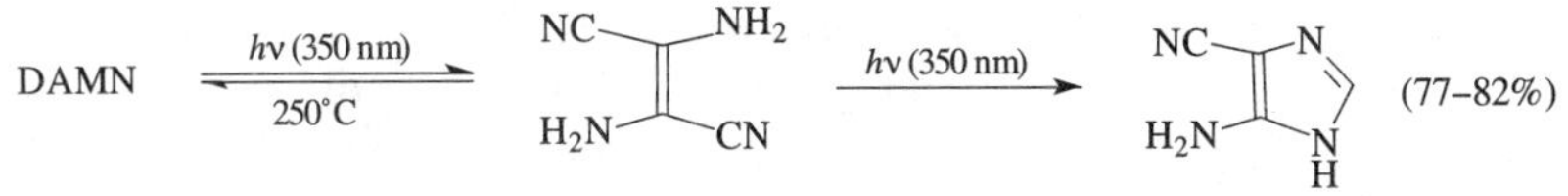

2.1. FORMATION OF THE 1,2 (OR 2,3) BOND

DAMN $\xrightleftharpoons[250°C]{h\nu\,(350\,\text{nm})}$ [NC, NH$_2$, H$_2$N, CN diaminomaleonitrile] $\xrightarrow{h\nu\,(350\,\text{nm})}$ (18) (77–82%)

Scheme 2.1.9

Tetrahydrobenzimidazole [57]

A 250 ml solution of 2-amino-1-cyclohexanecarbonitrile (10^{-2} mol l^{-1} in THF) is irradiated at 254 nm (24 h). Column chromatography (isopropylamine–methanol–benzene, 1:3:96) gives an 81% yield of tetrahydrobenzimidazole. After recrystallization from benzene–diethyl ether the m.p. is 146–148°C.

Suitably substituted 1,2-diamines and derivatives can by cyclized into 2*H*-imidazoles either by photolytic or thermal treatment in the presence of base. The 2*H*-imidazole products, however, cannot always be thermally isomerized to the 1*H*-isomers. The migratory aptitudes of 2-substituents have been shown to lie in the order methyl<ethyl<phenyl<benzyl [59–62].

Ring closure of *N*-(1-cyanoalkyl)alkylidene *N*-oxides (these compounds are nitrones, cf. Section 4.2) is induced by nucleophilic attack by the thiophenate ion (Scheme 2.1.10). The reactions are accelerated by the addition of small amounts of piperidine, but inhibited by temperatures above the melting points of the nitrones. In these reactions, 4-phenylthioimidazoles (**23**) are formed usually in 80–90% yields. Deficiencies in the general process include the problems of first making the nitrones, and the fact that other thiolate nucleophiles (methyl, ethyl, benzyl) are much less successful, giving only low yields of 4-alkylthioimidazoles [63].

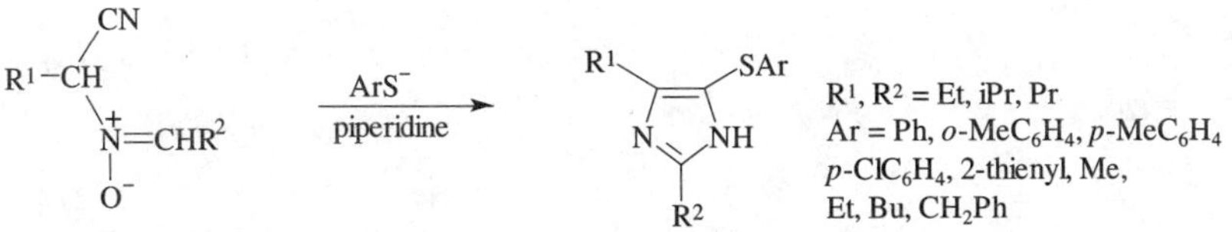

$R^1, R^2 = $ Et, iPr, Pr
Ar = Ph, *o*-MeC$_6$H$_4$, *p*-MeC$_6$H$_4$
p-ClC$_6$H$_4$, 2-thienyl, Me,
Et, Bu, CH$_2$Ph

(23)

Scheme 2.1.10

2,4(5)-Dialkyl-5(4)-arylthioimidazoles (**23**)

The *N*-(1-cyanoalkyl)alkylideneamine *N*-oxide (10 mmol), the thiol (11 mmol) and piperidine (0.2 mmol) are warmed in a stoppered vessel below the melting point of the nitrone. Within 2–4 days the product crystallizes. Disulfide and excess thiol are separated by refluxing with hexane, and the products are recrystallized in 80–90% yields from ethyl acetate or aqueous ethanol. Products are soluble in ethanol, ethyl acetate, acetone, diethyl ether, chloroform and hexane, but scarcely soluble in cold benzene, hexane, light petroleum and water.

When *N,N*-disubstituted arylhydrazones are treated with trifluoracetic anhydride in lutidine solution at room temperature the trifluoroacetyl-hydrazones (24) are formed in 60–98% yields. Thermal rearrangement of (24) leads to mixtures of oxadiazines and 4(5)-trifluoromethylimidazoles. High temperatures and more polar solvents suppress the formation of the six-membered rings, while solvent variation can also alter the proportions of the imidazole isomers (25) and (26) [64–66]. Thus, cyclization in refluxing toluene (under nitrogen, 48 h) can lead to varying ratios of 5- (25) and 4-trifluoromethylimidazole (26) isomers from 3:1 to 100% (25). The reaction is much more rapid if (24) is heated in a sealed tube without any solvent (5 min to completion), but as the temperature is raised, the proportion of the 4-isomer (26) and other by-products increases. Thermal cyclization in the presence of silica gel (with carbon tetrachloride under nitrogen) gives much more of the 4 isomer (~1:1) and virtually no oxadiazine. In heptane or toluene the ratio of (25) to (26) becomes 1:2 or 1:3. With no solvent at all, the ratio is about 1:3. (See Table 2.1.5 and Scheme 2.1.11.) The fact that the imidazole isomers are readily separable by column chromatography means that the process has some synthetic potential, particularly because the hydrazones are not difficult to make, and they can be acylated by other acylating or aroylating agents, e.g. the benzoyl analogue of (24) cyclizes after refluxing for 24 h in toluene to give an 87% yield of 1-methyl-4,5-diphenylimidazole [67, 68].

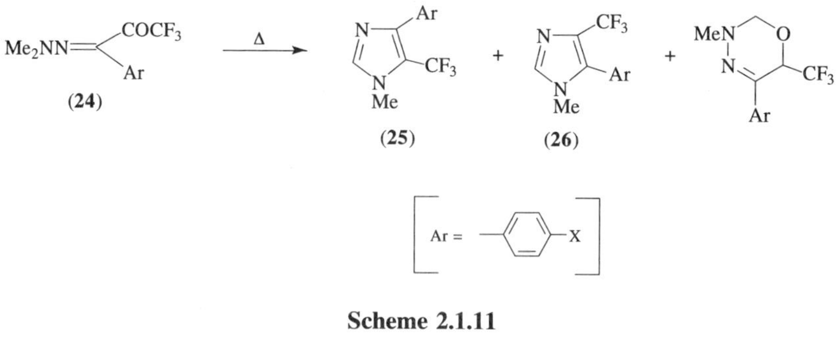

Scheme 2.1.11

Although aliphatic aldehyde dimethylhydrazones could not be trifluoro-acetylated, it is possible with t-butylmethylhydrazones. These give rise to

2.1. FORMATION OF THE 1,2 (OR 2,3) BOND

TABLE 2.1.5
Ratios of (25) to (26) formed by thermal cyclization of (24)[a]

X	(25):(26) ratio
H	23:77
Me	25:75
MeO	38:63
Cl	32:68
NO_2	37:63

[a]SiO_2 (no solvent), N_2, 80°C, 24 h.

5-trifluoromethylimidazoles only when refluxed in toluene, and under all other thermal regimes tried [64, 67, 69].

1-t-Butyl-4-isopropyl-5-trifluoromethylimidazole

A solution of the trifluoroacetylhydrazone (1 mmol) in toluene (20 ml) is refluxed (16 h) under nitrogen. Removal of the solvent gives a residue which is purified by Kugelrohr distillation ($b_{1.5}$ 110°C) to give the product (0.10 g, 81%) as a yellow oil.

In the presence of base, 2-isocyanoalkanenitriles [70] react with alcohols, thiols or hydrogen sulfide to form imidazoles which have oxygen or sulfur groups in the 4(5) position. The reactions appear to proceed via the 4*H*-imidazoles, which rearrange to the 1*H*-isomers [71]. There are similarities to the synthesis of 4-methoxyimidazole made, in 60% yield, by refluxing the *N*-cyanomethylimino ester of formic acid with sodium methoxide in methanol [72].

5-Methoxy-4-methylimidazole [71]

To solution of 2-isocyanopropanenitrile (1.6 g, 20 mmol) in methanol (15 ml) is added potassium carbonate (0.2 g), and the mixture is stirred at room temperature (20 h). After filtering, and removing the solvent under reduced pressure, the crude product is crystallized from cyclohexane (1.6 g, 70%), m.p. 88–89°C. Similarly prepared are 5-ethoxy-4-methyl- (62% from cyclohexane, m.p. 94°C), 5-methoxy-4-isopropyl- (75% from benzene, m.p. 96–97°C), 5-ethoxy-4-isopropyl- (51% from benzene, m.p. 122–123°C), 5-ethylthio-4-methyl- (58% from cyclohexane, m.p. 88°C), 5-benzylthio-4-methyl- (61% from benzene, m.p. 156°C), 5-ethylthio-4-isopropyl- (55% from cyclohexane, m.p. 124°C), and 5-benzylthio-4-isopropylimidazoles (45% from benzene, m.p. 133°C).

The rather uncommon 4- and 5-imidazolethiols are accessible mainly by nucleophilic substitution reactions of suitably activated 4(5)-halogenoimidazoles [73], from N-(1-cyanoalkyl)alkylidene N-oxides (see earlier in this section), and by a few other ring-synthetic or rearrangement processes (see Section 8.4). We can add to this a dehydration–cyclization reaction of the thionamides (**27**) (Scheme 2.1.12). The process is promoted by trimethylsilyl triflate and triethylamine, and provides a relatively mild and regiocontrolled multistep synthesis. The starting materials (**27**) are quite readily prepared in a multistep sequence starting from an aldehyde, primary amine and sodium cyanide, and there is no need to purify the reaction intermediates. The total time required for the whole sequence is around 1–2 days, and the chemicals are sufficiently inexpensive to make large-scale synthesis viable. Formic acid can also be used to promote the cyclization in cases where the products are relatively robust unfunctionalized thiols, but the silicon reagent is preferred. When the product has a 2-substituent it is necessary to raise the reaction temperature to around 25°C. It is easier to isolate the products (**28**) if they are "masked" as disulfides or sulfides. Treatment with hydrogen peroxide in aqueous alkali gives the disulfides; sequential reaction with sodium borohydride in ethanol and benzyl bromide gives the S-benzyl derivatives [74]. Table 2.1.6 lists some examples.

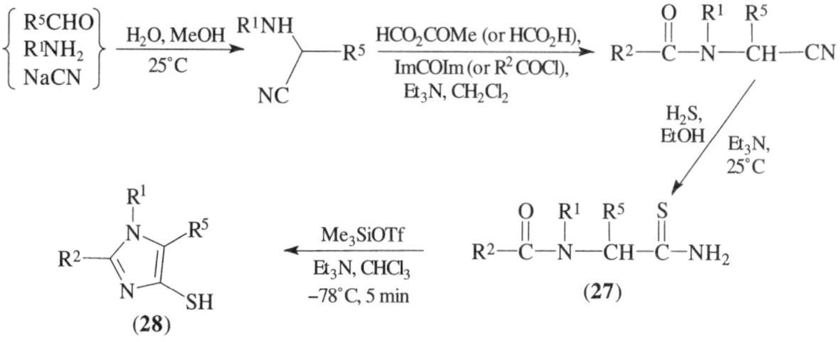

Scheme 2.1.12

4-Benzylthio-1-methyl-5-[(2-tetrahydropyranyl)oxy]methylimidazole

The thionamide (**27**) (R^1 = Me, R^2 = H, R^5 = CH_2OTHP) (0.15 g, 0.6 mmol) and triethylamine (0.25 g, 2.5 mmol) in dichloromethane (5 ml) are cooled to −78°C, when trimethylsilyl trifluoromethanesulfonate (0.40 g, 1.8 mmol) is added dropwise down the wall of the flask. The mixture is stirred at −78°C (5 min), then treated with methanol (5 ml) and sodium fluoride (0.5 g) before warming to 25°C and stirring (2 h). The volatiles are removed under

2.1. FORMATION OF THE 1,2 (OR 2,3) BOND

TABLE 2.1.6
Imidazole-4-thiols (28) prepared from thionamides (27)

R^1	R^2	R^5	Yield (%)	
			(27)	(28)[a]
Me	H	H	36	60
Bu	iPr	iPr	70	57
Bu	Ph	iPr	86	46
Me	H	CH_2OCH_2Ph	64	42
Me	H	$(CH_2)_2OCH_2Ph$	50	84
Me	H	CH_2OTHP[b]	52	69

[a] Based on (27).
[b] THP, tetrahydropyranyl.

reduced pressure, and the residue, dissolved in ethanol, is treated sequentially with sodium borohydride (0.5 g) and benzyl bromide (0.137 g, 0.8 mmol). After stirring (0.5 h) at 25°C the mixture is diluted with water and extracted with dichloromethane. The dried (MgSO$_4$) extracts are then concentrated *in vacuo* and chromatographed on silica gel eluted with 8% methanol in dichloromethane. The product (0.13 g, 69%) is isolated as a colourless oil.

Photochemical cyclization of nitrile ylides (**29**) derived from suitably substituted 2*H*-azirines gives imidazoles (Scheme 2.1.13) [75, 76].

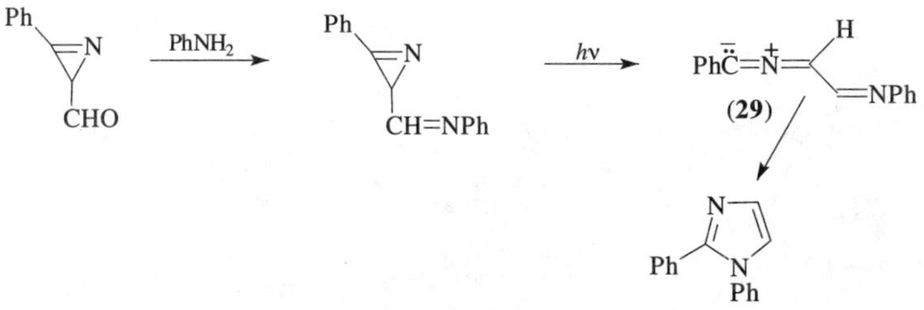

Scheme 2.1.13

2.1.2 Benzimidazoles

One of the most common approaches to benzimidazole synthesis (but less common for uncondensed imidazoles) utilizes an arylene (or alkylene) 1,2-diamino compound in which one of the amino functions has been acylated or thioacylated. The starting material has to be functionalized in such a

2. RING SYNTHESIS BY FORMATION OF ONE BOND

manner that an aromatic product can be formed when cyclization occurs. The methods are formally related to reactions of *o*-arylenediamines with appropriate carbonyl compounds (see Section 3.1.2), and a number of reviews have discussed these reactions in detail [77–83]. Typically, an *o*-(*N*-acylamino)- or *o*-(*N*-aroylamino)-arylamine (**30**) is cyclized by uncatalysed thermolysis, or with aqueous or ethanolic acid, or phosphoryl chloride. Heating under reflux in ethanolic alkali is also successful in some instances.

2-Methylbenzimidazole [84]

After heating *o*-aminoacetanilide (1 g) for 2 h in dry xylene (30 ml), cooling, and filtering the solid, the title compound is isolated (0.88 g, 100%) and recrystallized from water, m.p. 176–177°C.

This approach can be extended to the use of more accessible *o*-nitroacylaminoarenes (**31**) which cyclize in such reducing media as tin and acetic or hydrochloric acids, tin(II) chloride and hydrochloric acid, iron and acetic acid, hydrogen–palladium–carbon, hydrogen–Raney nickel–hydrochloric acid, sodium dithionite or sodium borohydride (Scheme 2.1.14). Yields are usually good, and often superior to those which are obtained from reaction between an *o*-aminoaniline and a carboxylic acid. The general method can be applied to the synthesis of 2,2′-bis-benzimidazoles [85], to 2-unsubstituted benzimidazoles by cyclization of an *o*-formamidoarylamine (**30**) (R^2 = H) [77, 82], and to 1-aminobenzimidazoles when *o*-acylaminophenylhydrazines (**30**) (R^3 = NH$_2$) are the substrates [86].

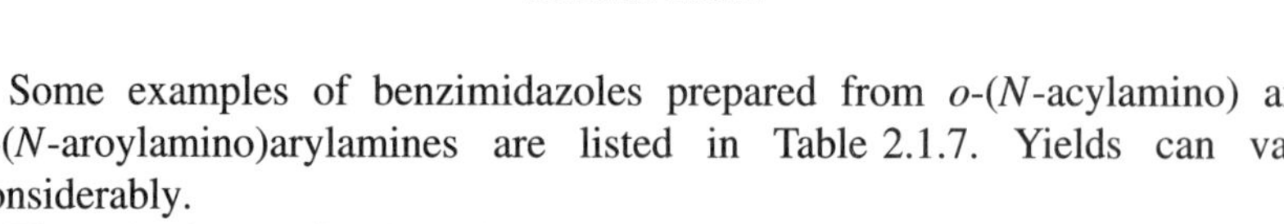

Scheme 2.1.14

Some examples of benzimidazoles prepared from *o*-(*N*-acylamino) and *o*-(*N*-aroylamino)arylamines are listed in Table 2.1.7. Yields can vary considerably.

The reductive cyclization of acylated *o*-nitroanilines (**31**) is a highly versatile process, since variation of both arylamine and acyl portions can lead to a wide variety of products. Thus, an acylated *N*-alkyl-*o*-nitroaniline gives a 1,2-disubstituted benzimidazole; an acylated *o*-nitrodiarylamine will lead

2.1. FORMATION OF THE 1,2 (OR 2,3) BOND

TABLE 2.1.7
Benzimidazoles (32) prepared by cyclization of various o-acylaminophenyl-hydrazines (30)

2 substituent	1 substituent	Reaction conditions	Yield (%)	Ref.
Me	H	Heat	100	84
Ph	H	4 M HCl or 2 M H_2SO_4	65	87
p-$NH_2C_6H_4$	H	4 M HCl or 2 M H_2SO_4	75	87
Me	Me	4 M HCl	2	88
Me	$PhCH_2$	4 M HCl	12	88
Me	o-$NO_2C_6H_4CH_2$	1:1 HCl	—	89
Me	o-$NO_2C_6H_4CH_2$	1:1 HCl	—	89
4-Thiazolyl	H	H_3PO_4	—	90
CHMe-tBu	H	$POCl_3$–pyridine	—	91

to 1-arylbenzimidazoles; and both 2-substituted and 1- and 2-unsubstituted benzimidazoles are accessible in moderate to good yields (40–80%) by judicious choice of substrate [82, 92–94].

Partial reduction of (31) can give benzimidazole N-oxides, and so careful choice of reducing agent and reaction conditions converts o-nitroanilides (including formanilides) into the oxides. Presumably the intermediate o-hydroxylaminoanilide spontaneously cyclizes before it can be further reduced. Synthetically useful procedures include hydrogenation in the presence of palladium or platinum with at least one molar proportion of acid present. The acid appears to catalyse the cyclization of the hydroxylamino species so that ring closure is more rapid than further reduction [82, 95]. Such reductive methods are especially appropriate to the synthesis of 1,2-disubstituted benzimidazole 3-oxides, and for cyclizations of nitroanilides which bear an electron-withdrawing substituent on the acyl carbonyl (e.g. (31) ($R^2 = CF_3$; $R^3 = H$). The use of sodium borohydride in the presence of platinum, palladium or Raney nickel [96] has been shown to give greater than 50% yields of benzimidazole N-oxides, but even this reagent is capable of reducing a nitro group to amino, and combined with product loss on work-up, yields of the N-oxides are often disappointing [97]. Sulfide reduction procedures (e.g. ammonium sulfide) seem better for 1-substituted benzimidazole 3-oxides than for benzimidazole 1-oxides [88, 92].

1,2-Dimethylbenzimidazole 3-oxide [92]

N-Methyl-o-nitroacetanilide [98] (20 g) is added to a solution of ethanol (200 ml) and ethanol saturated with ammonia at 0°C (100 ml). Hydrogen sulfide is bubbled through this solution for about 2 h. After standing overnight at room temperature, the brown solution is concentrated *in vacuo* to about 100 ml, and the precipitated sulfur is removed by filtration and washed with ethanol. The combined filtrate and washings are rotary evaporated to an oil which is dissolved in hot acetone and allowed to crystallize. The white crystals (80% yield) are recrystallized from acetone or ethyl acetate, m.p. 65–70°C (as dihydrate).

When an *o*-nitroaniline is acylated by ethyl chloroformate and then catalytically reduced, thermolysis of the reduction product (**33**) gives a 1-substituted 2-benzimidazolone (**34**) (Scheme 2.1.15) [99]. Presumably the carbamates (**33**) eliminate ethanol as they cyclize, and so the reactions bear similarities to those which proceed through isocyanates (see Scheme 2.1.18). In the presence of magnesium chloride, which appears to activate the "urea" carbonyl group to solvolysis and condensation, some benzimidazolones are converted into 2-alkyl- and 2-arylbenzimidazoles [100].

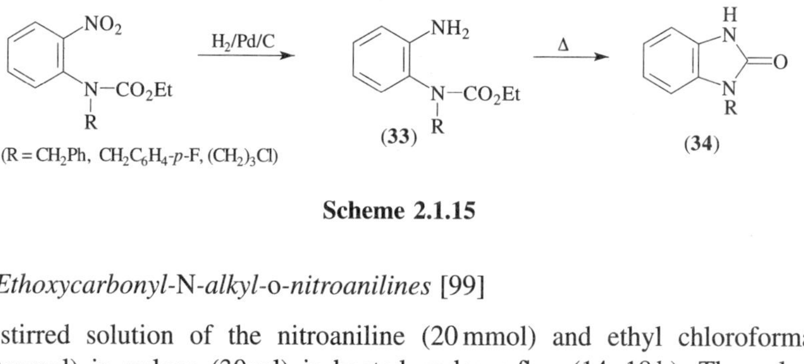

Scheme 2.1.15

N-Ethoxycarbonyl-N-alkyl-o-nitroanilines [99]

A stirred solution of the nitroaniline (20 mmol) and ethyl chloroformate (60 mmol) in xylene (30 ml) is heated under reflux (14–18 h). The solvent is removed *in vacuo* to give the pure carbamates as oils in quantitative yields. They can be used without further purification, but analytical samples can be obtained by preparative thin layer chromatography using petroleum ether–ethyl acetate as the eluent. The products have $\nu_{max} \sim 1710 \, cm^{-1}$.

N-Ethoxycarbonyl-N-alkyl-o-phenylenediamine (**33**) [99]

A mixture of the nitro compound (10 mmol) and 10% palladium on carbon (500 mg) in methanol (50 ml) is magnetically stirred under hydrogen at atmospheric pressure (2–4 h). The catalyst is removed by filtration, and the filtrate is concentrated under reduced pressure to give the pure amine (**33**) as an oil in almost quantitative yield. Again the products can be used directly, or purified

2.1. FORMATION OF THE 1,2 (OR 2,3) BOND

by the use of a short silica gel column eluted with petroleum ether-ethyl acetate. The products have $\nu_{max} \sim 3450-3460$ and $1685-1690\,cm^{-1}$.

1-Alkyl-1,3-dihydro-2H-benzimidazolin-2-ones (**34**) [99]

The amine (**33**) (10 mmol) is heated at 210°C (60-90 min) in a Kugelrohr apparatus in the presence of calcium chloride. The oily residue (**34**) which solidifies on cooling is recrystallized from the appropriate solvent as follows (R, solvent, m.p. listed): CH_2Ph, ethanol, 200-201°C; $CH_2C_6H_4-p$-F, isopropanol, 178-179°C; $(CH_2)_3Cl$, ethyl acetate-hexane, 118-120°C. Yields lie in the range 80-85%.

If the substrate is an acylated *o*-aminophenylhydrazine (**30**) (R_3 = NHR), the cyclization gives rise to 1-aminobenzimidazoles (**35**) in good yields (Scheme 2.1.16). Such *o*-aminoarylhydrazines can be made readily by catalytic reduction of the nitro precursors [101]. If the *o*-nitroarylhydrazine (**36**) is acetylated before catalytic reduction the product is an N'-acetyl-N-(2-aminophenyl)hydrazine (**37**). Under acidic conditions (**37**) cyclizes to 1-amino-2-methylbenzimidazoles (**35**) (R = H, R^2 = Me) via acetyl group migration, e.g. 1-amino-2-methylbenzimidazole (62%). Similarly prepared is 1-amino-2-phenylbenzimidazole (55%) [86]. (See also Section 3.1.2.)

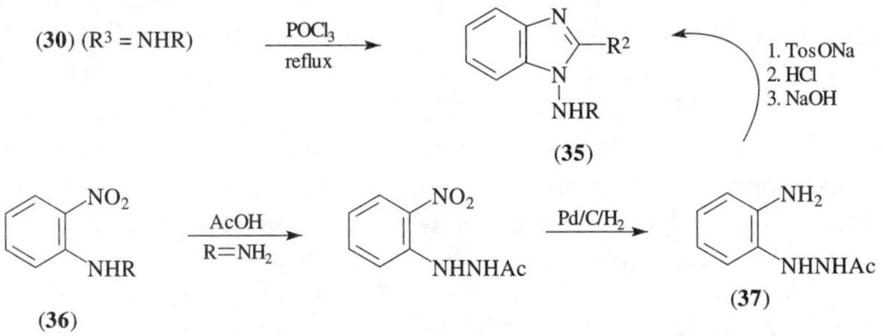

Scheme 2.1.16

Similarly, aldehyde *o*-aminophenylhydrazones, available by reduction of the nitro precursors, are converted in the absence of oxygen into 2-substituted benzimidazoles. The reaction is acid catalysed and, although complicated experimentally, gives good yields (67-87%), providing a useful alternative when the 2 substituent may be sensitive. When oxygen is not excluded, benzotriazines form instead [102]. In addition, only fair yields of 1-aminobenzimidazoles are obtained by acid-catalysed cyclization of *N*-(*o*-nitroanilino)-substituted aliphatic amines. Benzotriazoles and benzotriazole

N-oxides are usually the major products. This type of reaction is believed to follow a [1,5]-sigmatropic rearrangement [103].

N-*(2-Aminophenyl)-1H-pyrrol-1-amine* [101]

To a solution of (**36**) (R = 1-pyrryl) (4.06 g, 20 mmol) in ethyl acetate (100 ml) is added 10% palladium on carbon (200 mg). The mixture is hydrogenated (4 h) at room temperature under 4 atm of hydrogen (4 l). After filtration of the catalyst, the solvent is removed under reduced pressure to give an oil which solidifies after vacuum distillation (b.p. 130–132°C, 0.15–0.18 torr). The solid is recrystallized from cyclohexane (2.8 g, 81%, m.p. 67–70°C).

N-*(2-Acetylaminophenyl)-1H-pyrrol-1-amine* [101]

Acetyl chloride (1.0 g, 10 mmol) in anhydrous THF (25 ml) is dropped with stirring into an ice-cooled solution of N-(2-aminophenyl)-$1H$-pyrrol-1-amine (1.73 g, 10 mmol) and triethylamine (1.0 g, 10 mmol) in the same solvent (50 ml). The mixture is kept at room temperature (1 h), then filtered and rotary evaporated. The solid residue crystallizes from aqueous ethanol (1.8 g, 84%, m.p. 172–174°C; v_{max} 3250, 3300 and 1650 cm^{-1}).

1-(1-Pyrryl)-2-methylbenzimidazole (**32**) *($R^2 = Me$, $R^3 = 1$-pyrryl)* [101]

A mixture of the above amine (2.15 g, 10 mmol) and phosphoryl chloride (15 ml) is heated under reflux (30 min), then poured on to crushed ice (100 g) and made alkaline with concentrated aqueous ammonia. Extraction with ethyl acetate (150 ml), followed by drying (Na$_2$SO$_4$), filtration, and removal of the solvent gives a residue which is chromatographed on an alumina column with chloroform as the eluent. The first fractions are collected and rotary evaporated to give the benzimidazole, which is recrystallized from aqueous ethanol (1.57 g, 80%, m.p. 142–144°C).

Occasionally, benzimidazoles are made by cyclization of diacylated *o*-arylenediamines (**38**) (Scheme 2.1.17). Ring closure of these compounds requires higher temperatures than for the monoacyl derivatives. They can, however, be cyclized in a melt, or in the presence of acid catalysts, and when conditions are anhydrous, 1-acylbenzimidazoles may be isolated. These latter compounds are "azolides", and as such are prone to ready cleavage by nucleophiles. Mixtures of products are likely to be formed if the two acyl groups are different. For example, thermolysis of N-acetyl-N'-benzoyl-*o*-phenylenediamine gave a 3:1 mixture of 2-methyl- and 2-phenyl-benzimidazoles [104]. A recent modification has been used to prepare 1-acetyl-2-methylbenzimidazole (**39**) in quantitative yield [105]. Compound (**39**) can,

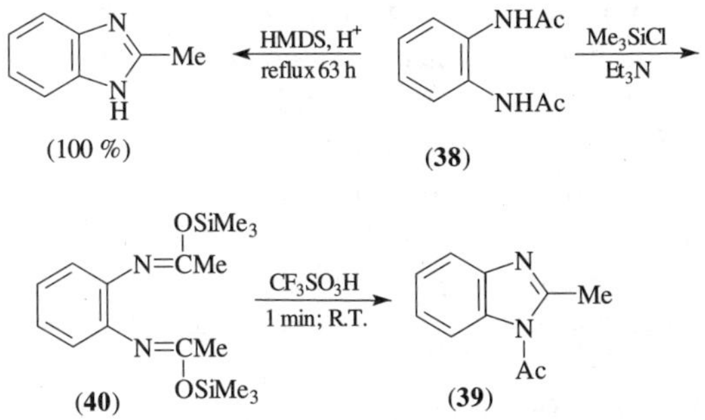

2.1. FORMATION OF THE 1,2 (OR 2,3) BOND

Scheme 2.1.17

of course, be made by direct acetylation of 2-methylbenzimidazole, but the general method may have applications.

1-Acetyl-2-methylbenzimidazole **(39)** [105]

A stirred solution of 1,2-di(acetylamino)benzene **(38)** (19.2 g, 0.1 mol), triethylamine (30.1 g, 0.298 mol), and toluene (300 ml) is heated under reflux while trimethylchlorosilane (37.9 ml, 0.298 mol) in toluene (50 ml) is added. After heating (4 h), the triethylamine hydrochloride is filtered off, and the solvent is removed under reduced pressure. Distillation *in vacuo* gives the silyl derivative **(40)** as a fraction, b_1 107°C (92%). Addition of triflic acid (0.033 g) to **(40)** (2.026 g, 0.006 mol) induces instant cyclization at room temperature. The 1-acetyl-2-methylbenzimidazole separates as a solid, and is recrystallized from diethyl ether in 100% yield, m.p. 80°C.

Related to the above synthesis is the process which converts the *o*-diamine into the diisocyanate **(41)** (Y = O) in the presence of excess phosgene (Scheme 2.1.18). When **(41)** (Y = O) reacts with water, methanol or aniline, the product is a benzimidazolone [81, 106]. The corresponding

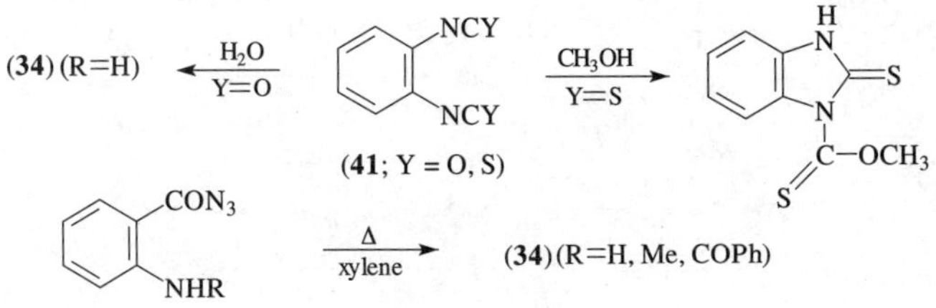

Scheme 2.1.18

diisothiocyanates (**41**) (Y = S) provide a source of benzimidazolethiones (>60% yields) when they are heated with alcohols [107].

Isocyanates are also formed when acid azides are heated, by means of a Curtius rearrangement proceeding via a nitrene. When alcohols are present, the products are carbamate esters. Hence, high yields (75–85%) of benzimidazolones can be obtained when *o*-aminoaroyl azides (**42**) are heated under reflux (~6 h) in acetic acid or xylene [62, 108]. Similarly, Hofmann rearrangement converts *o*-amidoanilines into *o*-cyanatoanilines, which can cyclize to form benzimidazolones [100].

Cyclization of *o*-azidoarylcarbamates to benzimidazolones (~30%) [109], and the action of thionyl chloride on *o*-arylazidocarboxamides to give 2-arylbenzimidazoles [110], are related reactions. Reductive cyclization of *o*-benzoquinonedibenzamide by triphenylphosphine gives 1-benzoyl-2-phenylbenzimidazole [111].

Schiff bases of *o*-amino- or *o*-nitroanilines can be cyclized (reductively in the latter case) to benzimidazoles often in rather better yields than when *o*-arylenediamines react with aldehydes (see Section 3.1.2). Thus, *N*-benzylidene *o*-nitroanilines (**43**) are converted into 2-arylbenzimidazoles by refluxing them with triethylphosphite in t-butylbenzene (Scheme 2.1.19). Yields usually lie in the range 30–50%, e.g. 2-phenylbenzimidazole is obtained in a yield of 47% [112]. Under rather less drastic base-catalysed conditions, such as potassium cyanide in methanol, the *N*-oxides can be obtained, e.g. 2-phenylbenzimidazole *N*-oxide (79%) [113]. The corresponding Schiff bases (**44**) of *o*-phenylenediamines are oxidatively cyclized using a variety of reagents (copper salts, active manganese dioxide, lead tetraacetate, nickel peroxide, barium manganate) to 2-arylbenzimidazoles (**45**) in yields between 15 and 90% (Scheme 2.1.19) [82, 114, 115]. Some examples are listed in Table 2.1.8.

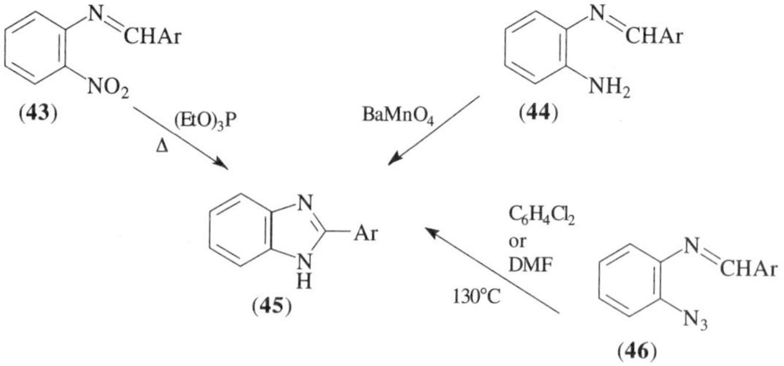

Scheme 2.1.19

2.1. FORMATION OF THE 1,2 (OR 2,3) BOND

TABLE 2.1.8
2-Arylbenzimidazoles (45) made from Schiff bases (44)[a]

2-Aryl substituent	Time (h)	Temperature (°C)	Yield (%)
Ph	5	R.T.	82
4-MeC$_6$H$_4$	9	R.T.	71
4-Me$_2$NC$_6$H$_4$	11	R.T.	70
4-NO$_2$C$_6$H$_4$	28	R.T.	80[b]
3-NO$_2$C$_6$H$_4$	5	10	60[c]

[a] BaMnO$_4$ as the oxidizing agent.
[b] 15% yield using MnO$_2$.
[c] 25% yield using MnO$_2$.

Barium manganate (which is also known to dehydrogenate imidazolines to imidazoles [116]) gives yields which are comparable with or better than those of other oxidizers, and it is an attractive alternative for other heterogeneous oxidizing agents. A reactant-to-oxidizer level of 1:10 in benzene solution is appropriate [115]. Using trichloroacetonitrile to form the Schiff base allows the general procedure to be adapted to prepare 2-aminobenzimidazoles [117].

Closely related synthetic methods thermolyse anils (46) derived from 2-azidoanilines, giving 2-arylbenzimidazoles in 48–96% yields (Scheme 2.1.19). These cyclizations possibly involve arylnitrenes, which form when (46) are heated in solvents such as nitrobenzene [62, 81, 118, 119]. The major problem which arises with this type of synthesis is the difficulty of making the Schiff bases. o-Azidoaniline decomposes at 65°C, and the azido group decreases the nucleophilicity of the *ortho* amino group, making it less reactive with aldehydes. Some of these problems can be overcome by carrying out low-temperature (~35°C) condensations in ethanol containing some acetic acid as a catalyst [118, 119], but the alternative methods which do not involve azidoanils seem more attractive for forming 2-arylbenzimidazoles.

Cyclization of o-aminoarylureas (or thioureas (47)) or, under reductive conditions, o-nitroaryl analogues can provide routes to benzimidazolones (or thiones) or 2-aminobenzimidazoles (49). Refluxing (48) for 5–10 min with iron in acetic acid gives a high yield of 5-nitrobenzimidazolone [120], and there are other examples [121]. Conversion of o-aminoarylureas or -thioureas into benzimidazolones or benzimidazolethiones is achieved by merely heating them (reactions may go through intermediate isocyanates) [122]. Oxidative methods, such as the use of mercury(II) chloride, oxide or acetate in ethanol, convert the o-aminothioureas (47) into 2-aminobenzimidazoles (49), with the best conditions using around 3 mol eq. of mercuric chloride in refluxing chloroform [123]. Such oxidative cyclizations possibly involve carbodiimide intermediates. Heating with either acids or bases also seems to convert the ureas or thioureas into 2-aminobenzimidazoles [82, 123–126]. Appropriate

2. RING SYNTHESIS BY FORMATION OF ONE BOND

o-aminothioureas (**47**) can be made by treating the *o*-diamines with alkyl isothiocyanates, and it is most convenient to transform them into (**49**) by heating them with around 8 eq. of an alkyl halide in ethanol (Scheme 2.1.20). Presumably *S*-alkylisothiouronium intermediates are involved under these reaction conditions. Yields range from 30 to almost 90% of both 2-alkyl- and 2-arylaminobenzimidazoles (e.g. (**49**), R, yield given: Ph, 76%; *o*-tolyl, 74%; Bu, 30%; PhCH$_2$, 50%) [122, 127].

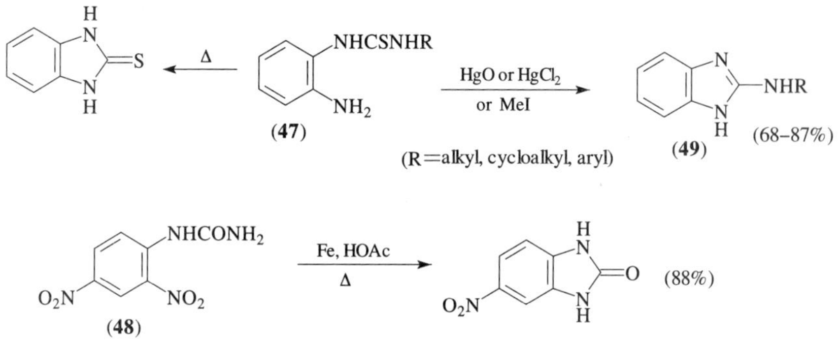

Scheme 2.1.20

2-Aminobenzimidazole (**49**) *(R = H)* [126]

Mercury(II) oxide (8.5 g) in ethanol (50 ml) is added at 70°C to *N*-(*o*-aminophenyl)thiourea (**47**) (R = H) (1.41 g). The mixture is stirred at 70–75°C (15 min), then three additional portions (4 g each) of mercury(II) oxide are added at 15 min intervals. The mixture is then filtered hot, and the filtrate is evaporated at 40°C to give 2-aminobenzimidazole (85%), m.p. 225–228°C.

2-Phenylaminobenzimidazole (**49**) *(R = Ph)* [122]

Iodomethane (1.14 g, 8 mmol) is added to a solution of (**47**) (R = Ph) (0.24 g, 1 mmol) in ethanol (15 ml), and the mixture is heated under reflux (9 h). Evolution of gas is observed as the mixture turns a deep violet colour. After addition of 10% hydrochloric acid (15 ml) and concentration to about 10 ml, the solution is filtered hot and allowed to cool. The crystals which separate are filtered, dissolved in water (25 ml), made alkaline with concentrated ammonia solution, and then repeatedly extracted with chloroform. The extracts, after washing with water and drying (MgSO$_4$), are rotary evaporated to give a yellow solid which is recrystallized from benzene (78%), m.p.195–197°C.

In *O*-benzoyl-2-aminobenzamide oxime (**50**) the two nitrogen atoms of the amidoxime, the amino nitrogen and C-1 of the aminophenyl group all compete

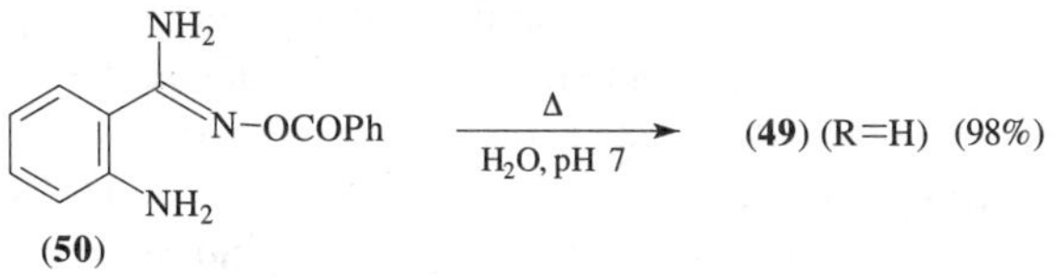

2.1. FORMATION OF THE 1,2 (OR 2,3) BOND

Scheme 2.1.21

for the role of nucleophilic centre (Scheme 2.1.21). Although this should lead to a mixture of thermolysis products, when the substrate is heated in aqueous medium the electron-rich C-1 attacks the oxime nitrogen. Beckmann rearrangement, followed by ring closure, gives an almost quantitative yield of 2-aminobenzimidazole. There is less selectivity, though, with the corresponding acetyl derivative, and in consequence the procedure may not have wide synthetic application [128, 129].

Although the most versatile methods of synthesis of 2-aminobenzimidazoles involve cyclizations of 2-substituted anilines with formation of the 1,2 and 2,3 bonds (see Section 3.1.2), the reduction of o-cyanoaminonitrobenzenes (**52**) offers a viable alternative route [83, 120, 130, 131]. Hydrogenation using a Raney nickel catalyst is frequently used, while iron–acetic acid is also a common reducing agent. The cyanamides (**52**) can be made from the aryl chlorides (**51**), or by direct reaction of the arylamine with cyanogen bromide (Scheme 2.1.22). Thus, 2,4-dinitrophenylcyanamide is converted by iron and acetic acid into 2-amino-5-nitrobenzimidazole in 55% yield [120]. A number of 2-amino-1-hydroxybenzimidazoles have been made by electrochemical reductive cyclization of 2-nitrophenylcyanamides (**52**) (R = H, alkyl, OH, NH_2, CN, SCN, SH). For example, 2-amino-1-hydroxybenzimidazole itself is isolated in 69% yield when (**52**) (R = H) is reduced at 1000 mV and 25°C in a mixture of sulfuric acid and methanol, using a platinum anode and mercury cathode (Scheme 2.1.22) [132]. This

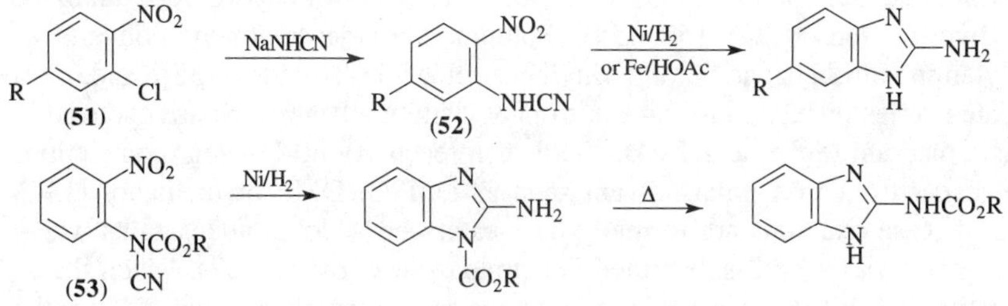

Scheme 2.1.22

latter process is, however, rather too exotic experimentally to supersede other approaches to 2-aminobenzimidazoles. Carbanilic acids (53), too, can be reduced directly to form alkyl esters of 2-aminobenzimidazole-1-carboxylic acids. These rearrange on heating to give the 2-carbamates [133].

There are a number of methods which cyclize o-amino- or o-nitroarylamines to benzimidazoles or benzimidazole N-oxides. These include cases where one of the amino groups is a secondary or a tertiary amine.

The most general route to benzimidazole N-oxides ring closes N-alkyl-o-nitroanilines which have the alkyl substituent activated by an electron-withdrawing group (e.g. acyl or nitrile). Many of these cyclizations are effected by base treatment [134–138], but acidic [139] and reductive [95, 140, 141] regimes have also been used, as have photochemical cyclizations of the corresponding o-nitroanilines [142, 143]. The synthesis of benzimidazole N-oxides has been reviewed [144]. 2-Substituted benzimidazole N-oxides (55) were first made in 1910 by reduction of o-nitroacylanilines (54) with ammonium sulfide (Scheme 2.1.23) [145]. Catalytic hydrogenation in acidic medium readily converts a range of compounds of type (54) into moderate to good yields of 1,2-disubstituted benzimidazole 3-oxides (55) (R = H, Me, Ph; R^2 = H, Me, Ph, Ar) in moderate to good yields [95, 138, 141, 146, 147].

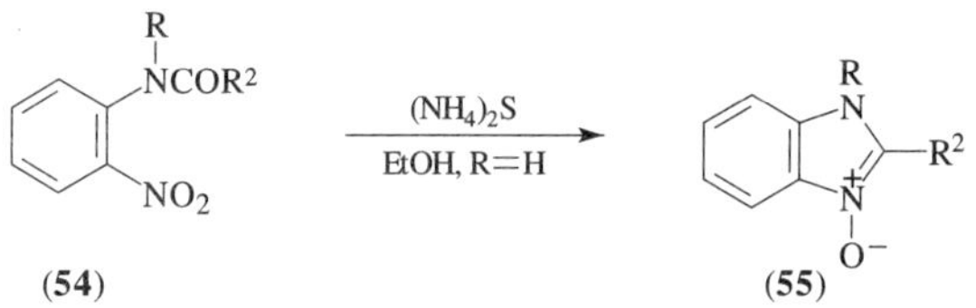

Scheme 2.1.23

In monosubstituted o-nitroanilines (56) (R = H) the activating group (R^2) can be acyl or aroyl [148], nitrile [149], ester [137], or even carboxylic acid provided that the solution is buffered. Weak electron-attracting groups such as phenyl require much stronger base treatment, e.g. N-benzyl-o-nitroanilines are only cyclized in the presence of hydroxide, alkoxide or hydride ions [134, 150–154]. Sodium ethoxide treatment converts o-allylaminonitrobenzene into 2-vinylbenzimidazole 1-oxide [138], and, in a related series of reactions, 2,4-dinitrophenylaminoalkenes (58) are cyclized in basic medium (Scheme 2.1.24). Yields can reach 70–80% when the reactions are carried out in a polar solvent such as DMSO, DMF or methanol [137]. All of these reactions are formally base-catalysed Aldol condensations which give good yields of 2-substituted benzimidazole N-oxides (57). When the 2-substituent is nitrile or ester these functions can be readily removed to give the 2-unsubstituted oxides (57) (R = R^2 = H) in 63–83% yields [97, 135]. Many

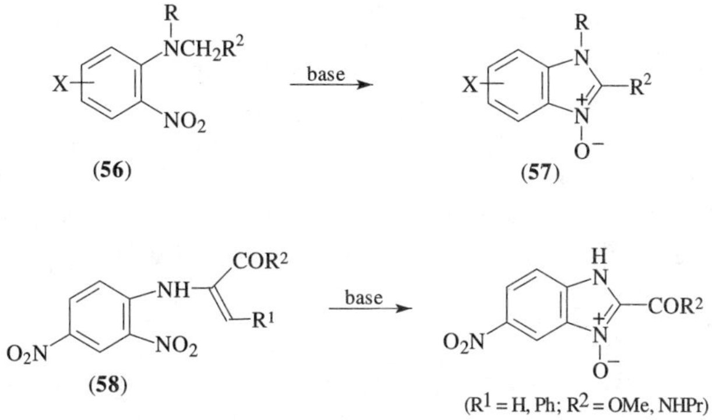

2.1. FORMATION OF THE 1,2 (OR 2,3) BOND

Scheme 2.1.24

TABLE 2.1.9
Benzimidazole N-oxides (57) (R = H) made by base treatment of various compounds (56) (R = H)

X	R^2	Base	Yield (%)	Ref.
H	Ph	NaOH	71	151
H	H	K_2CO_3	68[a]	97
H	CO_2Et	K_2CO_3	31	97
H	$CH=CH_2$	NaOEt	90	138
H	CN	Na_2CO_3	77	148
5-Me	COPh	KOH	—	149
5-OMe	CN	K_2CO_3	51	97
5-Me	CN	K_2CO_3	53	97
5-Cl	CN	K_2CO_3	58	97
5-F	CN	K_2CO_3	71	97
6-F	CN	K_2CO_3	70	97
4-NO_2	CN	K_2CO_3	34	97
5-Me	CO_2Et	K_2CO_3	46	97
5-OMe	CO_2Et	K_2CO_3	68	97
5-NO_2	CO_2Me	$NaHCO_3$	79	137
5-NO_2	CO_2Et	K_2CO_3	56	97
5-Me	H	K_2CO_3	72[a]	97
5-OMe	H	K_2CO_3	75[a]	97
5-Cl	H	K_2CO_3	76[a]	97
5-F	H	K_2CO_3	57[a]	97

[a]Yield based on the 2-cyano derivative.

such reactions take place under mildly basic conditions, e.g. using a peptide with a terminal 2,4-dinitrophenylglycine residue [155]. Some examples are listed in Table 2.1.9. N-Cyanomethyl-o-nitroanilines are readily made from o-nitroanilines by treatment with formaldehyde and potassium cyanide in acetic acid in the presence of zinc chloride.

N-Cyanomethyl-o-nitroaniline (**56**) *(R, X = H; R^2 = CN)* [97]

Glacial acetic acid (125 ml) containing concentrated sulfuric acid (eight drops) is added with stirring to a mixture of *o*-nitroaniline (6.9 g, 0.05 mol), paraformaldehyde (4.5 g, 0.15 mol of CH_2O), potassium cyanide (9.75 g, 0.15 mol) and anhydrous zinc chloride (5.25 g, 0.38 mol). The mixture is heated and stirred at 50°C (8 h), then poured into iced water before filtering. The solid is washed well with water and recrystallized from ethanol to give the product (6.73 g, 76%), m.p. 136–138°C.

2-Cyanobenzimidazole N-oxide (**57**) *(R, X = H; R^2 = CN)* [97]

Potassium carbonate (1.22 g) is added to a suspension of *N*-cyanomethyl-*o*-nitroaniline (3.08 g) in hot ethanol (170 ml), and the mixture is heated under reflux (4 h). The solvent is removed under reduced pressure and the residue dissolved as well as possible in water. The mixture is filtered, and the filtrate is acidified with hydrochloric acid to precipitate the pale yellow *N*-oxide (1.48 g, 54%), m.p. 232–234°C, when recrystallized from aqueous ethanol. 2-Carbethoxybenzimidazole *N*-oxides are prepared in a similar manner.

Hydrolysis of the 2-cyano or 2-carbethoxy group [97]

The nitrile or ester is heated under reflux with concentrated hydrochloric acid (20–25 ml per gram of substrate) for 4 h. The *N*-oxide hydrochloride which separates from the cooled solution is recrystallized, then decomposed by dissolving it in aqueous ammonia (d. 0.88; ~40 ml per gram of hydrochloride), and the solution is concentrated at 50°C under reduced pressure until crystallization commences. After cooling, the *N*-oxides are filtered.

In view of the observation that *N-o*-nitrophenyl derivatives of glycine and other α-amino acids could be converted into benzimidazolones by the action of heat, and the assumption that *N*-oxides were intermediates in the thermolysis [156], it was thought that flash vacuum pyrolysis with a very short reaction time might allow isolation of the *N*-oxides. This approach, however, did not turn out to be as synthetically efficient as the base treatment method [97]. Indeed, heating such compounds in sand at 200°C is probably of more use for making benzimidazoles or benzimidazolones [156]. Benzimidazole *N*-oxides can be made from acid-catalysed thermal or photochemical reactions of *N,N*-dialkyl-*o*-nitroanilines, but not from purely thermal reactions.

Photolysis of *N*-(2,4-dinitrophenyl) derivatives of amino acids gives mixtures of 4-nitro-2-nitrosoanilines and 5-nitrobenzimidazole *N*-oxides. In acidic media the *N*-oxides predominate with typical yields in the range 30–80% [80, 143, 157]. A related reaction which also appears to proceed

2.1. FORMATION OF THE 1,2 (OR 2,3) BOND

via an *o*-nitrosoanil is the almost quantitative photolytic conversion of *o*-nitrophenylaziridine derivatives into 2-phenylbenzimidazole *N*-oxides [153]. There are other similar examples [142].

Base-induced conversion of *N,N*-disubstituted-*o*-nitroanilines (59) into benzimidazole *N*-oxides may occur provided that one of the carbon substituents can be removed subsequently (e.g. cyano or arylsulfonyl). The reaction products, however, may well differ from those obtained from the monosubstituted analogues, e.g. cyclization of *N*-alkyl(or aryl)-*N*-cyanomethyl-*o*-nitroanilines gives not the 2-cyanobenzimidazole *N*-oxides, but the 2-"hydroxy" derivatives [148]. These processes do not seem to have great synthetic utility.

When *N,N*-dialkyl-*o*-nitroanilines are heated or photolysed in acidic media, good yields of benzimidazole *N*-oxides (60) can be obtained (Scheme 2.1.25). The thermal reactions usually require high temperatures (above 100°C) and extended heating times (often 12–48 h), but the reactions are versatile and simple experimentally. Refluxing in constant-boiling hydrochloric acid is usually successful, giving yields in the range 30–70% even with *N,N*-cycloalkyl derivatives [158–160]. Table 2.1.10 lists some examples.

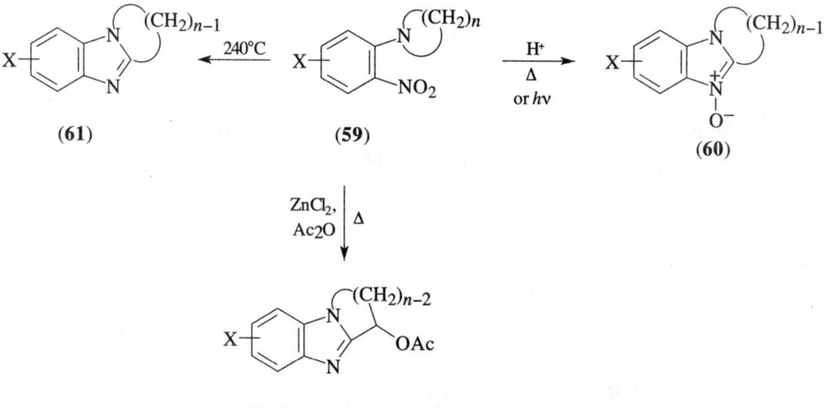

Scheme 2.1.25

Similar photolytic cyclizations which are usually carried out in methanolic hydrogen chloride solution can yield either benzimidazoles or the *N*-oxides, depending on the natures of the tertiary amino group and the ring substituents [161]. Under more vigorous reaction conditions (or in the presence of reducing agents), *N,N*-dialkyl(or aralkyl)-*o*-nitroanilines are converted into benzimidazoles (61) rather than the *N*-oxides (Scheme 2.1.25).

TABLE 2.1.10
Tricyclic benzimidazole 3-oxides (60) made by heating various (59) in acid[a,b]

X	n	Temperature (°C)	Time (h)	Yield (%)
H	5	110	20	51
5-Cl	5	110	20	70
5-NO_2	5	110	20	32
H	6	160	7	61

[a] From refs [159, 160].
[b] 1-Methylbenzimidazole 3-oxide is made similarly from o-nitro-N,N-dimethylaniline, 150°C, 12 h, 47% yield.

Reductive conditions have utilized iron(II) oxalate [162, 163], tin and hydrochloric acid (which gives only low yields) [160], and triethylphosphite [164, 165]. It is, however, much more convenient and efficient to merely heat the substrates. This can be done in sand at around 240°C, dispersed on glass beads, or in a high-boiling solvent such as polyphosphoric acid or diphenyl ether [62, 80–82, 160]. (See also similar thermal cyclizations of primary amino analogues in basic or reductive media, or thermally.) The reaction products are 1,2-disubstituted benzimidazoles (61) and, again, the N,N-dialkyl group can be a carbocyclic ring. Cyclizations are particularly smooth if the nitroarene contains an electron-withdrawing substituent *para* to the dialkylamino function. In the absence of such a substituent, or if there is an electron donor group present in the benzene ring, the reaction times have to be increased (1–4 h) and lower yields are to be expected. The reactions are believed to proceed through benzimidazole N-oxide intermediates. Some typical examples are listed in Table 2.1.11.

When o-nitroarylamines are treated with zinc chloride in boiling acetic anhydride, a variety of products can be formed. N,N-Dimethyl-o-nitroaniline forms 1-acetyl-3-methylbenzimidazolone in 48% yield, the N,N-diethyl

TABLE 2.1.11
Benzimidazoles (61) prepared by thermal cyclization[a] of (59)[b]

X	n	Time (h)	Yield (%)
H	5	4	62
5-CO_2H	5	0.5	88
5-Me	5	3	55
5-Cl	5	0.75	70
5-NO_2	5	0.5	82
5-NHAc	5	1	40
5-CF_3	5	0.5	65

[a] 240°C in sand.
[b] Ref. [166].

2.1. FORMATION OF THE 1,2 (OR 2,3) BOND

analogue gives 2-acetoxy-1-ethylbenzimidazole (65%), and N,N-cycloalkyl-o-nitroanilines (**59**) are similarly converted into α-acetoxy products (**62**) (68–87%) (Scheme 2.1.25) [167]. All of these reactions are Lewis acid catalysed, and presumably proceed through 3-acetoxy intermediates which can rearrange (see also refs [168, 169]).

When N,N-dimethyl-o-aminoaniline (**63**) reacts with acetic anhydride the primary amino group is initially acetylated (Scheme 2.1.26). Boiling the amide (**64**) with acetic anhydride induces cyclization to form 1,2-dimethylbenzimidazole. This reaction is quite general in that the N,N-dialkyl group can be part of a ring [160, 170]. Yields can range from 10 to 90%, with ring size affecting the reaction rates; steric factors can inhibit the cyclization. A related process is probably involved when sulfuryl chloride converts (**63**) at room temperature into 4,5,6,7-tetrachloro-1-methylbenzimidazole in 50% yield [171]. Acylated N,N-dialkyl-o-aminoanilines (**65**) can also be converted into benzimidazoles in hot polyphosphoric acid or under oxidative conditions (Scheme 2.1.26).

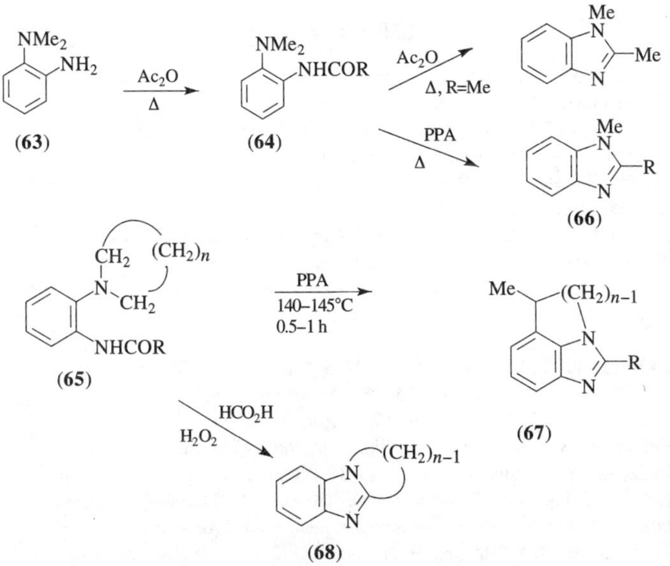

Scheme 2.1.26

With polyphosphoric acid the N,N-dimethyl substrates (**64**) (R = H, Me, Ph) are converted into 1,2-disubstituted benzimidazoles (**66**) (R = H (50%), Me (97%), Ph (53%)) (Scheme 2.1.26). The N,N-cycloalkyl analogues (**65**) are subject to a skeletal rearrangement, which generates the tricyclic products (**67**). These reactions are most successful with acetyl derivatives (**65**) (R = Me) and with piperidino compounds (**65**) ($n = 3$). Yields vary between 10 and 90%, with the top yield for (**65**) ($n = 3$, R = Me) [170].

A rather different cyclization process occurs when (**65**) is treated with trifluoroperacetic [172] or other peracids (Scheme 2.1.26) [173]. The acylamines appear to react most cleanly, with performic acid giving yields of (**68**) in the range 43–91% for a wide range of compounds. The *o*-substituted anilides (**65**) are made by treatment of *o*-chloronitrobenzene (1 mol) with the appropriate amine (2.1 mol) followed by reduction, and acylation with formic acid, acetic anhydride or benzoyl chloride [173].

Oxidative cyclization of acylated N,N-*diakyl*-o-*aminoanilines* (**65**) [173]

The acyl compounds (2 g) are heated with a mixture of 98% formic acid (12 ml) and 30% hydrogen peroxide (6 ml) at 100°C (10–15 min). The solution changes colour to yellow or brown during this period. Dilution with water and neutralization with concentrated ammonia solution precipitates the benzimidazole (**68**). Extraction of the filtrate with chloroform can increase the yield.

REFERENCES

1. O. Wallach and E. Schulze, *Ber. Dtsch. Chem. Ges.* **14**, 420 (1881).
2. M. R. Grimmett, *Adv. Heterocycl. Chem.* **12**, 103 (1970).
3. K. Hofmann, in *The Chemistry of Heterocyclic Compounds. Imidazole and its Derivatives, Part 1* (ed. A. Weissberger). Interscience, New York, 1953, p. 119.
4. E. S. Schipper and A. R. Day, in *Heterocyclic Compounds* (ed. R. C. Elderfield). Wiley, New York, 1957, Vol. 5, p. 194.
5. P. M. Kochergin, *J. Gen. Chem. USSR (Engl. Transl.)* **34**, 2758 (1964).
6. G. E. Trout and P. R. Levy, *Recl. Trav. Chim. Pays-Bas, Belg.* **84**, 125 (1965).
7. G. E. Trout and P. R. Levy, *Recl. Trav. Chim. Pays-Bas, Belg.* **85**, 765 (1966).
8. E. F. Godefroi, C. A. M. van der Eycken and P. A. J. Janssen, *J. Org. Chem.* **32**, 1259 (1967).
9. V. S. Korunskii, P. M. Kochergin and V. S. Shlikhunova, *Pharm. Chem. J. (Engl. Transl.)* **24**, 374 (1990); (*Chem. Abstr.* **113**, 132 086 (1990)).
10. P. Karrer and C. Gränacher, *Helv. Chim. Acta* **7**, 763 (1924).
11. C. Gränacher, V. Schelling and E. Schlatter, *Helv. Chim. Acta* **8**, 873 (1925).
12. T. Benincori, E. Brenna and F. Sannicolo, *J. Chem. Soc., Perkin Trans. 1* 675 (1993).
13. T. Kato, A. Takada and T. Ueda, *Chem. Pharm. Bull.* **22**, 984 (1974).
14. S. Furuya, K. Omura and Y. Furukawa, *Chem. Pharm. Bull.* **36**, 1669 (1988).
15. R. W. Begland, D. R. Hartter, F. N. Jones, D. J. Sam, W. A. Sheppard, O. W. Webster and F. J. Weigert, *J. Org. Chem.* **39**, 2341 (1974).
16. L. A. Reiter, *Tetrahedron Lett.* **26**, 3423 (1985).
17. L. A. Reiter, *J. Org. Chem.* **52**, 2714 (1987).
18. E. M. Beccalli, A. Marchesini and T. Pilati, *Synthesis* 127 (1991).
19. J. L. LaMattina, R. T. Suleske and R. L. Taylor, *J. Org. Chem.* **48**, 897 (1983).
20. R. K. Griffith and R. A. DiPietro, *Synthesis* 576 (1983).
21. I. Antonini, G. Cristalli, P. Franchetti, M. Grifantini and S. Martelli, *Synthesis* 47 (1983).
22. J. M. Kokosa, R. A. Szafasz and E. Tagupa, *J. Org. Chem.* **48**, 3605 (1983).
23. C. A. Lipinski, T. E. Blizniak and R. H. Craig, *J. Org. Chem.* **49**, 566 (1984).

REFERENCES

24. A. C. Veronese, G. Vecchiati, S. Sferra and P. Orlandini, *Synthesis* 300 (1985).
25. L. A. Reiter, *J. Org. Chem.* **49**, 3494 (1984).
26. B. Iddon, *Heterocycles* **23**, 417 (1985).
27. K. L. Kirk, *J. Heterocycl. Chem.* **22**, 57 (1985).
28. A. J. Carpenter and D. J. Chadwick, *Tetrahedron* **42**, 2351 (1986).
29. F. Benedetti, S. Bozzini, S. Fattuta, M. Forchiassin, G. Pitacco and C. Russo, *Gazz. Chim. Ital.* **121**, 401 (1991).
30. I. Lantos, W.-Y. Zhang, X. Shui and D. S. Eggleston, *J. Org. Chem.* **58**, 7092 (1993); W. R. Vaughan and D. R. Carlson, *J. Am Chem. Soc.* **84**, 769 (1962).
31. A. Lawson and H. V. Morley, *J. Chem. Soc.* 1695 (1955).
32. A. Lawson and H. V. Morley, *J. Chem. Soc.* 566 (1957).
33. W. Steglich and G. Höfle, *Chem. Ber.* **102**, 883 (1969).
34. W. Steglich and G. Höfle, *Angew. Chem. Int. Ed. Engl.* **8**, 981 (1969).
35. G. Höfle, W. Steglich and H. Vorbrüggen, *Angew. Chem. Int. Ed. Engl.* **17**, 569 (1978).
36. N. Engel and W. Steglich, *Liebigs Ann. Chem.* 1916 (1978).
37. J. Lepschy, G. Höfle, L. Wilschowitz and W. Steglich, *Liebigs Ann. Chem.* 1753 (1974).
38. Y. Mizuno and Y. Inoue, *J. Chem. Soc., Chem. Commun.* 124 (1978).
39. J. D. Bower and G. R. Ramage, *J. Chem. Soc.* 2834 (1955).
40. N. J. Cusack, G. J. Litchfield and G. Shaw, *J. Chem. Soc., Chem. Commun.* 799 (1967).
41. J. P. Ferris, S. S. Badesha, W. Y. Ren, H. C. Huang and R. J. Sorcek, *J. Chem. Soc., Chem. Commun.* 110 (1981).
42. J. P. Ferris, B. Devadas, C. Huang and W. Ren, *J. Org. Chem.* **50**, 747 (1985).
43. O. Moriya, H. Minamide and Y. Urata, *Synthesis* 1057 (1984).
44. W. K. Anderson, D. Bhattacharjee and D. M. Houston, *J. Med. Chem.* **32**, 119 (1989).
45. Y. Ohtsuka, *J. Org. Chem.* **41**, 629 (1976).
46. H. Bredereck and G. Schmotzer, *Liebigs Ann. Chem.* 95 (1956).
47. D. W. Woodward, US Patent 2,534,331 (1950); *Chem. Abstr.* **45**, 5191 (1951).
48. R. F. Shuman, W. E. Shearin and R. J. Tull, *J. Org. Chem.* **44**, 4532 (1979).
49. F. J. Weigert, US Patent 3,778,446 (1973); (*Chem. Abstr.* **80**, 59 941 (1974)).
50. M. H. Elnagdi, S. M. Sherif and R. M. Mohareb, *Heterocycles* **26**, 497 (1987).
51. Y. Ohtsuka, *ACS Congr. Abstr.* 393 (1979).
52. H. H. Lau and U. Schöllkopf, *Liebigs Ann. Chem.* 2093 (1982).
53. U. Schöllkopf, P. H. Porsch and H. H. Lau, *Liebigs Ann. Chem.* 1444 (1979).
54. J. P. Ferris and L. E. Orgel, *J. Am. Chem. Soc.* **88**, 1074 (1966).
55. B. Bigot and D. Roux, *J. Org. Chem.* **46**, 2872 (1981).
56. J. P. Ferris and J. E. Kuder, *J. Am. Chem. Soc.* **92**, 2527 (1970).
57. J. P. Ferris and R. W. Trimmer, *J. Org. Chem.* **41**, 19 (1976).
58. J. P. Ferris, R. S. Narang, T. R. Newton and V. R. Rao, *J. Org. Chem.* **44**, 1273 (1979).
59. M. P. Sammes and A. R. Katritzky, *Adv. Heterocycl. Chem.* **35**, 375 (1984).
60. D. Pooranchand and H. I. H. Junjappa, *Synthesis* 547 (1987).
61. D. Armesto, W. M. Horspool, M. Apoita, M. G. Gallego and A. Ramos, *J. Chem. Soc., Perkin Trans. 1*, 2035 (1989).
62. V. P. Semenov, A. N. Studenikov and A. A. Potekhin, *Chem. Heterocycl. Compd. USSR (Engl. Transl.)* **15**, 467 (1979).
63. M. Masui, K. Suda, M. Yamauchi and C. Yijima, *J. Chem. Soc., Perkin Trans. 1* 1955 (1972).
64. Y. Kamitori, M. Hojo, R. Masuda, T. Fujitani, S. Ohara and T. Yokoyama, *J. Org. Chem.* **53**, 129 (1988).
65. Y. Kamitori, M. Hojo, R. Masuda, T. Yoshida, S. Ohara, K. Yamada and N. Yoshikawa, *J. Org. Chem.* **53**, 519 (1988).

66. Y. Kamitori, M. Hojo, R. Masuda, T. Fujitani, S. Ohara and T. Yokoyama, *Synthesis* 208 (1988).
67. Y. Kamitori, M. Hojo, R. Masuda, S. Ohara, K. Kawasaki, Y. Kawamura and M. Tanaka, *J. Heterocycl. Chem.* **27**, 487 (1990).
68. Y. Kamitori, M. Hojo, R. Masuda, Y. Kawamura and X. Fang, *Heterocycles* **31**, 2103 (1990).
69. Y. Kamitori, M. Hojo, R. Masuda, K. Kawasaki, S. Takata and J. Ida, *Synthesis* 467 (1990).
70. K. Hantke, U. Schöllkopf and H.-H. Hausberg, *Liebigs Ann. Chem.* 1531 (1975).
71. U. Schöllkopf and K. Hantke, *Liebigs Ann. Chem.* 1602 (1979).
72. R. S. Hosmane, *Tetrahedron Lett.* **25**, 363 (1984).
73. S. Kulkarni and M. R. Grimmett, *Aust. J. Chem.* **40**, 1415 (1987).
74. A. Spaltenstein, T. P. Holler and P. B. Hopkins, *J. Org. Chem.* **52**, 2977 (1987).
75. A. Padwa, *Acc. Chem. Res.* **9**, 371 (1976).
76. A. Padwa, J. Smolanoff and A. Tremper, *Tetrahedron Lett.* 29 (1974).
77. J. B. Wright, *Chem. Rev.* **48**, 397 (1951).
78. K. Hofmann, in *The Chemistry of Heterocyclic Compounds. Imidazole and its Derivatives, Part 1* (ed. A. Weissberger). Interscience, New York, 1953, p. 258.
79. A. F. Pozharskii, A. D. Garnovskii and A. M. Simonov, *Russ. Chem. Rev. (Engl. Transl.)* **35**, 122 (1966).
80. P. N. Preston and G. Tennant, *Chem. Rev.* **72**, 627 (1972).
81. P. N. Preston, *Chem. Rev.* **74**, 279 (1974).
82. P. N. Preston, *Chemistry of Heterocyclic Compounds. Benzimidazoles and Congeneric Tricyclic Compounds.* Interscience-Wiley, New York, 1981.
83. R. Rastogi and S. Sharma, *Synthesis* 861 (1983).
84. C. H. Roeder and A. R. Day, *J. Org. Chem.* **6**, 25 (1941).
85. V. V. Korshak, A. L. Rusanov, D. S. Tugushi and S. N. Leont'eva, *Chem. Heterocycl. Compd. USSR (Engl. Transl.)* **9**, 232 (1973).
86. R. A. Abramovitch and K. Schofield, *J. Chem. Soc.* 2326 (1955).
87. V. K. Shchel'tsyn, A. Ya. Kaminski, T. P. Shapirovskaya, I. L. Vaisman, V. F. Andrianov and S. S. Gitis, *Chem. Heterocycl. Compd. USSR (Engl. Transl.)* **9**, 103 (1973).
88. S. von Niementowski, *Chem. Ber.* **43**, 3012 (1910).
89. I. Ganea and R. Taranu, *Stud. Univ. Babes-Bolyai, Ser. Chem.* 95 (1966); (*Chem. Abstr.* **67**, 32 648 (1967)).
90. M. Finotto, Ger. Patent 2,062,265 (1972); (*Chem. Abstr.* **77**, 61 997 (1972)).
91. R. S. Atkinson, J. Fawcett, D. R. Russell and G. Tughan, *J. Chem. Soc., Chem. Commun.* 832 (1986).
92. S. Takahashi and H. Kano, *Chem. Pharm. Bull.* **11**, 1375 (1963).
93. S. M. Hussain, A. M. El-Reedy and S. A. El-Sherabasy, *J. Heterocycl. Chem.* **25**, 9 (1988).
94. Australian Patent 257,959 (1965); (*Chem. Abstr.* **68**, 12 971 (1968)).
95. J. W. Schulenberg and S. Archer, *J. Org. Chem.* **30**, 1279 (1965).
96. Shionogi and Co. Ltd., French Patent 1,555,336 (1969); (*Chem. Abstr.* **72**, 43 679 (1970)).
97. I. W. Harvey, M. D. McFarlane, D. J. Moody and D. M. Smith, *J. Chem. Soc., Perkin Trans. 1* 681 (1988).
98. M. A. Phillips, *J. Chem. Soc.* 2820 (1929).
99. V. Gomez-Parra, F. Sanchez and T. Torres, *Monatsh. Chem.* **116**, 639 (1985).
100. C. H. Senanayake, L. E. Fredenburgh, R. A. Reamer, J. Liu, R. D. Larsen, T. R. Verhoeven and P. J. Reider, *Tetrahedron Lett.* **35**, 5775 (1994).
101. G. Stefancich, M. Artico, F. Corelli and S. Massa, *Synthesis* 757 (1983).
102. R. Cerri, A. Boido and F. Sparatore, *J. Heterocycl. Chem.* **16**, 1005 (1979).

103. D. W. S. Latham, O. Meth-Cohn and H. Suschitzky, *J. Chem. Soc., Chem. Commun.* 41 (1973).
104. M. Z. Girshovich and A. V. El'tsov, *J. Org. Chem. USSR (Engl. Transl.)* **10**, 542 (1974).
105. B. Rigo, D. Valligny and S. Taisne, *Synth. Commun.* **18**, 167 (1988).
106. W. J. Schnabel and E. Kober, *J. Org. Chem.* **34**, 1162 (1969).
107. D. Griffiths, R. Hull and T. P. Seden, *J. Chem. Soc., Perkin Trans. 1* 2608 (1980).
108. R. K. Smalley and T. E. Bingham, *J. Chem. Soc. (C)* 2481 (1969).
109. R. K. Smalley and A. W Stocker, *Tetrahedron Lett.* **25**, 1389 (1984).
110. R. K. Smalley and A. W. Stocker, *Chem. Ind. (London)* 222 (1984).
111. M. Sprechter and D. Levy, *Tetrahedron Lett.* 4957 (1969).
112. J. I. G. Cadogan, R. Marshall, D. M. Smith and M. J. Todd, *J. Chem. Soc. (C)* 2441 (1970).
113. R. Marshall and D. M. Smith, *J. Chem. Soc. (C)* 3510 (1971).
114. R. N. Leyden, M. S. Loonat, E. W. Neuse, B. H. Sher and W. J. Watkinson, *J. Org. Chem.* **48**, 727 (1983).
115. R. G. Srivastava and P. S. Venkataramani, *Synth. Commun.* **18**, 1537 (1988).
116. J. L. Hughey, S. Knapp and H. Schugar, *Synthesis* 489 (1980).
117. F. M. Abdelrazek, Z. E. Kandeel, K. M. H. Hilmy and M. H. Elnagdi, *Chem. Ind. (London)* 439 (1983).
118. L. O. Krbeckek and H. Takimoto, *J. Org. Chem.* **29**, 3630 (1964).
119. J. H. Hall and D. R. Kamm, *J. Org. Chem.* **30**, 2092 (1965).
120. J. Schulze, H. Tanneberg and H. Matschiner, *Z. Chem.* **20**, 436 (1980); (*Chem. Abstr.* **94**, 156 822 (1981)).
121. A. V. El'tsov, V. S. Kuznetsov and M. B. Kolesova, *J. Org. Chem. USSR (Engl. Trans.)* **1**, 1126 (1965).
122. A. M. M. E. Omar, *Synthesis* 41 (1974).
123. A. M. M. E. Omar, M. S. Ragab, A. M. Farghaly and A. M. Barghash, *Pharmazie* **31**, 348 (1976).
124. G. Depost, R. Salle and B. Sillion, *C. R. Acad. Sci. Ser. C* **275**, 697 (1972).
125. R. J. Stedman, US Patent 3,455,948 (1969); *Chem. Abstr.* **71**, 81 369 (1969).
126. A. W. Chow, US Patent 3,468,888 (1970); *Chem. Abstr.* **72**, 3489 (1970).
127. F. Janssens, M. Luyckx, R. Stokbroekx and J. Torremans, US Patent 4,219,559 (1980); (*Chem. Abstr.* **94**, 30 579 (1981)).
128. D. Korbonits and P. Kolonits, *J. Chem. Res. (S)* 209 (1988).
129. D. Korbonits and K. Horvath, *Heterocycles* **37**, 2051 (1994).
130. J. C. Watts, German Patent 2,204,479 (1973); (*Chem. Abstr.* **79**, 115 592 (1973)).
131. E. I. Du Pont de Nemours and Co., French Patent 2,170,981 (1973); (*Chem. Abstr.* **80**, 70 806 (1974)).
132. H. Schilling, K. Trautner, P. Gallien and H. Matschiner, German Patent 149,520 (1981); (*Chem. Abstr.* **96**, 142 855 (1982)).
133. J. C. Watts, British Patent 1,351,883 (1974); (*Chem. Abstr.* **81**, 105 512 (1974)).
134. G. W. Stacy, T. E. Wollner and T. R. Oakes, *J. Heterocycl. Chem.* **3**, 51 (1966).
135. J. M Harvey, M. D. McFarlane, D. J. Moody and D. M. Smith, *J. Chem. Soc., Perkin Trans. 1* 1939 (1988).
136. H. D. McFarlane, D. J. Moody and D. M. Smith, *J. Chem. Soc., Perkin Trans. 1* 691 (1988).
137. A. E. Luetzow and J. R. Vercellotti, *J. Chem. Soc. (C)* 1750 (1967).
138. I. N. Popov and O. V. Kryshtalyuk, *Chem. Heterocycl. Compd. (Engl. Transl.)* **27**, 802 (1991).
139. R. Fielden, O. Meth-Cohn, D. Price and H. Suschitzky, *J. Chem. Soc., Chem. Commun.* 772 (1969).

140. S. Takahashi and H. Kano, *Chem. Pharm. Bull.* **11**, 1375 (1963).
141. S. Takahashi and H. Kano, *Chem. Pharm. Bull.* **14**, 1219 (1966).
142. R. J. Pollitt, *J. Chem. Soc., Chem. Commun.* 262 (1965).
143. D. J. Neadle and R. J. Pollitt, *J. Chem. Soc., Chem. Commun.* 1764 (1967).
144. G. V. Nikitina and M. S. Pevzner, *Chem. Heterocycl. Compd. (Engl. Transl.)* **29**, 127 (1993).
145. S. von Niementowski, *Ber. Dtsch. Chem. Ges.* **43**, 3012 (1910).
146. Merck and Co., Netherlands Patent 6,517,256 (1966); (*Chem. Abstr.* **66**, 2568 (1967)).
147. P. L. Puigdellivol and E. Goday Baylina, Spanish Patent 2,006,167 (1989); (*Chem. Abstr.* **113**, 59 183 (1990)).
148. D. B. Livingstone and G. Tennant, *J. Chem. Soc., Chem. Commun.* 96 (1973).
149. J. D. Loudon and G. Tennant, *J. Chem. Soc.* 4268 (1963).
150. J. Machin, R. K. Mackie, H. McNab, G. A. Reed, A. J. G. Sagar and D. M. Smith, *J. Chem. Soc., Perkin Trans. 1* 394 (1976).
151. G. W. Stacy, B. V. Ettling and A. J. Papa, *J. Org. Chem.* **29**, 1537 (1964).
152. G. De Stevens, A. B. Brown, D. Rose, H. I. Chernov and A. J. Plummer, *J. Med. Chem.* **10**, 211 (1967).
153. H. W. Heine, G. J. Blosick and G. B. Lowrie, *Tetrahedron Lett.* 4801 (1968).
154. G. Gal and M. Sletzinger, US Patent 3,265,706 (1966); *Chem. Abstr.* **65**, 13 724 (1966).
155. L. A. Ljublinskaya and V. M. Stepanov, *Tetrahedron Lett.* 4511 (1971).
156. R. S. Goudie and P. N. Preston, *J. Chem. Soc. (C)* 1139 (1971).
157. D. J. Neadle and R. J. Pollitt, *J. Chem. Soc. (C)* 2127 (1969).
158. R. Fielden, O. Meth-Cohn and H. Suschitzky, *J. Chem. Soc., Perkin Trans. 1* 696 (1973).
159. R. Fielden, O. Meth-Cohn, D. Price and H. Suschitzky, *J. Chem. Soc., Chem. Commun.* 772 (1969).
160. O. Meth-Cohn and H. Suschitzky, *Adv. Heterocycl. Chem.* **14**, 211 (1972).
161. R. Fielden, O. Meth-Cohn and H. Suschitzky, *Tetrahedron Lett.* 1229 (1970).
162. R. H. Smith and H. Suschitzky, *Tetrahedron* **16**, 80 (1961).
163. R. A. Abramovitch, B. A. Davis and R. A. Brown, *J. Chem. Soc. (C)* 1146 (1969).
164. H. Suschitzky and M. E. Sutton, *J. Chem. Soc. (C)* 3058 (1968).
165. R. Garner, G. V. Garner and H. Suschitzky, *J. Chem. Soc. (C)* 825 (1970).
166. H. Suschitzky and M. E. Sutton, *Tetrahedron Lett.* 3933 (1967).
167. R. K. Grantham and O. Meth-Cohn, *J. Chem. Soc. (C)* 70 (1969).
168. S. Takahashi and H. Kano, *Chem. Pharm. Bull.* **12**, 783 (1964).
169. S. Takahashi and H. Kano, *Chem. Pharm. Bull.* **14**, 1219 (1966).
170. O. Meth-Cohn and H. Suschitzky, *J. Chem. Soc.* 2609 (1964).
171. J. Martin, O. Meth-Cohn and H. Suschitzky, *Tetrahedron Lett.* 4495 (1973).
172. M. D. Nair and R. Adams, *J. Am. Chem. Soc.* **83**, 3518 (1961).
173. O. Meth-Cohn and H. Suschitzky, *J. Chem. Soc.* 4666 (1963).

2.2 FORMATION OF THE 1,5 (OR 3,4) BOND

Reactions of this type are almost entirely confined to cyclizations of suitably substituted amidines, ureas or thioureas. Amidines give 2-substituted

2.2. FORMATION OF THE 1,5 (OR 3,4) BOND

imidazole or benzimidazole products, but ureas and thioureas lead only to products with oxygen or sulfur groups at C-2. The methods are not especially popular because of the problems associated with synthesis of a range of starting materials. Nevertheless, amidines are usually made by treating imidates (imidic esters, imino ethers) with ethanolic ammonia or with amines [1]. The imitates are prepared by passing dry hydrogen chloride gas into a solution of a nitrile in anhydrous alcohol (Scheme 2.2.1). Ureas are usually made from reaction of potassium isocyanate, phosgene or an alkyl isocyanate with a primary amine, but have more limited applicability because they do not allow variation of the imidazole 2-substituent.

$$R-C\equiv N + R^1OH \xrightarrow{HCl} RC(OR^1)=NH \cdot HCl \xrightarrow{R^2NH_2} RC(NHR^2)=NH$$

Scheme 2.2.1

2.2.1 Imidazoles

The reaction between a suitable imidate and an α-aminoaldehyde or α-aminoacetal to form an amidine, which cyclizes to an imidazole, rests largely on the availability of the aminoaldehydes from α-amino acids, which are readily reduced using the Akabori method [2]. Dimethyl or diethyl acetals frequently replace the aldehydes in these reactions [3, 4]. Table 2.2.1 lists some 2,5-disubstituted imidazoles prepared (ultimately) from amino acids. It is not possible to introduce a range of substituents at both the 4- and 5-positions by this method unless the amino acid is converted into a ketone rather than an aldehyde (see Section 2.1.1) (Scheme 2.2.2).

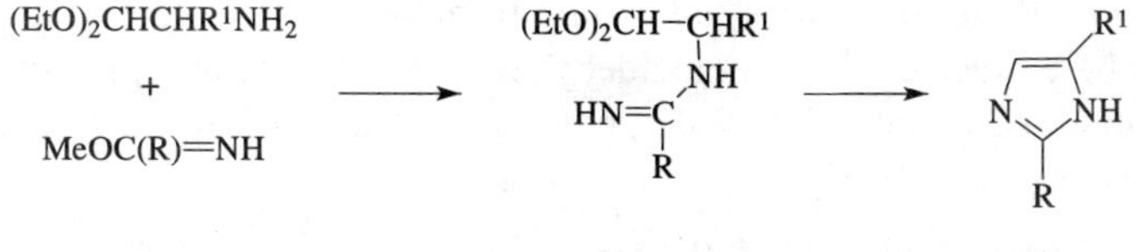

$$(EtO)_2CHCHR^1NH_2 + MeOC(R)=NH \longrightarrow (EtO)_2CH-CHR^1-NH-C(R)=NH \longrightarrow \underset{R}{\overset{R^1}{\text{imidazole}}}$$

(1)

Scheme 2.2.2

General method [5]

The amino acid (0.05 mol) is heated in ethanol (100 ml) saturated with dry HCl. After removal of the solvent under reduced pressure, the residue of the amino acid ester hydrochloride is dissolved in ice-cold water (100 ml), and 2.3% sodium amalgam (200 g) is added slowly with stirring, the pH being

TABLE 2.2.1
Imidazoles (1) prepared from amidines derived from α-amino acids[a]

Amino acid	R	R^1	Yield (%)
Glycine	Ph	H	18
Alanine	Ph	Me	46
Alanine	C$_5$H$_{11}$	Me	47
Aminomalonic	Ph	CO$_2$Et	—
α-Aminobutyric	Ph	Et	53
Aspartic	Ph	CH$_2$CO$_2$Et	37
Glutamic	Ph	(CH$_2$)$_2$CO$_2$Et	46
Serine	Ph	CH$_2$OH	10
α-Phenylglycine	Ph	Ph	31
Phenylalanine	Ph	CH$_2$Ph	33
Tyrosine	Ph	CH$_2$C$_6$H$_4$–p-OH	34
α-p-OH-phenylglycine	Ph	C$_6$H$_4$–p-OH	42

[a]Ref. [5].
[b]Based on amount of amino acid.

maintained around 4 by addition of 5 M hydrochloric acid, and the temperature between 0 and −5°C by addition of powdered dry ice. The resulting solution, after filtration, is evaporated under reduced pressure, and the residue is taken up in ethanol and filtered. The solution of α-aminoaldehyde is mixed with the imidate (0.05 mol), and the pH is adjusted (if necessary) to 7 by the addition of acetic acid. After 1 h at room temperature the mixture is heated at 100°C (3–4 h), effectively distilling off most of the ethanol.

Trituration of the residue with dilute hydrochloric acid, and cooling to 0°C, gives the imidazole hydrochloride salts, which are frequently crystalline. The free bases can be obtained by making the salts alkaline with aqueous sodium hydroxide followed by extraction with ether. The ether is removed to give the products (1). Some free bases precipitate from aqueous alkaline solution and need not be extracted. The aminoaldehydes from aspartic and glutamic acids are best distilled at around 1 mmHg pressure to purify them.

Preparation from α-aminoacetals [3, 4]

The acetal (0.01 mol) is heated with the imidate (0.01 mol) in glacial acetic acid (0.02 mol) at 100°C (3 h). Aqueous sodium hydroxide is added to form a dark-coloured basic solution which is extracted with diethyl ether, and the extracts are dried and rotary evaporated to give a residue (an oil in the case of α-benzamidinoethanal diethyl acetal). Heating with 5 M hydrochloric acid (4 ml) for 0.5 h, addition of water, and evaporation under reduced pressure again gives a residue which, after being dissolved in water, is extracted with ether to remove any amides (e.g. benzamide in the case of benzimidate).

2.2. FORMATION OF THE 1,5 (OR 3,4) BOND

Basification with aqueous alkali and re-extraction with ether, drying, and removal of the solvent gives the imidazole. Recrystallization may be necessary. Prepared in this way have been 2-pentyl- (75%), 2-phenyl- (85%), 1-methyl-2-pentyl- (33%) and 1-methyl-2-phenyl- (42%) imidazoles.

The general approach of amidine cyclization has been applied to the synthesis of a variety of 2-substituted imidazoles. Aminoacetaldehyde dimethyl and diethyl acetals are readily available commercially, and the N-substituted derivatives can be made with little difficulty, providing access to 1-substituted imidazoles on reaction with a suitable imidate. Thus, methyl β-hydroxypropanimidate (2), prepared from 3-hydroxypropanenitrile, and methanolic HCl, condenses with an aminoacetaldehyde acetal to give the amidine hydrochloride (3), which ring closes when heated in acidic medium to form the 1-substituted 2-hydroxyethylimidazole (4) (Scheme 2.2.3) [6]. The reaction has been adapted to the preparation of 2-arylimidazoles [5, 7–11],

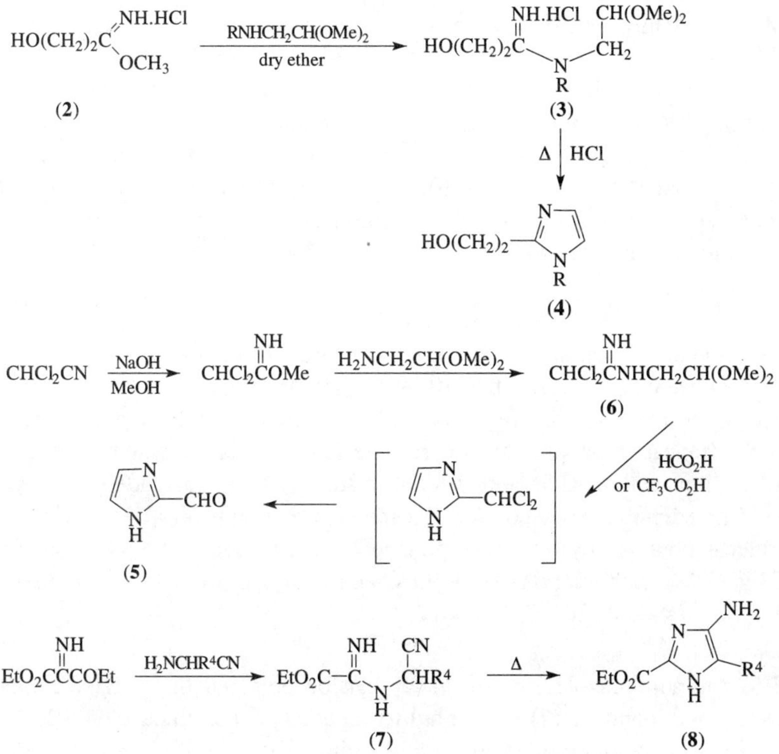

Scheme 2.2.3

2-alkyl- or 2-substituted alkylimidazoles [12-15], imidazole 2-carbaldehyde (5) [16] and 5-aminoimidazole-2-carboxylates (8) [17]. Concentrated sulfuric acid induces cyclization of alkyl or aryl amidines, but if the alkyl group is large, dilute acid suffices [14, 15].

The aldehyde (5) can be made in high yield from the amidine (6) via the unisolated 2-dichloromethylimidazole (Scheme 2.2.3). Dichloroacetonitrile is converted into its imidate with methanolic sodium methoxide, and then into the amidine with aminoacetaldehyde dimethylacetal. When (6) is heated in formic acid it is converted almost quantitatively into the 2-carbaldehyde; the use of trifluoroacetic acid at reflux gives around 60% of (5). Using essentially the same method, imidazole-2-carboxylic acid and its ethyl ester can also be made very efficiently when trichloroacetonitrile and the acetal are used to prepare the amidine [16].

N-(2,2-Dimethoxyethyl)dichloroacetamidine (6) [16]

A 1 M sodium methoxide solution (20 ml, 20 mmol) is added dropwise to a stirred solution of dichloroacetonitrile (13.7 g, 124.5 mmol) in anhydrous methanol (20 ml) cooled in a dry ice-acetone bath at −78°C. After 1.5 h at −78°C and 1 h at room temperature the solvent is removed under reduced pressure and at ambient temperature to give the crude imidate (15 g) as an oil, ν_{max} (CHCl$_3$) 3330 and 1670 cm^{-1}. A mixture of the imidate and aminoacetaldehyde dimethyl acetal (11.1 g, 105.7 mmol) is heated in an oil bath at 80°C (internal temperature; 1.45 h). The crude product is taken up in ethyl acetate and purified by column chromatography on Florisil using ethyl acetate to elute the crystalline product (6) (20.1 g, 88%), m.p. 85-89°C.

Imidazole-2-carbaldehyde (5) [16]

A solution of the amidine (6) (5.0 g, 23.2 mmol) in 95-97% formic acid (10 ml) is heated in an oil bath at 70-80°C (20 h). After removal of the solvent *in vacuo*, benzene is added to the residue, and the mixture is evaporated to dryness (repeat three times). The residue is dissolved in water (9 ml), and solid sodium carbonate is added to raise the pH to 8. The aldehyde precipitates immediately. After cooling the mixture overnight in a refrigerator the product is obtained by filtration and dried in a vacuum dessicator to give (5) (2.21 g, 99%). Sublimation at 80-90°C and 2 mmHg pressure gives the pure aldehyde, m.p. 204-205°C.

4(5)-Aminoimidazole-2-carboxylates can be prepared by initial formation of the α-amidinonitrile (7) from ethyl formimidate and aminoacetonitrile. The amidinonitrile can then be cyclized by heating. The overall process, however, also occurs merely by heating the imidate (or thioimidate) and α-aminonitrile

2.2. FORMATION OF THE 1,5 (OR 3,4) BOND

together and, as such, might be classified as a synthesis involving the formation of both 1,5 and 2,3 bonds. The products are ethyl 4(5)-aminoimidazole-2-carboxylates (**8**) [17]. Similar reactions cyclize other amidinonitriles in hot acetic anhydride in the presence of 1 eq. of sodium acetate (which gives acetylaminoimidazoles), or, when heated in pyridine, the products are 4(5)-amidinoimidazoles. 4-Aminoimidazoles are never noted for their stability, and they are usually isolated as their amides through reaction with acetic anhydride or ethyl chloroformate [7]. An improved synthesis of 5-amino-1-hydroxyimidazole-4-carboxamide cyclizes the amidine formed when N'-benzyloxy-N,N'-dimethylformamidine is treated at room temperature with 2-amino-2-cyanoacetamide. The amidine, formed in 63% yield, is ring closed, again in 63% yield, by treatment at 60°C with boron trifluoride etherate in DMF. The initial benzyl ether product can be catalytically reduced in almost quantitative yield to the 1-hydroxy compound [18].

Guanidines are also subject to acid-catalysed cyclization, which converts them into 2-aminoimidazoles. Cyanamide converts appropriate aminoacetaldehyde dialkyl acetals (**9**) into guanidines (**10**), which cyclize in hot aqueous acid (Scheme 2.2.4) [3, 19, 20]. Similarly, S-methylthioureas are converted via guanidines into mixtures of 2-arylamino- and 1-aryl-2-aminoimidazoles [19].

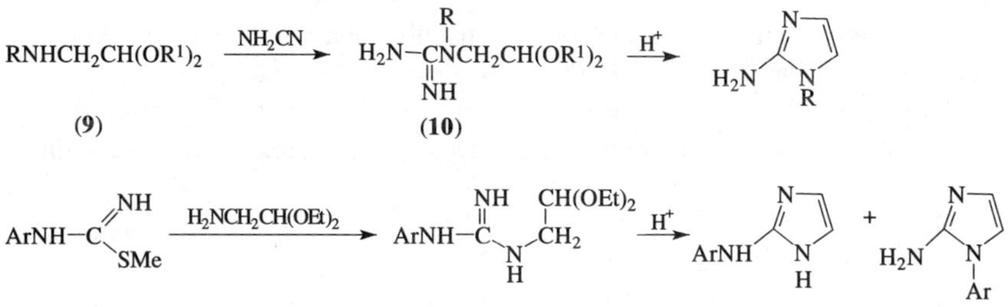

Scheme 2.2.4

3-Amino-1,2,4-oxadiazoles (**11**) are excellent sources of amidines. When they are treated with 1,3-dicarbonyl derivatives they are converted into enaminones (**12**), which cyclize to imidazoles in the presence of bases (Scheme 2.2.5). This reaction utilizes the well-known general attack of a nucleophilic group in the side chain at N-2 of the oxadiazole ring. Yields of 2-acylamino-4-acylimidazoles usually lie between 60 and 80% [21]. The free amino derivatives are readily isolated after acid hydrolysis.

General method for cyclization of an enaminone (**12**) [21]

The enaminone (**12**) (0.01 mol) and sodium ethoxide (0.01 mol) are mixed in dry DMF (50 ml) and heated at 110°C (2.5–3 h). After removal of the solvent

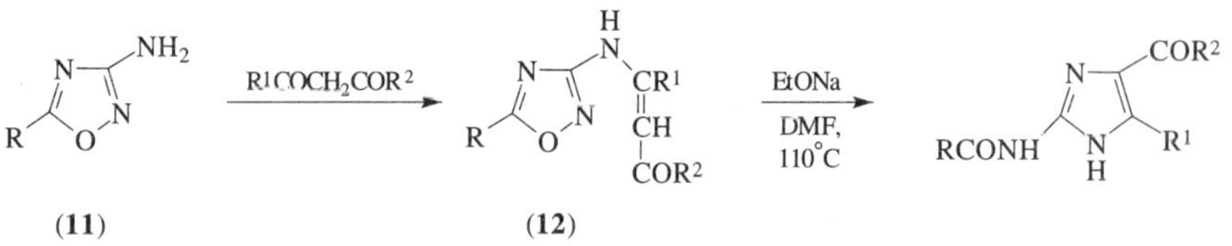

Scheme 2.2.5

under reduced pressure, the residue is dissolved in a minimum amount of water containing a few millilitres of 10% sodium hydroxide. Neutralization of the solution with acetic acid precipitates the product, which is recrystallized from solvents such as ethanol, petroleum ether, benzene, and ethanol–water (1:1). Substituent groups R, R^1, R^2 can by alkyl or aryl, while R^2 can also be ethoxy.

Benzamidine combines with 2-amino-3-phenacyl-1,3,4-oxadiazolium bromides to give 1-acylamino-2-benzimidoylamino-4-arylimidazoles. Yields are only moderate (14–43%), but the reaction works for a variety of 4-arylimidazoles [22]. Reactions of N-methyl-N-(N'-phenylbenzimidoyl)aminoacetonitrile (**13**) under acidic conditions lead to imidazolium salts which have amino (**14**) or amido (**15**) groups in the 4-position (Scheme 2.2.6). The 4-amino salt (**14**) undergoes Dimroth rearrangement to the 4-phenylaminoimidazole (**16**); direct conversion of (**13**) into (**16**) also occurs in warm alkali [8]. A Claisen rearrangement of the adduct (**17**), which forms from interaction of an arylamidoxime and a propiolate ester, provides a method

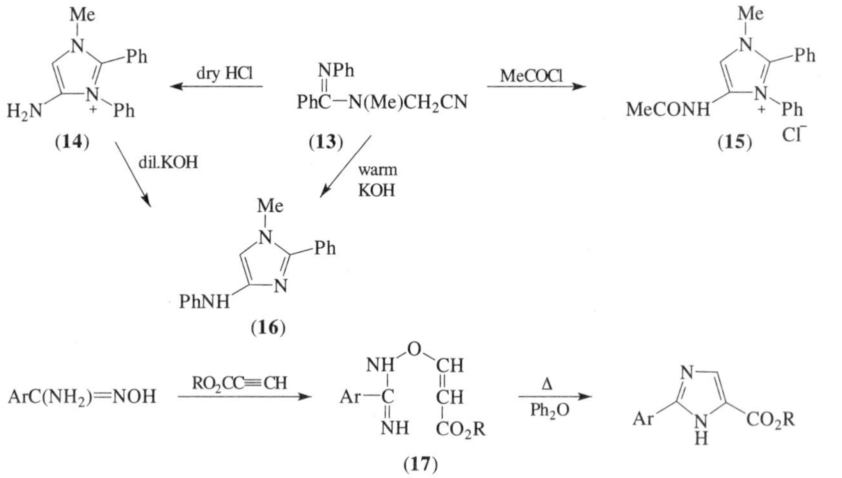

Scheme 2.2.6

2.2. FORMATION OF THE 1,5 (OR 3,4) BOND

TABLE 2.2.2
Imidazoles made by cyclization of amidines or guanidines

R^1	R^2	R^4	R^5	Yield (%)	Ref.
H	$(CH_2)_2OH$	H	H	40	12
Me	$(CH_2)_2OH$	H	H	65	12
H	$ArOCH_2$	H	H	5–87	13
H	Ph	NHAc	H	49	7
Ac	Ph	NHAc	H	15	7
H	Ph	$NHCO_2Et$	H	39	7
H	Ph	NHAc	Me	70	7
H	CO_2Et	H	NH_2	91	17
H	CO_2Et	CO_2Et	NH_2	92	17
H	CO_2Et	$CONH_2$	NH_2	92	17
H	CHO	H	H	99	16
H	CD_3	H	H	66	15
OH	H	$CONH_2$	NH_2	63[a]	18
$2,6-Me_2C_6H_3$	Ph	CF_3	F	58	9
$2,4,6-Me_3C_6H_2$	Ph	CF_3	F	63	9
$2,6-Me_2C_6H_3$	$2-ClC_6H_4$	CF_3	F	65	10
H	CO_2Et	H	H	78	16
H	CO_2H	H	H	94	16
$(CH_2)_2NMe_2$	NH_2	H	H	—	3
H	NMe_2	H	H	83	20
H	Ph	CO_2Me	H	72	23
H	Ph	CO_2Et	H	70	23
H	$4-ClC_6H_4$	CO_2Me	H	61	23

[a] O-benzyl derivative.

of making 2-arylimidazole-4-carboxylates in good yields (Scheme 2.2.6) [23, 24]. Table 2.2.2 lists imidazoles made by ring closure of amidines or guanidines.

An electron-deficient carbon bearing a trifluoromethyl group becomes part of an imidazole ring when Schiff bases (**18**) made from amidines and hexafluoroacetone are reduced by tin(II) chloride (Scheme 2.2.7). The 5-fluoro-4-trifluoromethylimidazoles (**19**) are isolated in 58–65% yields [9–11]. The fluoro substituents in the 5-position are readily displaced by nucleophiles such as alkoxy or cyanide, thereby extending the versatility of these syntheses [9].

When 4H-1,3-benzothiazine-4-thiones react with propargylamine, imidazoles are formed in 53–72% yields via amidine intermediates. This approach is, however, a little exotic to compete with the more common methods [25].

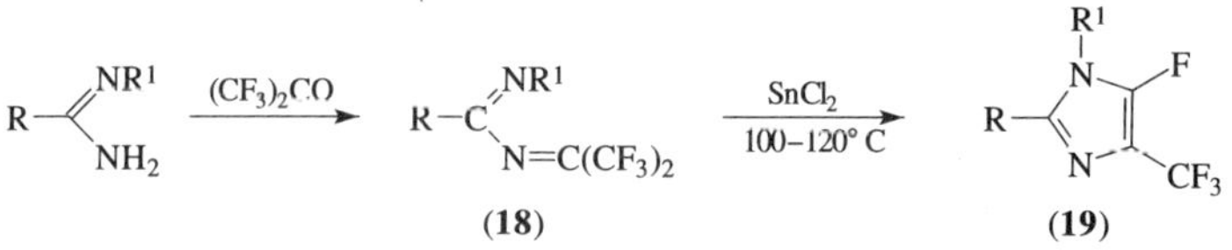

48   2. RING SYNTHESIS BY FORMATION OF ONE BOND

$$R-C{\overset{NR^1}{\underset{NH_2}{\diagdown}}} \xrightarrow{(CF_3)_2CO} R-C{\overset{NR^1}{\underset{N=C(CF_3)_2}{\diagdown}}} \xrightarrow[100-120°C]{SnCl_2} \underset{(19)}{\text{imidazole}}$$

(18)

Scheme 2.2.7

The great versatility of DAMN in imidazole synthesis has already been discussed (see Section 2.1.1). It also has applications under the present heading because the reagent can be converted into amidinium salts (**20**) by nitrilium triflates. In the presence of varying strength bases and temperatures (**20**) are transformed into 5-aminoimidazoles, substituted in the 4-position by nitrile (**21**), cyanoformimidoyl (**22**) or carboxamido (**23**) (Scheme 2.2.8). The amidinium salts (**20**) are isolated as mixtures of *cis* and *trans* isomers. The yields quoted for (**21**)-(**23**) have not been optimized, and their low recoveries reflect poor solubility in the usual solvents used for extraction [26, 27]. Since nitrilium salts with $R^1 = H$ are unstable, it is not possible to prepare similar imidazoles which have no 2-substituent by this method.

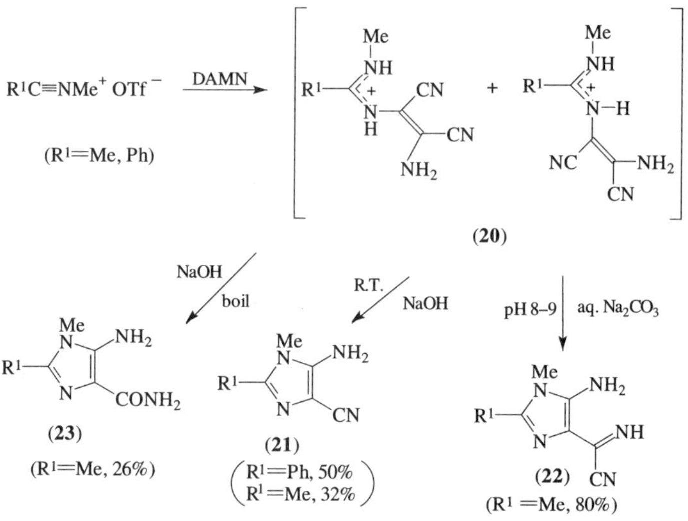

Scheme 2.2.8

Reaction of DAMN with formamidine gives the formamidine (**24**), which can either lose ammonia to form imidazole-4,5-dinitrile (**25**), or isomerize and then cyclize with elimination of HCN to form 4-aminoimidazole-5-nitrile (**26**) (Scheme 2.2.9) [28, 29]. The latter compound is of value for its

2.2. FORMATION OF THE 1,5 (OR 3,4) BOND

conversion in the presence of excess formamidine into adenine; with urea it gives guanine [30]. Formamidine acetate in refluxing ethanol, however, only gives a 2% yield of the formamidine (**24**) [30], and it is preferable to prepare it from the imidate (**27**) which is formed in high yield (80–85%) when DAMN is refluxed in dry dioxane with triethyl orthoformate (Scheme 2.2.9) [27]. At low temperatures in the presence of aniline hydrochloride as a catalyst the imidate reacts with ammonia to give the formamidine (**24**) in 95% yield. Aqueous ammonia converts (**24**) into (**26**); acetic acid leads to the dinitrile (**25**); barium hydroxide is best for formation of 4-cyanoformimidoyl-5-aminoimidazole (**28**). The base chosen to produce (**28**) is quite critical. If it is too weak the reaction will be too slow, and the product is likely to decompose. Saturated aqueous sodium carbonate or 1 M sodium hydroxide cause rapid cyclization, but base subsequently attacks the product to eliminate HCN and form (**26**). It is best to treat an ethanolic solution of (**24**) with solid barium hydroxide, which gives a 72% yield of (**28**) [27, 29].

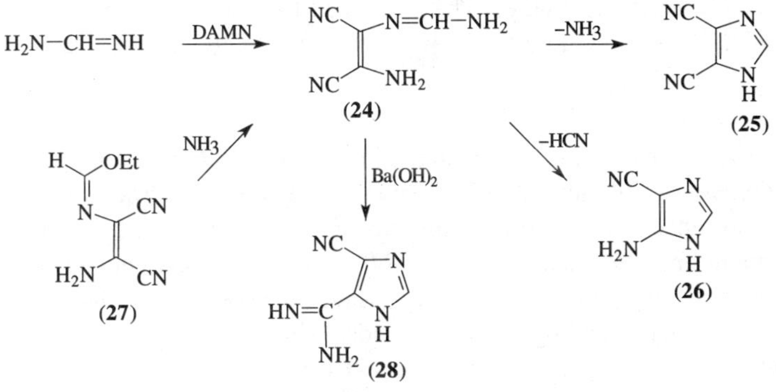

Scheme 2.2.9

Ethyl (Z)-N-(2-amino-1,2-dicyanovinyl) formimidate (**27**) [27]

A mixture of DAMN (2.0 g, 18.5 mmol) and triethyl orthoformate (2.74 g, 18.5 mmol) in dioxane (31.5 ml) is heated in a flask fitted with a short Vigreux column, distillation head, condenser and receiver. Ethanol mixed with dioxane is removed continuously until the temperature at the distillation head reaches 90°C (~17 ml, ~10 min). The clear brown solution remaining is allowed to cool overnight before the addition of hexane (16 ml), which precipitates dark brown crystals. Complete precipitation is ensured by the addition of further hexane. The filtered solid is dissolved in the minimum volume of hot diethyl ether, filtered to remove a dark brown solid impurity, then cooled to give (**27**) as white needles (2.55 g, 84%), m.p. 132.5°C (dec.).

(Z)-N-(2-Amino-1,2-dicyanovinyl) formamidine (24)

Ammonia is bubbled (30 min) through a cold suspension of (27) (4.0 g, 24.36 mmol) in dry chloroform (90 ml) containing a catalytic amount of anilinium chloride. When all of the imidate has dissolved, the solution is allowed to warm to room temperature and stirred (19 h), during which time the product separates as an off-white solid. After filtering, washing with dry ether–dry chloroform, and drying *in vacuo*, (24) is obtained (3.11 g, 95%), m.p. > 300°C.

5-Amino-4-(cyanoformimidoyl)imidazole (28)

Solid barium hydroxide dihydrate (8.0 g) is added to a suspension of the amidine (24) (3.0 g, 22.2 mmol) in 95% aqueous ethanol (270 ml), and the mixture is stirred vigorously (~50 min), till TLC shows that all of the (24) has been consumed; the solution turns deep yellow. Ether (300 ml) is added, and carbon dioxide is bubbled through the solution (10 min). The precipitated barium carbonate and unchanged barium hydroxide are filtered off and the solid is washed with ether. The combined filtrate and washings are rotary evaporated below 30°C to give the product as a pale green solid (2.17 g, 72%), m.p. > 300°C (dec.).

This general reaction can be adapted to prepare analogous 1-arylimidazoles by treatment of the imidate (27) with an arylamine at room temperature in the presence of a little anilinium chloride (Table 2.2.3). Yields of arylformamidines are usually around 60%, but vary between 35 and 95%. Electron-attracting groups in the aryl ring decrease the nucleophilicity of the amino group, leading to low yields. The reaction provides the first general method of preparation of 5-amino-1-arylimidazole-4-carbonitriles (and the 1-aryl analogues of (26) [31, 32]). The former are isolated when the arylformamidines are treated with ethanolic potassium hydroxide; the latter result when ethanolic or ethyl acetate solutions of the arylformamidines are

TABLE 2.2.3
1-Aryl-5-aminoimidazole-4-nitriles made from (27) [32]

1-Aryl group	Yield (%)	1-Aryl group	Yield (%)
Ph	90	4-PhCH$_2$OC$_6$H$_4$	95
4-MeOC$_6$H$_4$	86	4-ClC$_6$H$_4$	70
2-MeOC$_6$H$_4$	75	4-FC$_6$H$_4$	86
3,4-(MeO)$_2$C$_6$H$_3$	89	4-NH$_2$C$_6$H$_4$	67
2,4-(MeO)$_2$C$_6$H$_3$	83	3-NH$_2$C$_6$H$_4$	75
4-MeC$_6$H$_4$	96	4-NO$_2$C$_6$H$_4$	80
2,4-Me$_2$C$_6$H$_3$	86	4-CNC$_6$H$_4$	85

2.2. FORMATION OF THE 1,5 (OR 3,4) BOND

treated with a few drops of 1,8-diazabicyclo[5.4.0]undec-7-ene (DBU) for 1-3 h.

5-Amino-1-aryl-4-(cyanoformimidoyl)imidazoles [32]

To a stirred solution of the formamidine (1.00 g) in either dry ethyl acetate, ethanol, or a 1:1 mixture of ethyl acetate and isopropanol is added DBU (10 drops, 50 µl), and the reaction is monitored by TLC (Camlab polygram G_{254} silica gel; 9:1 chloroform-ethanol). The solid dissolves, and after 1-3 h the product precipitates as an off-white to pale yellow product, which is filtered off, washed with diethyl ether or light petroleum, and dried under vacuum to give the title compounds in 53-96% yields.

5-Amino-1-aryl-4-cyanoimidazoles [32]

An aqueous solution of 1 M potassium hydroxide (1 ml) is added to a suspension of the formamidine (1 mmol) in ethanol (1 ml). The mixture is stirred at room temperature (~1 h). The precipitated product is filtered, washed with water, a few drops of ethanol and, finally, with diethyl ether before drying under vacuum. (See Table 2.2.3.)

Similarly, 1-aralkyl analogues can be made by base-induced cyclization of N-aralkyl formamidines related to (**24**) (e.g. 5-amino-1-benzyl-4-cyanoimidazole (77%)) [33].

The imidate (**27**) reacts at 100°C with excess triethyl orthoformate to give 4,5-dicyano-1-ethylimidazole after vacuum distillation. The reaction becomes economically competitive with other reactions which alkylate 4,5-dicyanoimidazole since the 1-alkyl-4,5-dicyanoimidazoles can be made in one-pot reactions merely by heating DAMN at 100°C with excess ortho ester. No co-solvent is necessary. Presumably, (**27**) forms first, then 4,5-dicyanoimidazole, then the 1-alkyl derivative, with the anion of 4,5-dicyanoimidazole probably being the reactive species. Imidazole itself is not N-alkylated by ortho esters; instead it induces alkoxide displacement to give amide acetals [34].

4,5-Dicyano-1-methylimidazole [34]

A mixture of DAMN (100 g, 0.925 mol) and trimethyl orthoformate (350 ml, 3.20 mol) is stirred and heated under partial reflux through a 25 cm Vigreux distillation column. Around 200 ml of distillate, mainly methanol and methyl formate, is collected, b.p. 50-70°C, over 140 min. The distillation temperature is allowed to rise to 102°C, and distillate is slowly collected over 2 h (the pot temperature is 115°C). The fractionating column is removed, and the bulk of the remaining ortho ester is distilled off. The residue is distilled using a Kugelrohr apparatus (1.5 mbar; oven temperature 130-140°C) to give the title compound as a colourless solid (117.3 g, 96%), m.p. 87-89°C. Similarly prepared are the following 4,5-dicyanoimidazoles (1- and 2-substituents, temperature (reaction time), yield, m.p. listed): 1-Et, 100 (1 h)-150°C (3 h), 97%, 67-68°C; 1,2-Me$_2$, 100-75°C (18 h) (with added triethylamine), 76%, 101-103°C. This method is slower and less efficient with tri-n-propyl, tri-n-butyl and triisopropyl ortho esters [34].

The cyclization of formamidines (**24**) has been extended to similar reactions of amidrazones (**29**), which give 1,5-diaminoimidazoles when treated with bases (Scheme 2.2.10). Again, depending on the basic strength, the 4-substituent is cyano or cyanoformimidoyl. The amidrazones are available in almost quantitative yield by treatment of the imidate (**27**) with hydrazine hydrate at room temperature [35].

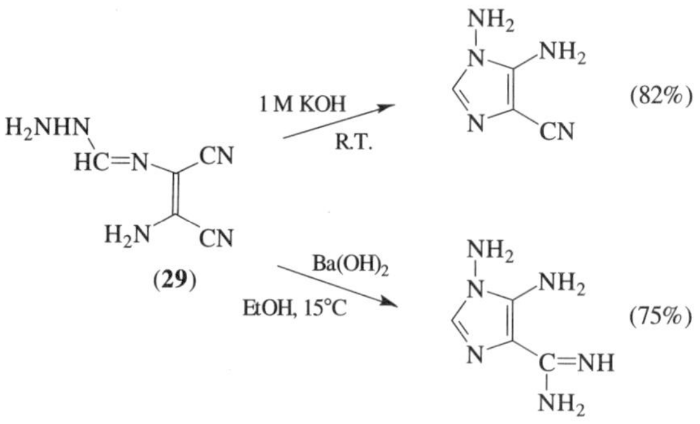

Scheme 2.2.10

(Z)-N^3-(2-Amino-1,2-dicyanovinyl)formamidrazone (**29**) [35]

Hydrazine hydrate (0.39 g, 0.38 ml, 7.72 mmol) is added at room temperature to a suspension of (**27**) (1.27 g, 7.72 mmol) in dry dioxane (8 ml). An immediate and slightly exothermic reaction gives a yellow solution, from which the product precipitates as yellow needles. Filtration, washing with diethyl ether, and drying *in vacuo* gives (**29**) (1.13 g, 98%), m.p. > 300°C (dec.).

1,5-Diamino-4-cyanoimidazole [35]

The formamidrazone (**29**) (1.01 g, 6.73 mmol) is dissolved in aqueous 1 M KOH (10 ml). A yellow solid soon precipitates. Filtration, washing with ether and drying as before gives the title compound (0.68 g, 82%), m.p. 214-215.5°C (dec.).

2.2. FORMATION OF THE 1,5 (OR 3,4) BOND

The utility of ureas and thioureas as substrates for making imidazoles is limited by the fact that the imidazole 2-substituent can only be an oxygen or sulfur function. Synthetic methods involving ureas and thioureas will also be discussed in Section 4.1, but some cyclizations of suitably functionalized species fall under the present heading. Appropriately substituted ureas and thioureas can be made from isocyanates and primary amines [36–38], from isocyanates and hydrazines [39] or thiocyanates and hydrazines [40], from α-aminonitriles and carbon dioxide [41] and by heating 1,3,4-oxadiazol-2-ones with amino acids [42]. Some of the substrates prepared in these ways, though, lead ultimately to reduced imidazoles such as hydantoins. Cyclizations are usually acid catalysed, but they can also be thermal [43].

Suitable ureas have been prepared by treating 1-hydroxylamino- or 1-alkoxylamino-2,2-diethoxyethane with cyanic acid or an arylisocyanate, or by the addition of hydroxylamine (or an alkoxyamine) to 2,2-diethoxyethylisocyanate. From the urea intermediates (30)–(32), 1-alkoxyimidazolin-2-ones (33) are formed in ~80% yields on acid treatment (Scheme 2.2.11). Catalytic reduction quantitatively converts the alkoxyimidazoles into the hydroxy analogues [37, 38, 44, 45].

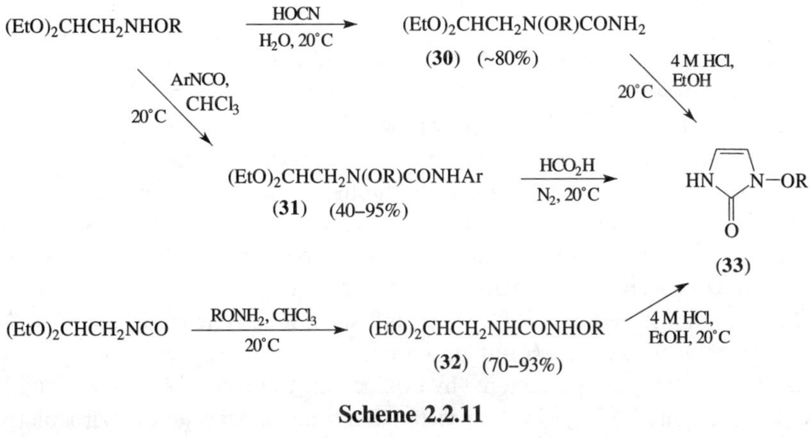

Scheme 2.2.11

3-Butynylurea and the corresponding thiourea cyclize in sulfuric acid to form 4,5-dimethylimidazolin-2-one and the analogous 2-thione [43].

Among the possible approaches to the preparation of 4-aminoimidazoles is the cyclization of α-cyanoalkyl cyanamides (34). When treated with anhydrous hydrobromic acid they give 4(5)-amino-2-bromoimidazoles (35) in a process which bears formal similarities to amidine cyclizations (Scheme 2.2.12). The instabilities of many 4(5)-aminoimidazoles necessitates their conversion into the more easily handled acetyl derivatives using acetic anhydride [46] (see also Section 3.1.1). Table 2.2.4 lists some examples.

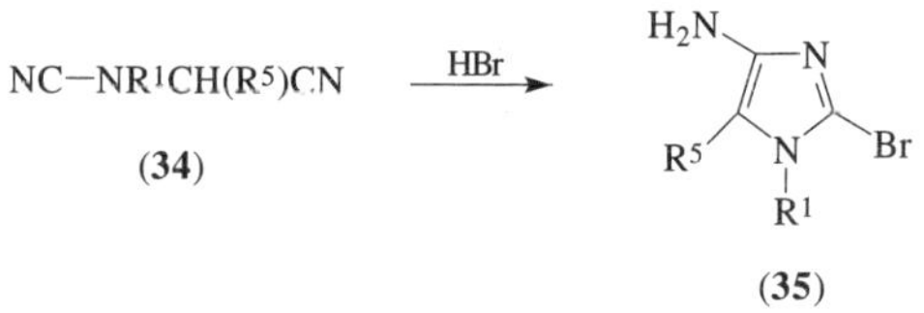

NC—NR¹CH(R⁵)CN $\xrightarrow{\text{HBr}}$

(34)

(35)

Scheme 2.2.12

TABLE 2.2.4
4(5)-Amino-2-bromoimidazoles (**35**) made by cyclization of **34** [46]

R¹	R⁵	Yield (%)	R¹	R⁵	Yield (%)
Me	H	33[a]	Ph	3-ClC₆H₄	55[a]
Et	H	53[a]	Ph	4-ClC₆H₄	50[a]
Bu	H	47[a]	Ph	2,4-Cl₂C₆H₃	68
Ph	H	82[a]	Ph	2-MeOC₆H₄	82
Ph	Ph	82	PhCH₂	H	68[a]
Ph	2-ClC₆H₄	10	CH₂CN	H	5–65[a]

[a]Isolated as the acetyl derivative.

The thermal ring closure of enecarbamoyl azides to imidazolones and reduced indazolones as minor products is unlikely to have compelling synthetic application [47]. The unusual cyclization of benzil monohydrazones to 4,5-diphenylimidazoles involves breaking and reorganizing an N—N bond. It is possible that this reaction proceeds through a diaziridine intermediate or a four-membered ring ylide. The reaction may have synthetic possibilities because yields of 4,5-diphenylimidazoles are 58–95%, and the benzil N-alkylhydrazones are readily made from benzil and 1,1-dialkylhydrazines [48].

β,γ-Acetylenic carbanilides cyclize and subsequently isomerize to 2(3H)-imidazolethiones and 2(3H)-imidazolones when treated respectively with hydrogen sulfide or potassium hydroxide in t-butanol. Although yields are almost quantitative, this approach does not appear to be particularly convenient [49]. Mixtures of imidazoles and pyrazoles are formed when phenacylethylene ketal hydrazones of benzaldehyde and acetaldehyde are treated with aluminium chloride in benzene [50].

2.2.2 Benzimidazoles

Under oxidative or acidic conditions, or merely by heating the reagents, appropriately functionalized aryl-amidines or -guanidines cyclize to 1-alkyl or 1-aryl-benzimidazoles [51, 52]. If guanidines are employed,

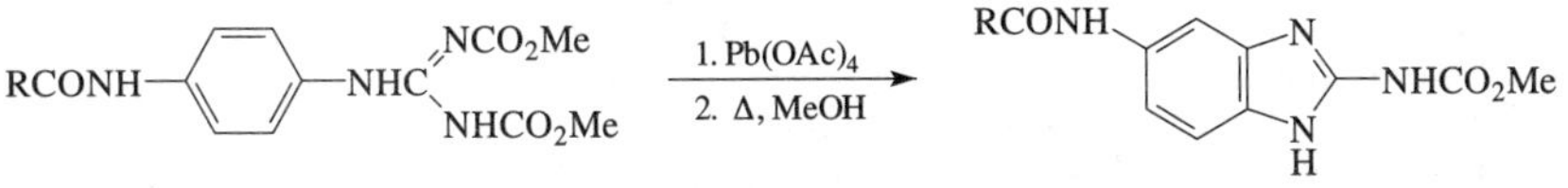

Scheme 2.2.13

2-aminobenzimidazoles (**36**) result [53–55]; *N*-arylbenzamidines give 2-phenylbenzimidazoles (77–83%); *N*-arylphenylacetamidines give 2-benzylbenzimidazoles (86–98%) (Scheme 2.2.13) [51].

Oxidation of N-*phenylphenylacetamidine; 2-benzylbenzimidazole* [51]

To a stirred solution of *N*-phenylphenylacetamidine (2.10 g, 0.01 mol) in dichloromethane (25 ml) and pyridine (8 ml) is added lead tetraacetate (4.88 g, 0.011 mol) is small portions over 15 min. After complete addition the mixture is heated in an oil bath at 80–90°C (1.5 h) before removal of the solvent under reduced pressure, and extraction of the residue with hot benzene. The benzene extracts are washed with water and dried (Na_2SO_4) before rotary evaporation of the solvent to give 2-benzylbenzimidazole (1.89 g, 90%), recrystallized from benzene, m.p. 182–183°C.

The appropriate amidines can be made from nitriles and metal amides, but reaction conditions appear to be critical with best yields reported using DMSO–THF. Alkylchloroaluminium amides, which can be generated from trimethylaluminium and ammonium chloride or amine hydrochlorides, have been shown to add efficiently to nitriles to give high yields (60–95%) of suitable amidines. Similarly, *N*-alkylcyanamides can be converted into guanidines, in an approach which appears to be useful and potentially quite general. Yields, however, are not always good [56].

N-Arylbenzimidazolones can be prepared in low to moderate yields from *N,N'*-diarylureas with sodium hypochlorite, e.g. 6-chloro-1-phenyl-2-benzimidazolone (49%) [57, 58]. Similarly, in a one-pot reaction, *N*-methoxy-*N'*-arylureas react with t-butyl hypochlorite to form unstable *N*-chloro-*N*-methoxy-*N'*-arylureas which cyclize in basic media to give 1-methoxy-2-benzimidazolones (56–83%) [59].

REFERENCES

1. A. McKillop, A. Henderson, P. S. Ray, C. Avendano and E. C. Molinero, *Tetrahedron Lett.* **23**, 3357 (1982).
2. A. Lawson, *J. Chem. Soc.* 1443 (1957).

3. F. Compernolle and N. Castagnoli, *J. Heterocycl. Chem.* **19**, 1403 (1982).
4. J. M. Bobbitt and A. J. Bourque, *Heterocycles* **25**, 601 (1987).
5. A. J. Lawson, *J. Chem. Soc.* 4225 (1957).
6. J. K. Lawson, *J. Am. Chem. Soc.* **75**, 3398 (1953).
7. M. Julia and H.-D. Tan, *Bull. Soc. Chim. Fr.* 1303 (1971).
8. K. T. Potts and S. Husain, *J. Org. Chem.* **36**, 3368 (1971).
9. K. Burger, R. Ottlinger, H. Goth and J. Firl, *Chem. Ber.* **115**, 2494 (1982).
10. K. Burger, K. Geith and D. Hübl, *Synthesis* 189 (1988).
11. K. Burger, D. Hübl and K. Geith, *Synthesis* 194 (1988).
12. J. K. Lawson, *J. Am. Chem. Soc.* **75**, 3398 (1953).
13. E. R. Freiter, L. E. Begin and A. H. Abdallah, *J. Heterocycl. Chem.* **10**, 391 (1973).
14. H. Jones, M. W. Fordice, R. B. Greenwald, J. Hannah, A. Jacobs, W. V. Ruyle, G. L. Walford and T. Y. Shen, *J. Med. Chem.* **21**, 1100 (1978).
15. M. Miyano and J. N. Smith, *J. Heterocycl. Chem.* **19**, 659 (1982).
16. E. Galeazzi, A. Guzman, J. L. Nava, Y. Liu, M. L. Maddox and J. M. Muchowski, *J. Org. Chem.* **60**, 1090 (1995).
17. A. McKillop, A. Henderson, P. S. Ray, C. Avendano and E. G. Molinero, *Tetrahedron Lett.* **23**, 3357 (1982).
18. M. R. Harnden, L. J. Jennings, C. M. D. McKie and A. Parkin, *Synthesis* 893 (1990).
19. T. Jen, H. Van Hoeven, W. Groves, R. A. McLean and B. Loev, *J. Med. Chem.* **18**, 90 (1975).
20. A. Dalkafouki, J. Ardisson, N. Kunesch, L. Lacombe and J. E. Poisson, *J. Heterocycl. Chem.* **32**, 5325 (1991).
21. M. Ruccia, N. Vivona and G. Cusmano, *Tetrahedron* **30**, 3859 (1974).
22. A. Hetzheim and G. Manthey, *Chem. Ber.* **103**, 2845 (1970).
23. N. D. Heindel and M. C. Chun, *Tetrahedron Lett.* 1439 (1971).
24. N. D. Heindel and M. C. Chun, *J. Chem. Soc., Chem. Commun.* 664 (1971).
25. P. Molina, A. Arques and M. V. Vinader, *Liebigs Ann. Chem.* 103 (1987).
26. B. L. Booth, R. D. Coster and M. F. J. R. P. Proenca, *J. Chem. Soc., Perkin Trans. 1* 1521 (1987).
27. M. J. Alves, B. L. Booth and M. F. J. R. P. Proenca, *J. Chem. Soc., Perkin Trans. 1* 1705 (1990).
28. A. W. Erian, *Chem. Rev.* **93**, 1991 (1993).
29. M. H. Elnagdi, S. M. Sherif and R. M. Mohareb, *Heterocycles* **26**, 497 (1987).
30. R. F. Shuman, W. E. Shearin and R. J. Tull, *J. Org. Chem.* **44**, 4532 (1979).
31. M. J. Alves, B. L. Booth, O. K. Al-Duaij, P. Eastwood, L. Nezhat, M. Fernanda, J. R. P. Proenca and A. S. Ramos, *J. Chem. Res. (S)* 402 (1993).
32. M. J. Alves, B. L. Booth, O. K. Al-Duaij, P. Eastwood, L. Nezhat, M. Fernanda, J. R. P. Proenca and A. S. Ramos, *J. Chem Res. (M)* 2701 (1993).
33. M. J. Alves, M. F. J. R. P. Proenca and B. L. Booth, *J. Heterocycl. Chem.* **31**, 345 (1994).
34. S. J. Johnson, *Synthesis* 75 (1991).
35. M. J. Alves, B. L. Booth, A. P. Freitas and M. F. J. R. P. Proenca, *J. Chem. Soc., Perkin Trans. 1* 913 (1992).
36. A. F. Rusakov, S. P. Epshtein, V. P. Tashchi, Yu. A. Baskalov and Yu. G. Putsykin, *Chem. Heterocycl. Compd. (Engl. Transl.)* **20**, 323 (1984).
37. O. Wong, N. Tsuzuki, M. Richardson, H. Rytting, R. Konishi and T. Higuchi, *Heterocycles* **26**, 3153 (1987).
38. V. Bock, W. Klötzer, N. Singewald and G. Strieder, *Synthesis* 1058 (1987).
39. S. P. Epshtein, A. F. Rukasov, V. P. Tashchi, Yu. G. Putsykin, Yu. A. Baskalov and T. G. Simonova, *Chem. Heterocycl. Compd. (Engl. Transl.)* **19**, 74 (1983).
40. N. Jacobsen and J. Toelberg, *Synthesis* 559 (1986).

2.3. FORMATION OF THE 4,5 BOND

41. R. A. O'Brien, J. J. Worman and E. S. Olson, *Synth. Commun.* **22**, 823 (1992).
42. Y. Saegusa, S. Harada and S. Nakamura, *J. Heterocycl. Chem.* **27**, 739 (1990).
43. C. Holzmann, B. Krieg, H. Lautenschläger and P. Konieczny, *J. Heterocycl. Chem.* **16**, 983 (1979).
44. G. V. Nikitina and M. S. Pevzner, *Chem. Heterocycl. Compd. (Engl. Transl.)* **29**, 127 (1993).
45. H. Hauser, W. Klötzer, V. Krug, J. Rzehak, A. Sandrieser and N. Singewald, *Sci. Pharm.* **56**, 235 (1988).
46. F. Johnson and W. A. Nasutavicus, *J. Org. Chem.* **29**, 153 (1964).
47. J. P. Chupp, *J. Heterocycl. Chem.* **8**, 557 (1971).
48. W. L. Collibee and J. P. Anselme, *Bull. Soc. Chim. Belg.* **95**, 655 (1986).
49. B. Devan and K. Rajagopalan, *Tetrahedron Lett.* **35**, 1585 (1994).
50. D. Scarpetti, K. Kano and J. P. Anselme, *Bull. Soc. Chim. Belg.* **95**, 1073 (1986).
51. S. Chaudhury, A. Debroy and M. P. Mahajan, *Can. J. Chem.* **60**, 1122 (1982).
52. T. Benincori and F. Sannicolo, *J. Heterocycl. Chem.* **25**, 1029 (1988).
53. S. Rajappa, R. Sreenivasan and A. V. Rane, *Tetrahedron Lett.* **24**, 3155 (1983).
54. T. Hisano, M. Ichikawa, K. Tsumoto and M. Tasaki, *Chem. Pharm. Bull.* **30**, 2996 (1982).
55. S. Rajappa, R. Sreenivasan and A. Khalwadekar, *J. Chem. Res (S)* 158 (1986).
56. A. D. Redhouse, R. J. Thompson, B. J. Wakefield and J. A. Wardell, *Tetrahedron* **48**, 7619 (1992).
57. L. Posnati, *Gazz. Chim. Ital.* **86**, 275 (1956).
58. M. L. Oftedahl, R. W. Radue and M. W. Dietrich, *J. Org. Chem.* **28**, 578 (1963).
59. J. Perronnet and J. P. Demoute, *Gazz. Chim. Ital.* **112**, 507 (1982).

2.3 FORMATION OF THE 4,5 BOND

This is an uncommon synthetic approach to imidazoles, being largely confined to examples in which an active methylene group cyclizes on to a nitrile carbon. There are, naturally, no such examples of benzimidazole synthesis.

Acyclic precursors (**5**) for these cyclizations can be made by alkylation of arylaminomethylene cyanamides (**1**) ($R^1 = SMe$) with α-halogenocarbonyl reagents [1]. Alternatively, condensation of 2-aminonitriles (or glycine esters) with *N*-cyanoacetamido esters (**2**) in the presence of triethylamine can be used to prepare suitable substrates (**6**) for cyclization. Subsequent base-catalysed ring closure gives 4-aminoimidazoles (**3**) and (**4**) in 40–90% yields [1, 2]. (See Table 2.3.1 and Scheme 2.3.1.) Such synthetic procedures provide useful alternatives in the preparation of 4-amino-5-cyanoimidazoles (see also the use of DAMN, Sections 2.1 and 2.2) in that the hetaryl *N*-substituent is put in place before cyclization in these latest examples. The *N*-cyanoamidine intermediates (**5**) and (**6**) can be isolated in good yields (40–85%), but they need

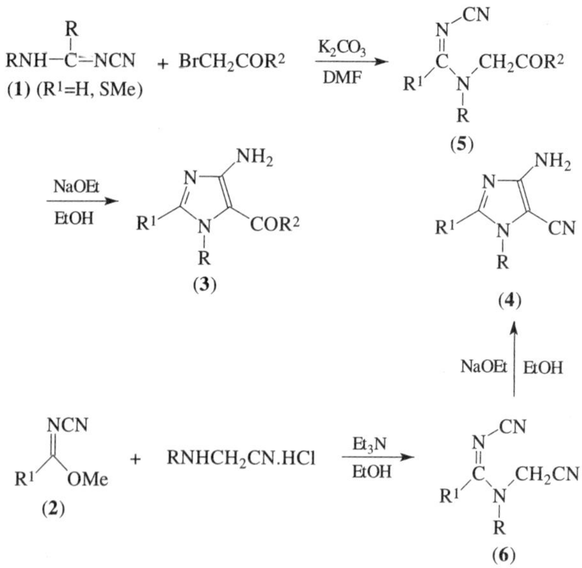

Scheme 2.3.1

TABLE 2.3.1
4-Aminoimidazoles (**3**) and (**4**) and precursors (**5**) and (**6**) made by alkylation of (**1**) and (**2**) by α-halogenocarbonyl reagents in basic medium

R	R^1	R^2	Products (yield, %)		Ref.
Ph	H	OEt	(**5**) (85)	(**3**) (85)[a]	1
Ph	H	Ph	(**5**) (84)	(**3**) (88)[a]	1
p-ClC$_6$H$_4$	H	OEt	(**5**) (79)	(**3**) (83)[a]	1
p-MeOC$_6$H$_4$	H	Ph	(**5**) (68)	(**3**) (84)[a]	1
Ph	MeS	Ph	(**5**) (62)	(**3**) (90)[a]	1
Ph	MeS	OEt	—[b]	(**3**) (68)[c]	1
Me	MeS	OEt	—[b]	(**3**) (45)[c]	1
CH$_2$CH=CH$_2$	MeS	OEt	—[b]	(**3**) (41)[c]	1
Me	Me	—	(**6**) (68)	(**4**) (79)[d]	2
H	Me	—	(**6**) (41)	(**4**) (41)[d]	2
Me	Me	—	(**6**) (49)	(**4**) (88)[d]	2
H	Me	—	(**6**) (70)	(**4**) (63)[d]	2

[a] Yield based on (**5**).
[b] (**5**) not isolated.
[c] Yield based on (**1**).
[d] Yield based on (**6**).

2.3. FORMATION OF THE 4,5 BOND

not be isolated in all cases. Consequently, some syntheses classified in this class could also be listed under 1,5 and 4,5 bond formations.

N-Ethoxycarbonylmethyl-N'-cyano-N-phenylformamidine **(5)** *(R = Ph, $R^1 = H, R^2 = OEt$)* [1]

A mixture of **(1)** (R = Ph) (1.5 g) and ethyl bromoacetate (1.7 g) with dry, powdered potassium carbonate (1.4 g) in absolute ethanol (10–15 ml) is warmed at 75–80°C for 30 min. The cooled solution is then stirred into 150 ml of water, treated with a little acetic acid, concentrated, and allowed to stand at room temperature. The product which separates is filtered and recrystallized from ethanol, giving **(5)** (R = Ph, $R^1 = H$, $R^2 = OEt$) (1.9 g, 85%), m.p. 84–86°C.

Ethyl 5-amino-2-phenylimidazole-4-carboxylate **(3)**; *(R = Ph, $R^1 = H$, $R^2 = OEt$)*

The above phenylformamidine (2.1 g) (*note*: the crude product can also be used in the cyclization step) is warmed in a sodium ethoxide solution made by dissolving sodium (0.5 g) in ethanol (15 ml). All of the phenylformamidine dissolves in 3–5 min. After 15 min a 2–3-fold volume of water is added to the solution, which is allowed to stand (1 h) to crystallize. Recrystallization from ethanol gives nearly colourless crystals (1.8 g, 85%), m.p. 97–99°C.

A similar reaction sequence condenses *N*-cyanoiminodithiocarbamic esters **(7)** with α-cyanoammonium salts (e.g. sarcosine nitrile sulfate) (Scheme 2.3.2). The isothiourea product **(8)** readily cyclizes to give 4-amino-5-cyanoimidazoles when heated with sodium ethoxide in DMF solution. It is possible to convert the dithiocarbamic esters directly into imidazoles if they are heated in DMF solution with sarcosine ethyl ester hydrochloride or methylaminoacetophenone in the presence of triethyamine [3]. These direct cyclizations are, however, examples of both 1,2 and 1,5 bond formation. (See also Section 3.2).

When carbanions are generated from the products of reaction between azirines and formamidine, a similar process also leads to imidazoles in moderate yields. Azirines are, however, rather too exotic as substrates to make this an appealing synthetic approach [4]. Treatment of perhydro-1,3-thiazine-2-thione with trifluoracetic anhydride gives an imidazothiazine [5].

Imidazoles can be made by 1,5-dipolar cyclization of suitable 2-azavinamidinium salts [6, 7]. Such salts **(10)** are readily made from amidines (or guanidines) and amido or ureido chlorides **(9)**, and in the presence of sodamide in liquid ammonia they cyclize to form electron-rich imidazoles

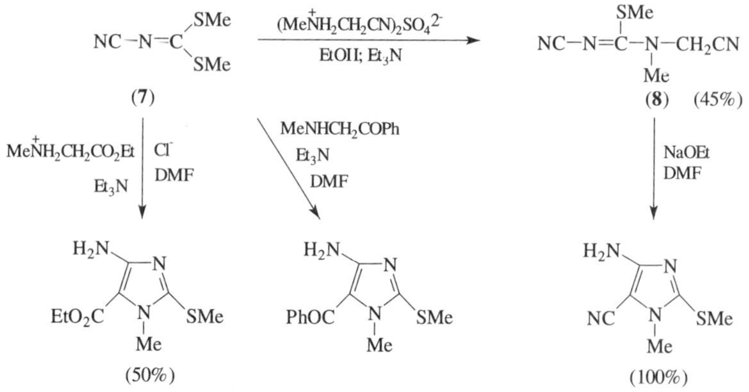

Scheme 2.3.2

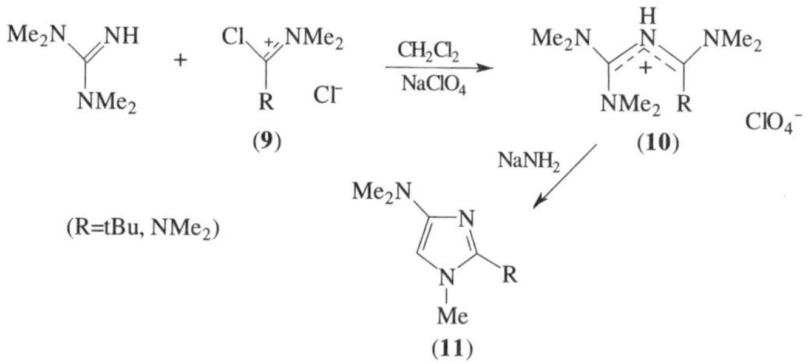

Scheme 2.3.3

(Scheme 2.3.3). The approach has been used to make imidazoles (11) with dimethylamino groups in the 4- and/or 2-positions. Such compounds are not well known (they are inclined to be sensitive to the effects of air and moisture, and their behaviour in electrophilic aromatic substitution reactions parallels that of *N,N*-dimethylaniline [8]), and the method is of value for that reason alone. Its wider application to imidazole synthesis seems unlikely.

N,N-*Dimethyl-(1-t-butyl-3,3-bis[dimethylamino]-2-aza-3-propenyliden)- ammonium perchlorate* (10) *(R = tBu)* [6]

To *N,N*-dimethyltrimethylacetamido chloride (9) (R = tBu) (6.23 g, 33 mmol) dissolved in dichloromethane (50 ml), an ethereal solution of tetramethylguanidine (7.79 g, 66 mmol) is added dropwise with stirring. After heating the mixture for 1 h under reflux, the solvent is removed under reduced pressure, and the residue taken up in a little water before addition of a saturated aqueous solution of sodium perchlorate. The precipitated perchlorate salt is isolated as colourless plates after recrystallization from ethanol (71%), m.p. 96–97°C. Similarly prepared in 93% yield is (**10**) (R = NMe_2).

1-Methyl-2,4-bis(dimethylamino)imidazole (**11**) *(R = NMe_2)* [6]

To a suspension of sodamide in liquid ammonia (200 ml) (add sodium (3.91 g, 170 mmol) in portions with vigorous stirring to the ammonia) is added (**10**) (R = NMe_2) (10.7 g, 34 mmol). After 45 min, pentane (150 ml) saturated with nitrogen is added, and the mixture is very gently warmed, or allowed to stand, until the ammonia has evaporated. The pentane solution is decanted, rotary evaporated and the residue is distilled using a Kugelrohr apparatus to give a pale yellow liquid, $b_{0.8}$ 70°C, in 70% yield. Similarly prepared is 2-t-butyl-1-methyl-4-dimethylaminoimidazole, $b_{0.8}$ 65°C.

REFERENCES

1. K. Gewald and G. Heinhold, *Monatsh. Chem.* **107**, 1413 (1976).
2. E. Edenhofer, *Helv. Chim. Acta* **58**, 2192 (1975).
3. R. Gompper, M. Gäng and F. Saygin, *Tetrahedron Lett.* 1885 (1966).
4. A. Kascheres, C. M. A. Oliviera, M. B. M. de Azevedo and C. M. S. Nobre, *J. Org. Chem.* **56**, 7 (1991).
5. C. W. Horzapfel and A. E. Von Platen, *S. Afr. J. Chem.* **40**, 153 (1987); *Chem. Abstr.* **108**, 21 788 (1988).
6. R. Gompper and C. S. Schneider, *Synthesis* 215 (1979).
7. R. Gompper, *Bull. Soc. Chim. Belg.* **92**, 781 (1983).
8. R. Gompper, P. Kruk and J. Schelble, *Tetrahedron Lett.* **24**, 3653 (1983).

–3–
Ring Syntheses Involving Formation of Two Bonds: [4 + 1] Fragments

3.1 FORMATION OF 1,2 AND 2,3 BONDS

3.1.1 Imidazoles

As will be discussed in Section 3.1.2 the most common approaches to benzimidazoles involve cyclizations of *o*-arylenediamines with suitable carbonyl species. This general strategy can be applied to uncondensed imidazoles by similar cyclization of 1,2-alkylenediamines (or 1,2-diaminoalkanes in the presence of a dehydrogenating agent). Such syntheses have been best adapted to the preparation of 2-alkyl- and 2-arylimidazoles. Thus, when a 1,2-diaminoalkane is treated with an alcohol, aldehyde or carboxylic acid at high temperature, and in the presence of a dehydrogenating agent, the products are 2-substituted imidazoles (Scheme 3.1.1). Suitable dehydrogenating agents include platinum and alumina, or palladium on carbon [1–6]. When there is no dehydrogenating agent present the products will be imidazolines or imidazolidines. The C-2 fragment can also be introduced by replacement of the carbonyl reagent with an iminoether hydrochloride, an amidine hydrochloride, a thioamide or a nitrile [7]. Apart from the high temperatures often needed to carry out the reactions, another problem is product purification. One method is to distil the product azeotropically with an aralkyl hydrocarbon whose boiling point is 10–40°C lower than that of the imidazole. 2-Methylimidazole can be purified by distillation with 1- or 2-methylnaphthalene, and then isolated by washing with toluene or pentane, solvents in which the azole is only sparingly soluble. On the laboratory scale, however, combinations of acidification and extraction methods are likely to be more convenient [8]. Reaction yields can be high, but the methods are experimentally complicated. The manufacture of imidazole itself uses a modification of the general approaches [3, 4]. When carbon disulfide is introduced as the C-2 fragment the products are imidazolidin-2-thiones [9].

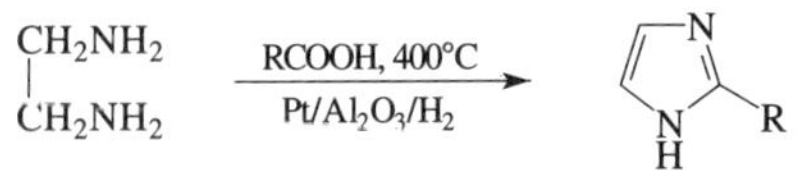

$$\begin{array}{c}CH_2NH_2\\|\\CH_2NH_2\end{array} \xrightarrow[Pt/Al_2O_3/H_2]{RCOOH, 400°C} \text{imidazole-2-R}$$

Scheme 3.1.1

A specific synthesis of 1,4- and 1,5-disubstituted imidazoles has been accomplished in about 70% yields by cyclization of 2-amino-3-methylaminopropanoic acid and 3-amino-2-methylaminopropanoic acid, respectively, with triethyl orthoformate (Scheme 3.1.2). The initial product is the 2-imidazoline, which needs to be aromatized by treatment with active manganese dioxide [10].

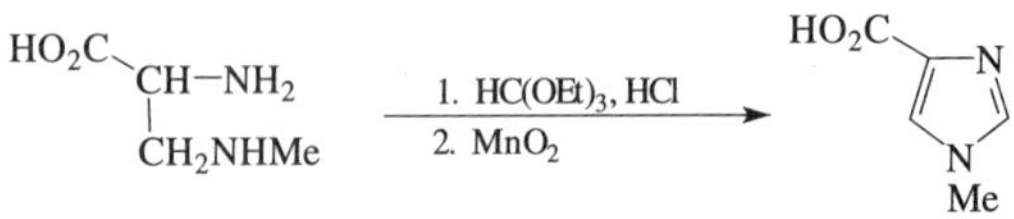

$$\begin{array}{c}HO_2C\\ \diagdown CH-NH_2\\|\\CH_2NHMe\end{array} \xrightarrow[2.\ MnO_2]{1.\ HC(OEt)_3,\ HCl} \text{1-methyl-imidazole-4-CO}_2H$$

Scheme 3.1.2

1-Methyl-2-imidazoline-4-carboxylic acid [10]

2-Amino-3-methylaminopropanoic acid hydrochloride (4.0 g) is stirred with freshly distilled triethyl orthoformate (40 ml) in the presence of 36% HCl (2.5 ml) at 90–100°C (12 h). During the initial few hours of heating, a low-boiling (56°C) distillate is collected, and fresh triethyl orthoformate (10 ml) is added after the first hour. The mixture is cooled, and the solid imidazole is filtered and washed with a little acetone (3.38 g, 79%). Recrystallization from absolute ethanol gives the pure product, m.p. 194–195°C.

Methyl 1-methyl-2-imidazoline-4-carboxylate [10]

To a suspension of the acid (11.1 g) in dry methanol (125 ml) is added 2 g of molecular sieve. Dry HCl is bubbled in rapidly until the solution is saturated, and then more slowly while the solution is refluxed (4 h). The excess HCl is removed in a stream of nitrogen, and the solution is neutralized with solid sodium bicarbonate. Most of the methanol is evaporated, the residue is dissolved in water, and the mixture is filtered, cooled in an ice bath, and adjusted to above pH 9 with cold sodium carbonate solution, before extracting with chloroform. The dried extracts are rotary evaporated to leave a pale yellow oil (81%). The crude product seems to be unstable when distilled, and is best dehydrogenated without purification.

3.1. FORMATION OF 1,2 AND 2,3 BONDS

Methyl 1-methylimidazole-4-carboxylate [10]

To a solution of the imidazoline (7.5 g) in chloroform (150 ml) is added active manganese dioxide* (30 g). The suspension is stirred at room temperature (16 h) then filtered, and the solid washed with hot chloroform. Evaporation of the combined chloroform fractions gives the imidazole ester (5.2 g). Sublimation at 80°C (0.05 mmHg) gives the pure product, m.p. 97–98°C.

*Active manganese dioxide is available commercially from a number of suppliers. It can also be prepared quite readily (see L. F. Fieser and M. Fieser, *Reagents for Organic Synthesis*, Wiley, Chichester, 1967, p. 637).

Another interesting example of the method involves making the 1,2-diamine by a Diels–Alder reaction. The 2-substituted imidazoles which result are isolated in yields of 40–79% based on the diamine (Scheme 3.1.3) [11].

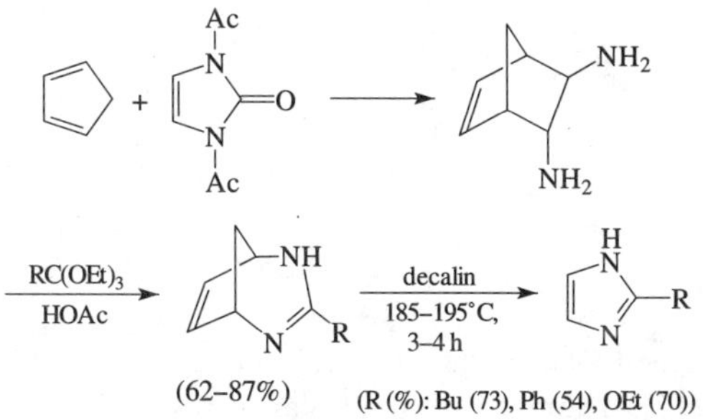

Scheme 3.1.3

It has been known for many years that 1,2-dibenzoylaminoethene (which forms when imidazole is benzoylated under Schotten–Baumann conditions) can be recyclized to 2-methylimidazole when heated with acetic anhydride at 180°C. Although some other acid anhydrides take part in this reaction, it is not synthetically viable for a wide range of 2-substituted imidazoles [12, 13].

When 3-arylamino-2-nitro-2-enones (**1**) condense with orthoesters in the presence of reducing agents, 4-methoxycarbonylimidazoles are produced, albeit in only moderate yields (Scheme 3.1.4) [14]. The synthetic utility of this method is probably limited to occasions when particular 4-acylimidazoles are desired (see also Sections 7.2.2 and 8.3).

A corollary of the foregoing reactions is the conversion of (**2**) into the 5-hydroxyimidazole (**3**) by treatment with an orthoester (Scheme 3.1.5) [15].

Suitable 1,2-diimino species can replace 1,2-diaminoalkanes or -alkenes in cyclizations of this general type. Thus, diazadienes (**4**), which are easily prepared from glyoxal and primary amines, can be converted into 2-substituted 1,3-dialkylimidazolium salts by treatment with dry HCl (Scheme 3.1.6)

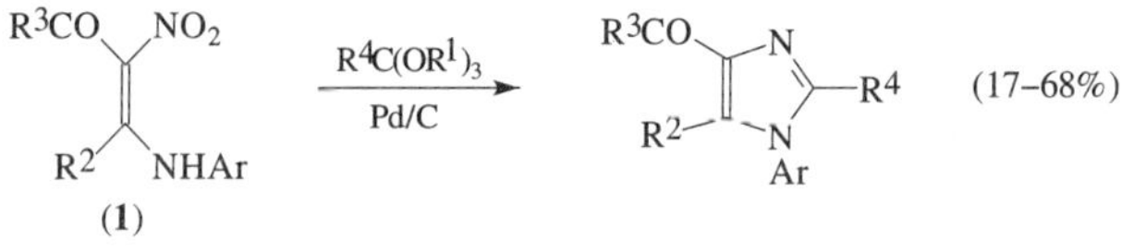

3. RING SYNTHESES INVOLVING FORMATION OF TWO BONDS

Scheme 3.1.4

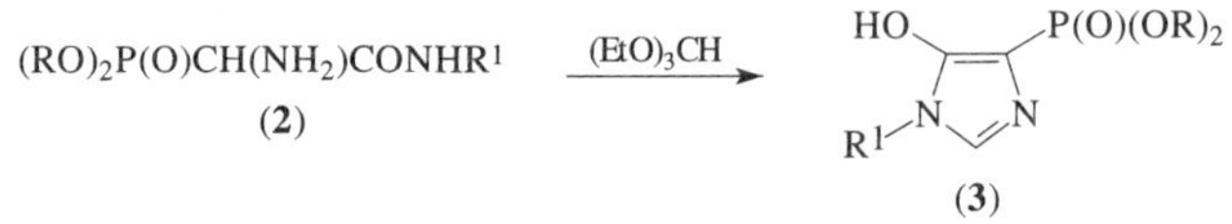

Scheme 3.1.5

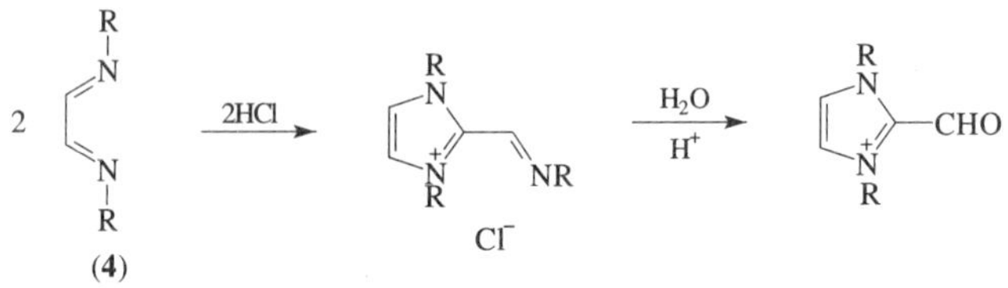

Scheme 3.1.6

[16]. Similar trimethylsilyldiimines are converted by phosgene into 2-imidazolinones [17]. There are also similarities in the reaction of cyanogen with aldehydes in the presence of HCl. The products are 2-substituted 4,5-dichloroimidazoles in yields of 36–72%. This reaction sequence has value because of the difficulty of access of 4,5-dichloroimidazoles, but it is restricted to aldehydes which cannot enolize and undergo alternative Aldol-type side reactions. In addition, basic heteroaromatic aldehydes form salts under the reaction conditions [18].

Synthetic methods based on the versatility of DAMN have already been discussed. Its conversion into amidinium salts by nitrilium triflates and cyclization of these products to form 5-aminoimidazoles, and its reactions with formamidine, or as its imidate with ammonia, amines, or orthoformate esters, have been classified as 1,5 bond formations (see Section 2.2.1). Oxidative cyclization of Schiff bases formed when DAMN is treated with appropriate carbonyl reagents (see Section 2.1) also provides versatile entry to a variety of imidazoles.

Under the current heading can be considered processes in which DAMN (or a DAMN derivative) is heated in alcoholic solution with an orthoester to give 4,5-dicyanoimidazole [19, 20]. Note that with excess orthoester the 1-alkyl-4,5-dicyanoimidazole is formed (see Section 2.2.1). With formic

3.1. FORMATION OF 1,2 AND 2,3 BONDS

TABLE 3.1.1
2-Substituted 4,5-dicyanoimidazoles prepared from DAMN

Reagent	Reaction conditions	2 substituent in product	Yield (%)	Ref.
$Cl_2C=NSO_2Me$	Δ, C_6H_6	$NHSO_2Me$	90	27
$Cl_2C=NSO_2Ph$	Δ, C_6H_6	$NHSO_2Ph$	79	27
$(EtO)_3CH$	135°C, PhOMe	H	83	27
HCO_2H	Δ, diglyme	H	76	22
ClCN	0–50°C, THF	NH_2	75	25, 26
$COCl_2$	50°C, NaOH, dioxane	"OH"	80	23
DMF	160°C, $POCl_3$	H	90	26
$Cl_2C=N$-tBu	R.T. 16 h, THF	NH-tBu	11	26
MeC(=NH)OEt	Δ, 9 h, PhOMe	Me	51	28

acid in refluxing xylene, DAMN is converted into 4-cyanoimidazole-5-carboxamide [21, 22], with phosgene in the presence of base the product is 4,5-dicyanoimidazolin-2-one [23, 24], and with cyanogen chloride or other isocyanide dichlorides, 2-amino-4,5-dicyanoimidazoles result [25, 26]. A further example is the reaction of DAMN with dichloromethylenesulfonamides [27]. With iminoether (imidic ester) hydrochlorides, DAMN is converted into 2-alkylimidazole-4,5-dinitriles [28]. Table 3.1.1 lists some examples.

In the reaction with formic acid, reaction conditions modify the products isolated. Refluxing for 15 min in dry benzene merely formylates one of the amino groups. Refluxing for 6 h in diglyme gives 4,5-dicyanoimidazole, whereas in refluxing xylene one of the nitrile groups is converted into an amido group.

The synthetic utility of DAMN is summarized in Scheme 3.1.7.

2-t-Butylamino-4,5-dicyanoimidazole [26]

To a solution of t-butyl isocyanide dichloride (8 g, 96 mmol) in THF (100 ml) is added DAMN (5.0 g, 46 mmol). The temperature rises to ~43°C, and a precipitate forms. After stirring (18 h), the precipitate redissolves, and the resulting dark solution is chromatographed on silica with chloroform to give the product (0.94 g, 11%). White crystals separate from ether–petroleum ether, m.p. 171–172°C.

4,5-Dicyano-2-sulfonylaminoimidazoles [27]

To a suspension of DAMN (2.7 g, 25 mmol) in dry benzene (100 ml) is added, dropwise with vigorous stirring at room temperature, a solution of an N-dichloromethylenesulfonamide (25 mmol) in benzene (100 ml). After refluxing (8–10 h), the reaction mixture is cooled, and the precipitate filtered, dried and recrystallized from acetonitrile (see also Table 3.1.1).

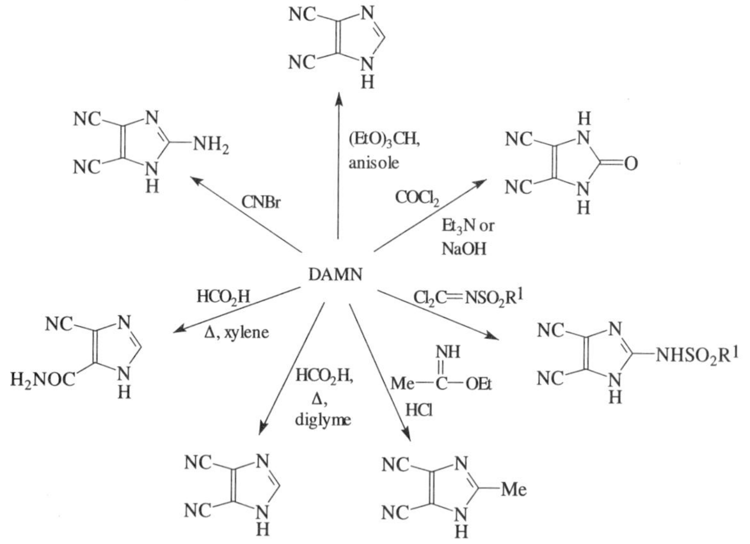

Scheme 3.1.7

4-Cyanoimidazole-5-carboxamide [22]

A mixture of DAMN (4 g), formic acid (4 ml) and xylene (150 ml) is stirred at room temperature (30 min), and then refluxed (6 h). The resulting dark solids (which adhere to the walls of the flask) are collected, washed with diethyl ether, and extracted repeatedly with hot water. Concentration of the aqueous extracts gives the above product (3.9 g, 77%), m.p. 273°C. From the residue some imidazole-4,5-dicarboxamide (0.12 g, 2.2%) is obtained.

Whereas diiminosuccinonitrile is converted by trifluoracetic anhydride into 2H-imidazoles, these sometimes rearrange at fairly low temperatures (<80°C) to 1,2-dialkyl-4,5-dicyanoimidazoles [17, 29, 30].

A synthetic strategy which leads to 2,4(5)-disubstituted imidazoles utilizes the reaction between 2-aminonitriles and aldehydes. The aldehyde provides the 2-substituent [31, 32]. Thus, 2-methylaminophenylacetonitrile reacts with benzaldehyde to give 1-methyl-2,5-diphenylimidazole (76%) [32]. With orthoformates the products are 2-imidazolin-4(5)-ones. Thus, when aminoacetonitrile hydrochloride is refluxed for 30 min with trimethyl orthoformate, some 2-imidazolin-4(5)-one is formed (although the reaction is not very efficient — the major product is N-formylacetonitrile) (Scheme 3.1.8) [33, 34]. A rapid method for making N-substituted α-aminonitriles under mild conditions in a one-pot reaction uses an aldehyde, a primary or secondary amine, and trimethylsilyl cyanide in methanol [35].

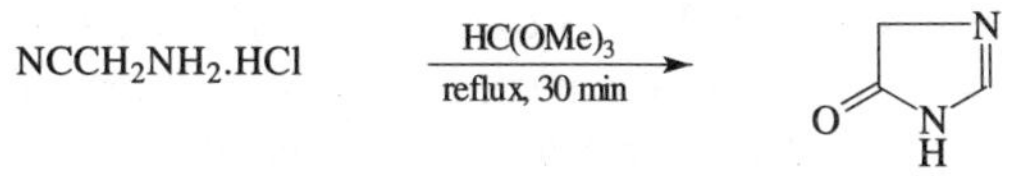

3.1. FORMATION OF 1,2 AND 2,3 BONDS

NCCH$_2$NH$_2$·HCl $\xrightarrow[\text{reflux, 30 min}]{\text{HC(OMe)}_3}$

Scheme 3.1.8

Condensation of aldehydes with 2-amino-2-phenylacetamidooxime (5) gives high yields (~80%) of 5-aminoimidazoles as their Schiff bases [36]. In a similar reaction, 4-amino-1-substituted imidazoles are formed in the reaction between the hydrochlorides of benzyloxycarbonyl or tosyl derivatives of aminoacetamidine and orthoformic esters (Scheme 3.1.9) [37].

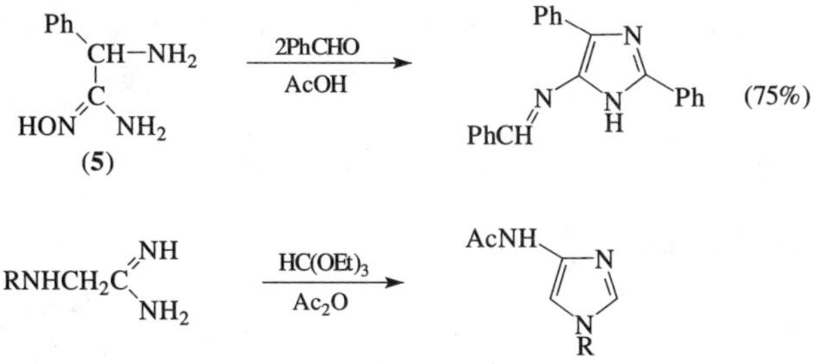

Scheme 3.1.9

Imidazole N-oxides and/or 1-hydroxyimidazoles are also accessible by ring-synthetic methods which fall into this general classification. Of most importance are the reactions of (E)-α-hydroxylamino oximes with aldehydes [38–41]. When the carbonyl compound is aryl or hetaryl, alternative pathways can result in the formation of imidazole N-oxides, pyrazine-1,4-dioxides or open-chain nitrones [42]. Methods of synthesis of α-hydroxylamino oximes have been surveyed [41], and their applications to imidazole synthesis have been reviewed [41, 43]. Suitable (E)-α-hydroxylamino oximes (6) react at room temperature with formaldehyde, or when warmed with acetaldehyde or chloroacetaldehyde, to give 1-hydroxy-3-imidazoline 3-oxides (7) in 40–100% yields (Scheme 3.1.10). Provided that the 2 and 5 positions of (7) bear at least one hydrogen atom, aromatization can take place. Treatment with ethanol saturated with hydrogen chloride gives 1-hydroxyimidazoles (8), which can be isolated either as hydrochloride salts or as the free bases [39]. Warming at 50–60°C with dilute alkali also "dehydrates" compounds of type (7) [44]. The isomeric 1-hydroxides (9) are formed in high yield from the acyloxy or aroyloxy derivatives of (7). Either heating *in vacuo* or saturation of an ethanolic solution with dry HCl is effective [39, 45, 46]. When oximes (6) are condensed with triethyl orthoformate by heating in ethanolic solution,

2-ethoxy-1-hydroxy-3-imidazoline 3-oxides (**7**) ($R^1 = OEt$) are formed in 60–65% yields [47, 48]. Further heating in ethanol or merely increasing the reaction time gives 2-ethoxy-1-hydroxyimidazoles (e.g. (**8**) ($R^1 = OEt$, $R = Me$, $R^2 = Ph$, 80%)). The isomer (**8**) ($R^1 = OEt$, $R = Ph$, $R^2 = Me$) can again be made by the acetylation–aromatization sequence above [47].

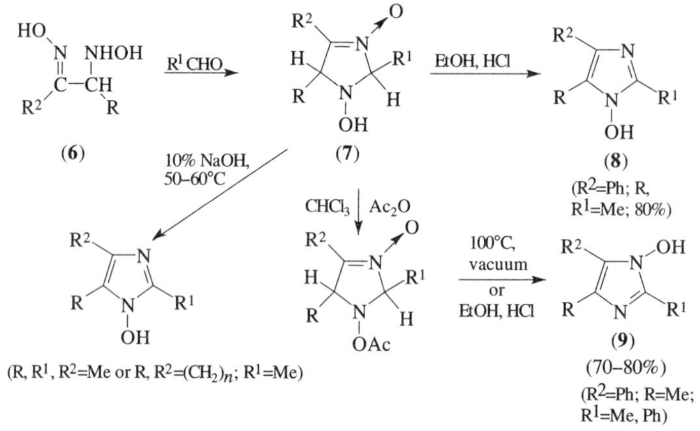

Scheme 3.1.10

Reaction of phosgene with the *anti* isomer of 1-amino-1-phenyl-2-propanone oxime gives 1-hydroxy-5-methyl-4-phenyl-3-imidazolin-2-one in 25% yield [49]. A review of *N*-oxide formation has appeared recently [50], and discussions of other approaches to the compounds are found in Section 4.1 and Chapter 5.

The Pinner salt (**10**), which is obtained by reaction of methanol and gaseous HCl with α-azidoacetonitrile [51], reacts sequentially with triethylamine, triphenylphosphine and either acid chlorides or isothiocyanates to give imidazoles in moderate yields (Scheme 3.1.11) [52].

General method [52]

A mixture of α-azidoacetonitrile (2.32 g, 28 mmol), dry methanol (0.90 g, 28 mmol) and dry CH_2Cl_2 (25 ml) cooled to 0°C is prepared. Through this stirred solution a stream of dry HCl is passed (3 h). Addition of dry diethyl ether precipitates the Pinner salt (**10**), which is rapidly filtered off and dried *in vacuo* at room temperature. To a suspension of (**10**) (1.50 g, 9.96 mmol) in dry CH_2Cl_2 (30 ml) triethylamine (1.00 g, 9.96 mmol) is added. After stirring at ambient temperature (1 h) the solution is cooled to 0°C before dropwise addition under nitrogen of a solution of triphenylphosphine (2.61 g, 9.96 mmol)

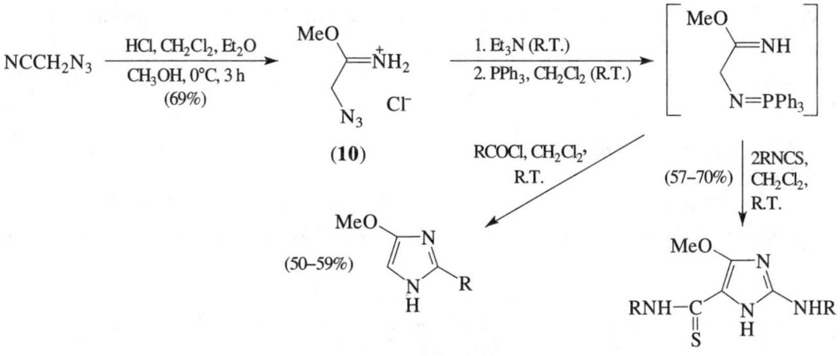

3.1. FORMATION OF 1,2 AND 2,3 BONDS

Scheme 3.1.11

in dry CH_2Cl_2 (15 ml). Stirring is continued (2 h) before addition of the appropriate aroyl halide (1 eq.) or isothiocyanate (2 eq.) in dry CH_2Cl_2. The reaction mixture is allowed to warm up to ambient temperature and stirred (15 h). The solvent is removed under reduced pressure, and the residue chromatographed on a silica gel column (40 × 3.5 cm, 70–230 mesh), eluting with ethyl acetate–diethyl ether (1:4).

3.1.2 Benzimidazoles

The most important synthetic methods for preparation of a wide range of benzimidazoles condense o-diaminobenzenes with carboxylic acids or derivatives. Benzimidazole itself can be made in greater than 80% yield merely by standing a mixture of o-phenylenediamine and formic acid at room temperature for 5 days. At around 100°C the process takes only 2 h, and it is applicable to a wide range of 2-substituted benzimidazoles. Careful choice of reaction conditions is, however, essential if good yields are to be obtained in all instances [53, 54].

The most widely used conditions (Phillips method [55]) involve heating the reagents together in the presence of hydrochloric acid, usually around 4 M concentration. As detailed below, it is more difficult to make 2-arylbenzimidazoles in this way, but application of more vigorous conditions is frequently successful, e.g. heating in a sealed tube at 180°C in the presence of hydrochloric acid or, better still, substitution of polyphosphoric acid for the hydrochloric acid. Other acidic catalysts have also been found to be effective [56].

The range of reaction conditions which has been used is quite wide, from merely heating the diamines with a carboxylic acid, to heating in the presence of acids such as hydrochloric acid [57–59], PPA [60–64] and boric acid [65], hexamethyldisilazane in the presence of a catalytic amount of triflic acid [66], heating with a titanium catalyst [67], heating at elevated pressure [68], or use of photolytic conditions [69]. Tables 3.1.2–3.1.4 list some typical examples.

2-Trifluoromethylbenzimidazole [74]

A mixture of *o*-phenylenediamine (5.0 g, 46 mmol), trifluoroacetic acid (5.7 g, 50 mmol) and 4 M HCl is heated under reflux (3 h). The cooled solution is

TABLE 3.1.2
Benzimidazoles made from reaction of aliphatic carboxylic acids and *o*-phenylenediamines

R	Reaction conditions[a]	Other substituents on benzimidazole	Yield (%)	Ref.
H	A	5,6-Me$_2$	85	70
H	E	4,7-(OMe)$_2$	88	58
H	A	4-CF$_3$	44	57
H	A	5-CF$_3$	35	57
H	E	4-F	80	71
H	A	5,6-(OMe)$_2$	93	58
H	E	4,7-(NO$_2$)$_2$	88	72
H	A	5-CO$_2$Et	58	73
Me	H	—	73	69
Me	C	4,7-(NO$_2$)$_2$	85	72
Me	A	4,7-(OMe)$_2$	90	58
Me	A	5,6-(OMe)$_2$	65	58
Et	H	—	62	69
CF$_3$	A	—	64	74
CF$_3$	A	4-CF$_3$	29	74
CF$_3$	A	5-CF$_3$	48	74
CF$_3$	E	4-F	83	71
Et	A	4,7-(OMe)$_2$	57	58
Pr	A	4,7-(OMe)$_2$	50	58
CH$_2$OH	A	4,7-(OMe)$_2$	72	58
CH$_2$Cl	A	1-(CH$_2$)$_2$OEt	47	75
tBu	F	—	60	76
CH$_2$SBu	A	—	49	73
CO$_2$H	J	—	62	77
Cl	J	—	30	77
CH=CH–CO$_2$H	J	—	96	77

[a] A — Δ, HCl (Phillips); B — HBO$_3$, 190-200°C; C — Ac$_2$O, Δ; D — PPA, Δ; E — 180-210°C; F — Δ, high pressure; G — Ph$_3$P(OTf)$_2$, R.T.; H — Photolysis; I — Δ, Ti(OBu)$_4$; J — HCl, acid resin.

3.1. FORMATION OF 1,2 AND 2,3 BONDS

TABLE 3.1.3
Benzimidazoles made from reaction of aromatic carboxylic acids and o-phenylenediamines

Ar	Reaction conditions[a]	Yield (%)	Ref.
C_6H_5	I	54	67
C_6H_5	G	85	78
C_6H_5	D	81	60
o-ClC_6H_4	I	65	67
m-ClC_6H_4	E	44	68
p-ClC_6H_4	I	69	67
o-BrC_6H_4	E	22	68
m-BrC_6H_4	E	35	68
p-BrC_6H_4	B	80	65
o-IC_6H_4	E	26	68
m-IC_6H_4	E	23	68
p-IC_6H_4	E	22	68
o-FC_6H_4	E	26	79
m-FC_6H_4	E	46	79
p-FC_6H_4	E	39	79
o-MeC_6H_4	G	95	78
m-MeC_6H_4	D	85	60
p-MeC_6H_4	B	58	65
o-$NH_2C_6H_4$	D	60	60
o-HOC_6H_4	D	29	60
o-HSC_6H_4	A	12	59
o-$HO_2CC_6H_4$	D	58	60
o-$NO_2C_6H_4$	I	95	67
p-$NO_2C_6H_4$	I	48	67
$3,4$-$Cl_2C_6H_3$	D	62	60

[a]See Table 3.1.2

neutralized with aqueous ammonia, and the precipitated product is filtered and recrystallized from ethanol (5.5 g, 64%), m.p. 209–210°C.

Most aliphatic acids enter the process easily, but more vigorous reaction conditions are necessary for aryl carboxylic acids or for sterically hindered alkanoic acids. Increased pressure can help counteract such problems, e.g. 2,2-dimethylpropanoic acid heated at 112°C in ethanol with o-phenylenediamine dihydrochloride for 24 h at 8 kbar pressure gives a 48% yield of 2-t-butylbenzimidazole [76]. Care has to be taken, though, that the ethanol concentration is not too high, as esterification of the carboxylic acid becomes a competing reaction; DMSO appears to be a suitable alternative

TABLE 3.1.4
Benzimidazoles made from reaction of other aromatic and heteroaromatic carboxylic acids and o-phenylenediamines

Carboxylic acid	Reaction conditions[a]	Other substituents on benzimidazole	Yield (%)	Ref.
1-naphthyl	B	—	77	65
2-pyridyl	D	5-OMe	33	61
3-pyridyl	D	—	11	60
3-pyridyl	D	5-OMe	9	61
3-pyridyl	D	5-Cl	45	61
3-pyridyl	D	7-Cl	45	61
4-pyridyl	D	5-OMe	61	61
4-pyridyl	D	5-Cl	40	61
4-pyridyl	D	5,6-Me$_2$	81	64
2-HS-3-pyridyl	A	—	14	59

[a]See Table 3.1.2.

solvent. This method is probably only of synthetic utility when the acids are sterically hindered. There has been a review of the applications of high pressure to such reactions [80]. Phillips-type reactions can also be carried out in the presence of a mixture of hydrochloric acid and an acid resin such as Dowex-50W-X8. Such reactions go at room temperature, much milder conditions than the usual refluxing in 4 M HCl [77]. If the o-diaminoarene has one of the amino groups substituted by an alkyl or aryl group, 1-substituted benzimidazoles are formed [72]. Under prolonged heating, α-amino acids react with o-phenylenediamine to give satisfactory yields of 2-aminoalkylbenzimidazoles. When the α-carbon is highly substituted, reactivity is reduced, and compounds such as glutamic acid naturally react at the γ-carboxyl group [81].

Yields of 2-arylbenzimidazoles have been shown to decrease in the order 2-NO$_2$ >2-Me>4-Cl>H>4-NO$_2$ >2-NH$_2$ for a range of phenylcarboxylic acids reacting at 185–250°C in the presence of stoichiometric amounts of hydrochloric, polyphosphoric or boric acid. In the HCl-catalysed reactions the yields increase in line with the pK_a values, i.e. associated with increasing polarization of the carbonyl bond [67].

5-Methoxy-1-methylbenzimidazole [82]

A solution of 3-amino-4-(methylamino)anisole (46.6 g, 0.27 mol) in 96% formic acid (50 ml, 1.27 mol) is refluxed (2 h). After addition of toluene (200 ml) and water (50 ml) the volatile liquids are removed under reduced pressure, the residue poured into water (300 ml), and extracted with ethyl acetate (2 × 200 ml). The combined extracts are washed with water (200 ml) before drying (MgSO$_4$), filtering and evaporating to give 5-methoxy-1-methylbenzimidazole (35.6 g, 81%). The brownish product of approximately 95% purity can be purified by filtration in dichloromethane solution through silica gel. Recrystallization from dichloromethane–hexane gives plates, m.p. 112°C.

2-t-Butylbenzimidazole [76]

A solution of o-phenylenediamine (1.0 g, 9 mmol) and 2,2-dimethylpropanoic acid (1.0 g, 10 mmol) in 75% (v/v) ethanol–water (6 ml) is heated at 107°C for 66 h under 8 kbar pressure (reactions are carried out in glass or Teflon precision-bore tubes of 5–30 ml capacity pressurized with a hydraulic press and electrically heated). Dilution with water and neutralization with ammonia (d. 0.88) gives the benzimidazole (0.73 g, 42%), which is recrystallized from aqueous methanol, m.p. 334°C.

2-(2'-Pyridyl)benzimidazole [61]

A mixture of o-phenylenediamine (3.24 g, 30 mmol), 2-picolinic acid (4.31 g, 35 mmol) and PPA (7 g) is heated at 170–180°C (3 h). The reaction mixture is poured into ice–water, neutralized with sodium bicarbonate and filtered. Further treatment of the filtrate with 5% sodium hydroxide solution produces additional crystals which are filtered, combined with the earlier crop, washed with water, dried, and recrystallized from benzene to give the product (4.7 g, 80%).

2-Methyl-5,7-dinitro-1-phenylbenzimidazole [72]

A solution of 4,6-dinitro-2-phenylaminoaniline (1.00 g, 3.65 mmol) in glacial acetic acid (7 ml) and acetic anhydride (5 ml) is refluxed (12–15 h). The cooled mixture is quenched with iced water to give a grey solid (1.08 g, 99%) which, on recrystallization from ethanol, gives off-white needles (0.80 g, 74%), m.p. 189°C.

The last synthesis can be adapted to prepare 2,2'-bi-benzimidazoles from dibasic acids [83].

General procedure [83]

A 1:2 mixture of the dibasic acid and o-phenylenediamine is placed in a two-necked flask equipped with a reflux condenser, a calcium chloride drying tube and a mechanical stirrer. PPA is added to make a thick paste, then sufficient extra PPA to make a smooth slurry. The temperature is raised slowly until the desired temperature is reached, maintained for 3–4 h, then the slurry is cooled to about 80°C and added slowly in a fine stream with rapid stirring to cold water. Stirring is continued (1 h) before filtering the solution and neutralizing with dilute aqueous ammonia to precipitate the product, which is filtered, dried and recrystallized (e.g. 2,2'-bi-benzimidazole is isolated in 85% yield using oxalic acid heated for 4 h at 240°C; analogous products have been made using malonic acid (85% yield, 250°C, 3.5 h), succinic acid (80% yield, 230°C, 3.5 h) and glutaric acid (90% yield, 200°C, 3 h)).

Recently, some modified reaction conditions have been proposed to improve yields, particularly for aromatic carboxylic acids. A "phosphonium anhydride" reagent (11) made from triphenylphosphine oxide and triflic anhydride appears to induce reaction at low temperatures in common solvents giving high yields of 2-arylbenzimidazoles in 30–60 min at room temperature (Scheme 3.1.12) [78].

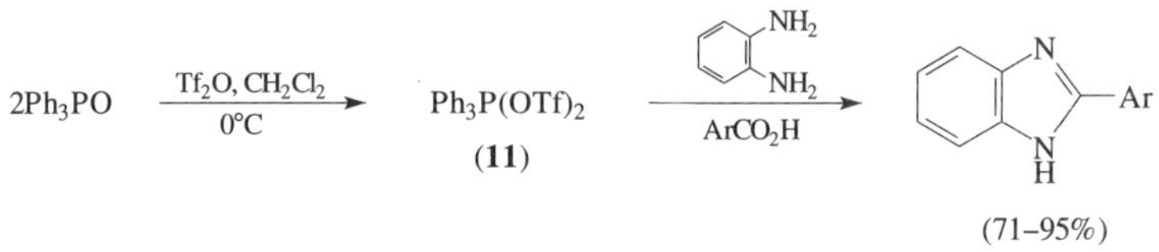

$$2Ph_3PO \xrightarrow{Tf_2O, CH_2Cl_2}_{0°C} Ph_3P(OTf)_2 \xrightarrow{\text{o-diaminobenzene}}_{ArCO_2H} \text{2-arylbenzimidazole}$$

(11)

(71–95%)

Scheme 3.1.12

2-Phenylbenzimidazole [78]

A solution of triflic anhydride (1.57 ml, 10 mmol) [84] in dichloroethane (30 ml) at 0°C is added to a solution of triphenylphosphine oxide (5.56 g, 20 mmol) or equivalent phosphinamide in dichloroethane. After appearance of a precipitate (usually in less than 15 min), a solution containing o-phenylenediamine (0.44 g, 4 mmol) and benzoic acid (0.61 g, 5 mmol) in dichloroethane (10 ml) is added dropwise. After stirring (0.5 h), the solution is washed with 5% sodium bicarbonate solution, dried (MgSO$_4$) and evaporated. The residue is passed through a short column packed with silica and eluted with hexane–ethyl acetate (3:1) to remove excess phosphine oxide. Evaporation gives 2-phenylbenzimidazole (0.66 g, 85%), m.p. 287°C.

An alternative procedure utilizes photo-oxidation of o-phenylenediamine in the presence of a carboxylic acid [69]. It is unlikely that this approach will supersede the more usual methods.

2-Ethylbenzimidazole [69]

A solution of o-phenylenediamine (1.08 g, 10 mmol) in glacial acetic acid (100 ml) is irradiated in a quartz flask at 253.7 nm (3 days) while air is slowly

3.1. FORMATION OF 1,2 AND 2,3 BONDS

bubbled through the solution. Evaporation of the solution gives a brown solid which is redissolved in hot water (200 ml), treated with charcoal, filtered and concentrated. The crystalline product (0.90 g, 62%) separates on standing. Similarly prepared is 2-methylbenzimidazole (73%).

A novel palladium-catalysed carbonylation of iodobenzene has recently been linked to base-induced coupling and cyclization with o-phenylenediamine, to give 2-arylbenzimidazoles without having to use an aryl carboxylic acid (Scheme 3.1.13). Provided that bases with pK_a values around 6.6 are used, the yields of 2-arylbenzimidazoles lie in the range 70–98%. This route is tolerant of a variety of functional groups and complements the classical approaches where the required benzoic acids are not readily available [85].

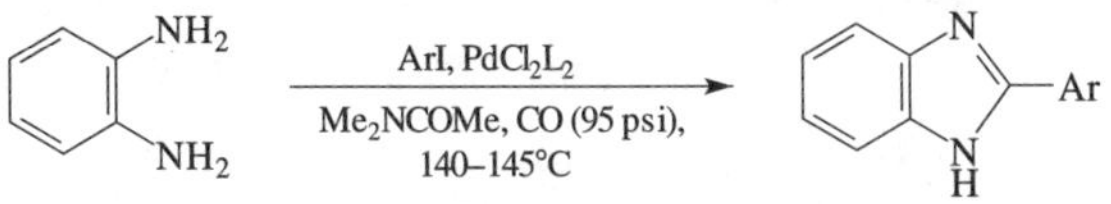

(Ar (%): Ph (75), p-MeC$_6$H$_4$ (81), p-MeOC$_6$H$_4$ (71), p-ClC$_6$H$_4$ (93), p-CNC$_6$H$_4$ (74))

Scheme 3.1.13

A wide range of acid derivatives can be substituted for carboxylic acids: esters [86–90], orthoesters [91, 92], nitriles [53, 60, 93], amides [60, 94, 95], imidates [96–100], acid chlorides (including phosgene) [101] and anhydrides (Scheme 3.1.14) [102, 103].

4-Hydroxybenzimidazole [91]

A mixture of 2,3-diaminophenol (1.24 g, 10 mmol), triethyl orthoformate (1.48 g, 10 mmol) and a catalytic amount of p-toluenesulfonic acid is heated at 120°C in a distillation flask. Ethanol is removed by distillation as it is formed, and final traces are removed by azeotroping with added toluene. The solid residue is recrystallized from ethanol to give 4-hydroxybenzimidazole (1.20 g, 90%), m.p. 182–183°C.

Orthoesters such as ethyl triethoxyacetate react easily when heated with o-phenylenediamines, to give 2-carbalkoxybenzimidazoles in good yields, e.g. 2-carbomethoxybenzimidazole (55%) and 1-phenyl-2-carbethoxybenz-imidazole (71%) [92]. This provides an alternative to the usual approaches involving oxidation of the 2-hydroxyalkyl group [104], alcoholysis of 2-trichloromethyl group [98], or cleavage of the aryl–silicon bond of 1-methyl-2-trimethylsilylbenzimidazole with ethyl chloroformate [105], all of which give low yields and are multistep reactions.

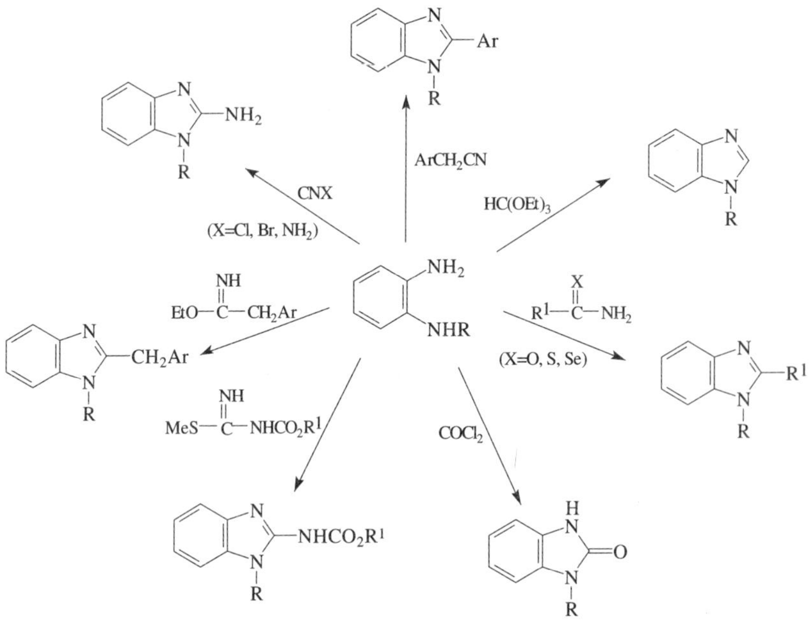

Scheme 3.1.14

When polyphosphoric or hydrochloric acids are used as the condensing media with *o*-diaminoarenes, the diamine often competes successfully with the carbonyl oxygen atom for the proton provided by the acid catalyst. For this reason there are advantages to replacing the carbonyl oxygen by the more basic imino group. In consequence, imino ethers (imidates) are used, often very successfully, in the reactions [54, 97, 106]. These imidates can be made quite readily from the corresponding nitriles [96], or they can be made *in situ* [93]. Whereas nitriles are converted into basic imidates in the presence of a catalytic amount of base, benzimidazoles form better when the imidinium ion is used [100]. The reactions will not work satisfactorily when there are strongly electron-withdrawing groups in the *o*-phenylenediamine ring.

Ethyl 2,4-dinitrophenylacetimidate [96]

To 2,4-dinitrophenylacetonitrile (20.7 g, 0.10 mol) in dry chloroform (200 ml) is added ethanol (6 ml). The solution is cooled to 0°C and then saturated with dry HCl during 30 min. After standing at 0°C (12 h) the colourless crystalline product is filtered and washed with dry chloroform giving the above imidate (as its hydrochloride) (25 g, 90%), m.p. 181°C.

3.1. FORMATION OF 1,2 AND 2,3 BONDS

2-(2',4'-Dinitrobenzyl)benzimidazole [96]

To a suspension of the above hydrochloride salt (6.94 g, 24 mmol) in methanol (15 ml) is added, with stirring, a solution of o-phenylenediamine (2.7 g, 25 mmol) in methanol (15 ml). After stirring for 30 min at 20–25°C a thick mass of crystals separates. The mixture is then boiled (1 h) and cooled, and the precipitate filtered and washed with water, to give the benzimidazole (6.31 g, 84%). Recrystallization from 50% aqueous ethanol gives colourless crystals, m.p. 202–203°C. Similarly prepared are 2-benzylbenzimidazole (73%) [93], 2-benzyl-5(6)-nitrobenzimidazole (65%) [107] and 2-p-chlorobenzylbenzimidazole (53%) [107]. It should be noted that such reactions of iminoesters with o-diamines are rapid only when carried out in an acidic medium.

2-[(2'-Carbomethoxyphenoxy)methyl]benzimidazole [100]

Under nitrogen, sodium metal (0.35 g, 15.0 mmol) is added to a stirred solution of (2-carbomethoxyphenoxy)acetonitrile (made from methyl salicylate) (3.0 g, 15.7 mmol) in dry methanol (50 ml) at room temperature. Stirring is continued (40 min) before addition of o-phenylenediamine dihydrochloride (2.72 g, 15.0 mmol) when the colourless solution turns yellow and a solid separates. After further stirring (2 h) the salts are filtered and the filtrate is treated with decolorizing charcoal. Water (about 70–75 ml) is added until the methanolic solution becomes cloudy. After standing at room temperature for several hours, the precipitate which separates is filtered, washed with water and air dried, to give the benzimidazole (3.66 g, 88%) as colourless plates, m.p. 159°C.

2-Trichloromethylbenzimidazoles [99]

To a cooled solution or suspension of the appropriate o-arylenediamine monohydrochloride (0.1 mol) in methanol, ethanol or 1,2-dimethoxyethane (200–300 ml), methyl trichloroacetimidate [108] (0.1 mol) is added slowly. When the resultant exothermic reaction has subsided, the mixture is maintained at room temperature for several hours. Aqueous quenching, or evaporation of the solvent, yields the products (substituent, solvent, yield, recrystallizing solvent listed): H, EtOH, 95%, HOAc; 5(6)-Cl, MeOH, 55%, xylene; 5,6-Cl$_2$, MeOH, 67%, MeOH; 5(6)-Me, (MeOCH$_2$)$_2$, 60%, benzene; 5,6-Me$_2$, (MeOCH$_2$)$_2$, 35%, benzene; 1-Me, MeOH, 90%, MeOH.

Under oxidative conditions, aldehydes will react with 1,2-arylenediamines. With an equimolar amount of any aromatic or (especially) heterocyclic aldehyde the initial product is a monoanil. In the presence of suitable oxidizing agents (e.g. heating in nitrobenzene) these anils cyclize to benzimidazoles (see also Section 2.1.2). In a number of instances, atmospheric oxygen is sufficient to achieve the desired oxidation, but copper acetate, mercuric oxide or chloroanil have also been used [65, 106, 109-111]. Sometimes it may only be necessary to heat the aldehyde with the diamine for a few minutes in nitrobenzene or o-dinitrobenzene to achieve cyclization (e.g. 2-(2′-furyl)benzimidazole (73%), 2-(2′-pyridyl)benzimidazole (85%), 2-benzylbenzimidazole (68%) and 2-isopropylbenzimidazole (61%)) [110]. When 2 mol or more of aldehyde is used, mixtures of 1,2-di- and 2-mono-substituted benzimidazoles can result, often along with a small amount of 2,3,4-triaryldiazepine [65, 81, 112, 113]. Thus, moderate yields of 1-substituted benzimidazoles have been isolated when 2 mol of formaldehyde in boiling hydrochloric acid reacts with an N-substituted o-phenylenediamine [114].

5(6)-Carbethoxy-2-(4′-hydroxyphenyl)benzimidazole [115]

A mixture of 4-carbethoxy-1,2-diaminobenzene (0.18 g, 1 mmol) and p-hydroxybenzaldehyde (0.12 g, 1 mmol) is heated at 140-150°C (24-36 h) in nitrobenzene (10 ml). Removal of the solvent under reduced pressure, followed by chromatography of the residue on silica gel or florisil, gives the benzimidazole (0.24 g, 85%). Reaction progress can be monitored by TLC with visualization under UV or by spraying with 2.5% methanolic phosphomolybdic acid and heating.

1-Methyl-4-nitrobenzimidazole [116]

3-Nitro-1,2-diaminobenzene (153 g, 1 mol) is dissolved in a mixture of ethanol (3 l) and concentrated hydrochloric acid (1 l). After addition of 40% aqueous formaldehyde (140 ml, 2 mol), the mixture is refluxed (2 h), allowed to cool to room temperature, and then neutralized with 20% aqueous sodium hydroxide. The precipitated product is collected, washed with water and dried before recrystallizing from toluene-heptane (2:1) to give the product (138 g, 77%) as a yellow crystalline solid, m.p. 168°C.

Reaction of 1,2-diaminoarenes with phosgene or urea gives 2-benzimidazolones (Scheme 3.1.14). The latter reagent has been preferred as it is less toxic. There are, however, a variety of other reagents which lead to the same products: carbonate esters [117], diethyl pyrocarbonate [118], N,N-diethylcarbamyl chloride [119], 1,1′-carbonyldiimidazole [120], cyanic acid [121] and carbon dioxide [122]. (See also reactions with ketones below.) Preston has reviewed these processes [54]. The reactions with urea (or

3.1. FORMATION OF 1,2 AND 2,3 BONDS

substituted ureas) are usually carried out at around 130°C, but a microwave oven can also be used [123]. High-temperature (~400°C) reactions with aryl isocyanates have been used recently, but do not appear to offer advantages over the earlier approaches [124]. A one-pot process utilizes the reductive carbonylation of an o-nitroaniline using thermal decomposition of ammonium formate as the carbonyl source. This is much less troublesome than using carbon monoxide, sulfur and water, but the process requires the use of an autoclave [125].

Benzimidazolone — method A [123]

A mixture of o-phenylenediamine (2.16 g, 20 mmol) and urea (1.32 g, 22 mmol) in N,N-dimethylacetamide (0.5 ml) and diethylene glycol (5 ml) contained in a 100 ml Erlenmeyer flask covered with a watch glass is irradiated in a microwave oven at 385 W (2 min) in an efficient fume hood. The mixture is allowed to cool to room temperature, and the precipitate is collected, washed with water (3 ml), acetone (5 ml) and ethanol (5 ml) before recrystallization from ethyl acetate. The yield is 91%, m.p. 308–309°C.

Benzimidazolone — method B [125]

In a 100 ml stainless steel autoclave, o-nitroaniline (2.76 g, 20 mmol), sulfur (2.57 g, 80 mmol), water (2.0 ml, 110 mmol), ammonium formate (6.30 g, 100 mmol), potassium carbonate (4.14 g, 30 mmol) and dimethylacetamide (40 ml) are placed along with a magnetic stirring bar. The reaction is carried out at 180°C for 3 h (maximum pressure 10 kg cm^{-2}). After extraction with 1-butanol, the product (2.52 g, 94%) is recrystallized from 1:1 aqueous methanol. Similarly prepared are 5-methyl- (27%), 5-methoxy- (36%) and 5-chlorobenzimidazolones (76%).

Analogous methods lead to benzimidazolethiones when o-phenylenediamines are treated with carbon disulfide, thiophosgene, 1,1'-thiocarbonyldiimidazole, thioureas, thiocyanates or potassium ethyl xanthate [123, 126–129]. The reactions with carbon disulfide take place in basic media, e.g. with KOH or pyridine. Again, microwave irradiation offers advantages [123].

1-Trifluoroacetyl-5-fluorobenzimidazolin-2-thione [130]

A mixture of 2-amino-4-fluorotrifluoroacetanilide (6.7 g, 0.03 mol), KOH (1.9 g) and carbon disulfide (2.3 g, 0.03 mol) in 95% aqueous ethanol (30 ml) and water (5 ml) are heated under reflux (5 h). Activated charcoal is cautiously added, the mixture is refluxed again (10 min) and filtered while hot. The filtrate, diluted with warm water (30 ml), is treated with stirring with a solution of acetic acid (2.4 ml) in water (5 ml). The product separates as white crystals, which are recrystallized from aqueous methanol (4.6 g, 66%). Similarly prepared are the 5-chloro (65%), 5-methyl (65%) and 5-methoxy derivatives (63%).

When o-phenylenediamine reacts with aliphatic or aliphatic–aromatic ketones the initially formed benzimidazolines can be thermally decomposed, losing a hydrocarbon fragment to yield 2-substituted benzimidazoles [81, 131]. While this method does not appear to have major synthetic importance, the analogous reaction with a β-diketone has some application [132]. Presumably it involves an acid-catalysed retro-Claisen condensation of the β-dicarbonyl compound. When acetylacetone is used, acetone is formed. The process, then, offers an unambiguous approach to compounds such as 2-methyl-4-nitrobenzimidazoles (not available by direct nitration). Neither the numbers of substituents on the arylenediamine nor their natures appear to affect yields or reaction times significantly.

2-Methyl-4-nitrobenzimidazole [132]

To a solution of 3-nitro-1,2-diaminobenzene (0.15 g, 1 mmol) in hot ethanol (15 ml) and 5 M HCl (4 ml) is added acetylacetone (0.2 g, 2 mmol). The mixture is heated under reflux (~10 min) (the reaction is monitored by TLC using dichloromethane–ethyl acetate (1:1)), cooled to room temperature and filtered. Further product is obtained by extraction of the filtrate with chloroform (3 × 15 ml). Recrystallization from 75% aqueous ethanol gives the product (0.13 g, 75%). Similarly prepared are a wide range of 2-methylbenzimidazoles in yields of 74–89%.

When the required 1,2-diaminoarenes are not readily available it is often possible to utilize o-nitroanilines as substitutes. They can be successively reduced or hydrogenated and then cyclized [82, 91, 133], or the cyclization and reduction processes can be combined. Suitable reducing agents are triethyl phosphite, iron pentacarbonyl, titanium(III) chloride, Raney nickel–hydrazine, palladium–carbon, bisulfite or dithionite, and metal–acid [54, 134, 135]. Thus, 4,5-dimethyl-2-nitroaniline heated at 90°C (1 h) with formic acid and sodium dithionite gives 5,6-dimethylbenzimidazole in 92% yield [136].

A recent, efficient, one-pot benzimidazole synthesis treats an o-nitroaniline with an allyl, benzyl or alkyl halide in the presence of sodium hydride as a base. The products are 1-alkyloxy-2-alkyl-, 1-benzyloxy-2-aryl- and 1-allyloxy-2-vinylbenzimidazoles in 75–98% yields. This novel sequence combines N-alkylation, heterocyclization and O-alkylations in the reaction vessel [137, 138].

3.1. FORMATION OF 1,2 AND 2,3 BONDS

General procedure

o-Nitroaniline (0.20 g, 1.45 mmol) is dissolved in THF (10 ml) and NaH (80% in oil, 1.45 mmol) is added at room temperature. The reaction mixture is warmed to gentle reflux with efficient magnetic stirring, alkyl halide (1.45 mmol) is added, and refluxing is continued (4 h). After cooling to room temperature, further NaH (1.45 mmol) is added, and heating is resumed (4 h), before addition of further alkyl halide (1.45 mmol) and heating again (4 h). The reaction is again cooled to ambient temperature, a third portion of NaH is added, heated (4 h), then cooled and quenched by the addition of brine (5 ml), and extracted with CH_2Cl_2 (3 × 40 ml). The combined organic extracts are washed with brine (25 ml), dried ($MgSO_4$), and the solvent removed *in vacuo*. The crude product is flash chromatographed on silica gel, eluting with hexane–ethyl acetate to give the pure benzimidazole N-alkoxide as a thick oil or coloured solid. In most cases the benzimidazole is the last to elute.

When arylenediamines are treated with β-ketoesters by heating in neutral solvents the products are benzimidazolones. In acidic media, 2-substituted benzimidazoles are formed predominantly [139–142]. Generally, in view of the yields reported and the mixtures which may result in these reactions, they do not offer much potential as alternatives to the more standard procedures. A recent modification appears more promising. Here, 2-substituted benzimidazoles can be made in 75–92% yields by reaction with a β-ketoester using solid mineral supports under microwave irradiation. The solid supports are "argiles" with surface acidities equivalent to those of concentrated nitric or sulfuric acid [140].

General method [140]

The diamine (3 mmol) in acetonitrile or THF (15 ml) is added with shaking to the "argile" (e.g. montmorillonite KSF or bentonite K10) at room temperature (~5 min). The solvent is then removed below 50°C. The solid impregnated with the diamine is then placed in a Pyrex tube (2 × 15 cm), and while stirring with a glass rod, ethyl acetoacetate or ethyl benzoylacetate (4 mmol) is added and the mixture is microwaved at 490 W (4 min). The benzimidazole products are extracted from the support with acetone, and ultimately recrystallized from xylene.

A number of unsaturated compounds of the type $X_2C{=}CHY$ where X = Cl, F, CN, SMe and Y = OR, Ar, NO_2, CF_3, SO_2Ph have been converted into benzimidazoles by reaction with o-phenylenediamines. An 80% yield of 2-nitromethylbenzimidazole is obtained when o-phenylenediamine reacts at 10°C in methanol with 1,1-dichloro-2-nitroethene [143]. With

1-nitro-2,2-bis-methylmercaptoethene in refluxing ethanol the yield rises to 89% [144]. Similarly prepared are 2-phenylsulfonylmethyl- (31%) [145], 2-chloronitromethyl- (76%) [146], 2-fluoro(trifluoromethyl)methyl- [147] and a variety of 2-arylbenzimidazoles [148]. Similarly, dimethyl N-aryldithiocarbonimidates (MeS)$_2$C=NAr (or the corresponding dichloro derivatives [149]) give good yields of 2-arylaminobenzimidazoles. A basic medium is required to promote nucleophilic displacement of the methanethiolate groups. In contrast to other methods, no desulfurizing agent is needed [150], but the reaction is not always successful [151]. The dimethyl N-aryldithiocarbonimidates can be made by a variety of methods [152]. 2-Sulfonylaminobenzimidazoles have previously been made by reaction of 2-aminobenzimidazoles with an appropriate sulfonyl chloride [153], from the reaction between -phenylenediamines and sulfonylguanidines [154], and from methyl N-sulfonyldithiocarbonates [155]. The second method needs high temperatures and gives only moderate yields, while the last mentioned is unsuitable for phenylenediamines with two or more substituents. Now one can use N-dichloromethylenesulfonamides (Cl$_2$C=NSO$_2$R) that are easily prepared by chlorination of the corresponding N-sulfonyldithiocarbonates [156], which react readily when heated with -phenylenediamines [157].

Dimethyl N-*aryldithiocarbonimidates* [150]

To a stirred solution of methyl N-aryldithiocarbamate (50 mmol) in DMF (50 ml) is added at room temperature 20 M NaOH (3 ml, 60 mmol). After 1 h, iodomethane (8 g, 50 mmol) is added dropwise, stirring is continued (4 h), and the mixture is then poured into water (400 ml) to give a suspension. Extraction with hexane (3 × 50 ml), followed by drying of the hexane layer (MgSO$_4$), filtration, and removal of the solvent gives the products, usually as yellow oils (yields 78–82%).

2-Phenylaminobenzimidazole [150]

To a vigorously stirred solution of -phenylenediamine (0.27 g, 2.5 mmol) in DMSO (3 ml) is added at room temperature 20 M aqueous NaOH solution (0.25 ml, 5 mmol). After 30 min a solution of N-phenyldithiocarbonimidate (0.49 g, 2.5 mmol) in DMSO (3 ml) is added dropwise, and the mixture is refluxed (7 h), cooled in an ice bath, and poured into water (300 ml). The precipitate is filtered, dried, and recrystallized from acetonitrile (0.26 g, 50%) m.p. 191–192°C. Similarly prepared are other 2-arylaminobenzimidazoles, including examples with substituents at C-5 and/or C-6 in yields which range between 55 and 80%.

2-Methylsulfonylaminobenzimidazole [157]

To a suspension of -phenylenediamine (2.70 g, 25 mmol) in benzene (100 ml) is added dropwise with stirring a solution of N-dichloromethylenemethylsulfonamide (4.10 g, 25 mmol). The mixture is refluxed (8–10 h) and cooled, and the precipitated product is isolated, dried, and recrystallized from DMF (3.60 g, 68%). Similarly prepared are a wide range of analogues in yields of 64–94%.

There are now a number of ways in which one can prepare 2-aminobenzimidazoles, and some of the most useful can be classified under this general heading. Both cyanogen chloride and cyanogen bromide react with -phenylenediamines, as does cyanamide. Cyanogen bromide is available commercially, or it can be generated *in situ*. Yields are usually high and the processes are quite versatile in terms of the -arylenediamines which will react (Table 3.1.5) [158–161].

2-Amino-1-benzylbenzimidazole [162]

A warm solution of sodium cyanide (17 g, 0.35 mol) in water (50 ml) is added to a stirred solution of bromine (16 ml, 0.30 mol) in water (5 ml) at 30°C. The mixture is stirred (2 h) before adding N-benzyl- -phenylenediamine (44 g, 0.41 mol) in methanol (100 ml) dropwise at ambient temperature. After the addition is complete the mixture is refluxed (2 h), cooled, and made basic with aqueous NaOH solution. The precipitated product (24.5 g, 50%) is recrystallized from diethyl ether, m.p. 150°C.

The cyanogen halide can be fruitfully replaced by cyanamide [163], especially if the -phenylenediamine is heated with cyanamide and

TABLE 3.1.5
2-Aminobenzimidazoles made from o-phenylenediamines

Reagent	Substituents	Yield (%)	Ref.
CNBr	5-Me	65	160
CNBr	5,6-Me$_2$	60	160
CNBr	4,6-Me$_2$	55	160
CNBr	5-Cl	60	160
CNBr	1-Ph	60	160
CNBr	1-Me	75	160
CNBr	1,5-Me$_2$	57	160
CNBr	1,5,6-Me$_3$	83	160
CNNH$_2$	H	92	164
CNNH$_2$	1-CH$_2$Ph	50	162

hydrochloric acid [164]. In a variation of the cyanamide method, cyanamide is reacted with chloroformate esters *in situ* to form cyanocarbamates, which combine with -phenylenediamines in the presence of bases (e.g. triethylamine, NaOH) or at elevated temperatures, to form benzimidazole-2-carbamates [165–166]. Cyanoguanidines (dicyandiamides) can similarly be converted into 2-guanidinobenzimidazoles [167].

2-Aminobenzimidazole [163]

A mixture of -phenylenediamine mono- -toluenesulfonate (9.2 g, 33 mmol) is heated (8 h) at 180°C with cyanamide (1.4 g, 33 mmol) until evolution of ammonia has ceased. The crystalline residue separates from ethanol to give 2-aminobenzimidazole -toluenesulfonate (4.6 g, 46%), m.p. 190.5–191.5°C. Trituration with aqueous NaOH, filtration and recrystallization from benzene–light petroleum gives the free base as plates, m.p. 222°C. Replacement of cyanamide with dimethylcyanamide gives 2-dimethylaminobenzimidazole, albeit in low yield.

2-Amino-1-(2′-pyridyl)benzimidazole [161]

A mixture of N-(2′-pyridyl)cyanamide (0.119 g, 1 mmol), -phenylenediamine (0.108 g, 1 mmol) and ethanol (10 ml) is heated under reflux (7 h). Upon cooling, the separated product is filtered and recrystallized from ethanol (0.070 g, 33%), m.p. 259–260°C.

2-Arylaminobenzimidazoles have also been made by refluxing -phenylenediamines with N,N-diphenylcarbodiimide, with N-arylcarbonimidates (see above) [168, 169] or with -methylisothiourea [169]. Carbodiimides can be efficiently made in 85–100% yields under mild conditions by treating a thiourea with methanesulfonyl chloride in the presence of triethylamine and 4-dimethylaminopyridine [170]. When -phenylenediamine is heated in DMF solution with a methyl N-aryldithiocarbamate the product is benzimidazoline-2-thione; at room temperature in the presence of a desulfurizing agent such as red mercuric oxide, moderate to good yields of 2-arylaminobenzimidazoles are formed [171].

General method [171]

To a suspension of -phenylenediamine (1.1 g, 10 mmol) and red HgO (2.2 g, 10 mmol) in DMF (20 ml) a solution of a methyl N-aryldithiocarbamate [152] (10 mmol) in DMF (10 ml) is added with vigorous stirring at room temperature. After 3 h the solution is poured into 3% aqueous HCl (600 ml), boiled (2 min) and hot-filtered, before cooling the filtrate in an ice bath and making it alkaline (pH 8) with concentrated ammonia. The precipitate is filtered, dried and

3.1. FORMATION OF 1,2 AND 2,3 BONDS

recrystallized from acetonitrile. Example products (aryl group, yield given): Ph, 85%; -MeC$_6$H$_4$, 78%; -CF$_3$C$_6$H$_4$, 50%.

There will be some occasions, however, when it is more convenient to prepare 2-aminobenzimidazoles by nucleophilic substitution, e.g. in the presence of copper(I) bromide 2-chlorobenzimidazole reacts with diethylamine to form 2-diethylaminobenzimidazole [172], and there are many other examples [169]. Table 3.1.5 lists some examples of 2-aminobenzimidazoles made from -arylenediamines.

When it is desired to make 1,2-diaminobenzimidazoles, the compounds can be prepared directly from acetyl 2-aminophenylhydrazines (**12**) and cyanogen bromide [173, 174] with subsequent hydrolysis of the 1-amido derivative (Scheme 3.1.15). It may, however, be more convenient to N-aminate a 2-aminobenzimidazole directly using -(2,4-dinitrophenyl)hydroxylamine or hydroxylamine -sulfonic acid [174, 175], which gives yields in the range 40–76%.

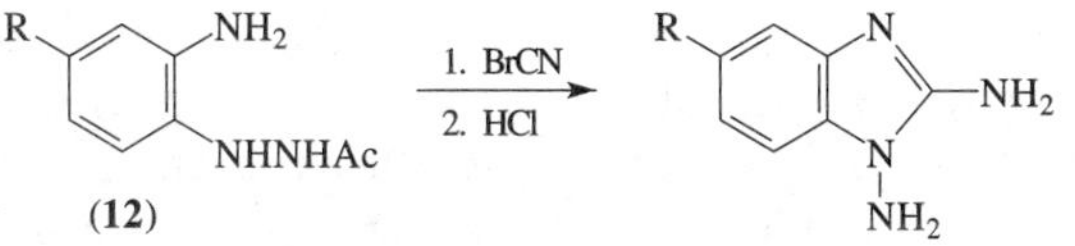

R	%
CF$_3$	56
Cl	59
Me	65

Scheme 3.1.15

1,2-Diamino-5-trifluoromethylbenzimidazole [174]

A solution of (**12**) (R = CF$_3$, 0.44 g, 1.9 mmol) in water (50 ml) is mixed with cyanogen bromide (1.16 g, 11 mmol) in water (10 ml). After stirring (2 h) at room temperature the solution is rotary evaporated to give a solid, 1-acetamidobenzimidazole hydrobromide monohydrate, which is recrystallized from acetonitrile (60% yield), m.p. 249–251°C. A solution of this product (0.63 g, 1.8 mmol) dissolved in 4 M HCl (4.5 ml) is heated under reflux (1 h), cooled and made basic with saturated sodium bicarbonate. The solid which separates is recrystallized from ethanol (0.35 g, 93%), m.p. 250–251°C.

Benzimidazolyl-2-carbamates can be made either by variation of the cyanamide method, by the use of cyanocarbamates (see above) or by using isothioureas, which can be generated *in situ* by treating thiourea or S-methylisothiourea with an alkyl chloroformate. Subsequent condensation with an -phenylenediamine at pH ~ 6 gives moderate to good yields [176–178]. The general process involves adding the alkyl chloroformate to an aqueous solution of the thiourea at room temperature, raising the temperature to 100°C for a short time, cooling and adding a further portion of alkyl chloroformate to complete formation of the isothiourea. Addition of the requisite amount of aqueous sodium hydroxide, then -phenylenediamine in acetic acid, followed by heating at 100°C for 2-3 h completes the cyclization. The yield of methylbenzimidazolyl-2-carbamate from this method is about 70% [177].

Other reagent types which condense with -phenylenediamines include trichloromethyl derivatives of aromatic and heteroaromatic compounds [179], N-1-chloroalkylpyridinium salts [180] and triazines [149]. Whereas reactions of aldehydes with -diamines are widely applicable, the work-up procedures may be tedious. It is often better to convert the aldehyde into an N-(1-chloroalkyl)pyridinium chloride; indeed, the process can be adapted to a one-pot synthesis, it uses mild conditions and generally gives high yields.

General procedure for preparation of benzimidazoles [180]

To a cooled (0°C) solution of thionyl chloride (0.9 ml, 12 mmol) in dichloromethane (10 ml) is added dropwise a solution of pyridine (1.0 ml, 12 mmol) in dichloromethane (6 ml). The aldehyde (10 mmol) is then added dropwise, and the mixture is allowed to come to room temperature. Formation of the N-(1-chloroalkyl)pyridinium chloride is confirmed by NMR. The diamine (30 mmol) is then added slowly and stirred overnight, before evaporation of the solvent. The residue is triturated with water to give the crude benzimidazole, e.g. 2-Ph (95%), 2-Pr (75%), 2-iPr (75%), 2-tBu (70%), 2-Ph$_2$CH (80%), 2- -MeC$_6$H$_4$ (80%), 2-(2'-furanyl) (80%), 2-(2'-thienyl) (75%) and 2- -NO$_2$C$_6$H$_4$ (70%).

A commercially important method of synthesis of benzimidazoles is carried out by heating mixtures of an -nitroaniline (or -dinitroarene) with an alcohol in the gas phase. Although yields are quite good, secondary reactions can occur (e.g. with -nitroanilines the 2-alkylbenzimidazole primary products are converted into 1,2-dialkylbenzimidazoles, and mixtures or products are obtained [181]). In contrast, -dinitroarenes react with alcohols in the presence of vanadium or copper and alumina (at around 320°C) to give 2-alkylimidazoles exclusively [182]. Such reactions are, however, experimentally difficult in the usual laboratory situation and offer no significant advantages. A more recent modification which treats an -arylenediamine with an alcohol in the presence of a ruthenium catalyst (RuCl$_2$(PPh$_3$)$_3$) seems promising. The catalyst has a dual role in that it oxidizes the alcohol and promotes cyclization [183].

Under this heading fall some approaches to the synthesis of benzimidazole N-oxides which are formed when -nitro- or -nitrosoanilines react with aldehydes. Undoubtedly these reactions involve cyclization of an initially formed

Schiff base, and as such they could be classified as a 1,2 bond formation. The acid-catalysed condensation between -nitrosoaniline and an aldehyde gives yields of 2-substituted benzimidazole 1-oxides in the range 50-65%, but the method is of limited utility because the nitrosoanilines are difficult to make and they are unstable [184, 185]. The *N*-oxides can also be made from -nitroanilines and aromatic aldehydes (ideally with the aldehyde:amine ratio slightly in excess of 2:1). Reactions are carried out in a boiling hydrocarbon solvent with azeotropic removal of water as it is formed. If higher temperatures are employed, or if the excess of aldehyde is greater, deoxygenation is likely to ensue [186, 187]. -Benzoquinone dioxime, readily prepared by reduction of benzofuroxan, reacts with aromatic (and some cinnamyl) aldehydes in the presence of strong acids to give 2-aryl-1-hydroxybenzimidazole 3-oxides. The yields are dependent on the strength of the acid used. Perchloric acid (70% aqueous) is best, but 48% aqueous hydrobromic and 4 M ethanolic hydrochloric acids both give high yields [188].

1-Hydroxybenzimidazole 3-oxides [188]

The aldehyde (50 mmol) and -benzoquinone dioxime (6.90 g, 50 mmol) are dissolved in ethanolic hydrochloric acid (3.5 M, 100 ml). The reaction mixture is then refluxed for several hours until the deep red colour disappears. After cooling, the hydrochloride is filtered off, suspended in ethanol (100 ml), water (200 ml) is added, and the mixture is neutralized with aqueous ammonia. The precipitate is filtered, washed in turn with water and ethanol, and dried. Prepared in this way are the following 1-hydroxybenzimidazole 3-oxides (2-substituent, yield, m.p. listed): Ph, 72%, 223-224°C; -Me$_2$NC$_6$H$_4$, 91%, 180-185°C; -HOC$_6$H$_4$, 65%, 227-228°C; -HOC$_6$H$_4$, 72%, 263-265°C; -NO$_2$C$_6$H$_4$, 80%, 224-225°C; CH=CHPh, 70%, 226-228°C; CH=CHC$_6$H$_4$- -NMe$_2$, 70%, 223-225°C; 2-thienyl, 83%, 183-185°C.

REFERENCES

1. S. E. Voltz, J. H Krause and W. E. Erner, US Patent 2,891,965 (1959); (*Chem. Abstr.* **54**, 1557 (1960)).
2. H. A. Green, US Patent 3,037,028 (1962); (*Chem. Abstr.* **57**, 12 501 (1962)).
3. H. A. Green, US Patent 3,255,200 (1966); (*Chem. Abstr.* **65**, 5467 (1966)).
4. H. Kroeper and J. Sand, Belgian Patent 661,322 (1965); (*Chem. Abstr.* **64**, 2094 (1966)).
5. M. Ya. Kraft, P. M. Kochergin and A. M. Tsyganova, *Khim.-Farm Zh.* **23**, 1246 (1989); *Chem. Abstr.* **112**, 216 784 (1990)).
6. V. M. Stoyanov, M. M. El'chaninov and A. F. Pozharskii, *Chem. Heterocycl. Compd. USSR (Engl. Transl.)* **27**, 1140 (1991).
7. E. S. Schipper and A. R. Day, in *Heterocyclic Compounds* (ed. R. C. Elderfield) Wiley, New York, 1957, Vol. 5, p. 194.

8. Air Products and Chemicals, Inc., French Patent 1,362,689 (1964); (*Chem. Abstr.* **61**, 13 317 (1964)).
9. S. G. Davies and A. A. Mortlock, *Tetrahedron Lett.* **32**, 4791 (1991).
10. P. K. Martin, H. R. Matthews, H. Rapoport and G. Thyagarajan, *J. Org. Chem.* **33**, 3758 (1968).
11. M. A. Eissenstat and J. D. Weaver, *J. Org. Chem.* **58**, 3387 (1993).
12. A. Windaus and W. Langenbeck, *Chem. Ber.* **55**, 3706 (1922).
13. M. R. Grimmett, *Adv. Heterocycl. Chem.* **12**, 103 (1970).
14. A. Gomez-Sanchez, F. J. Hidalgo and J. L. Chiara, *J. Heterocycl. Chem.* **24**, 1757 (1987).
15. G. L. Matevosyan and P. M. Zavlin, *Chem. Heterocycl. Compd. USSR (Engl. Transl.)* **26**, 599 (1990).
16. M. Zettlitzer, H. tom Dieck, E. T. K. Haupt and L. Stamp, *Chem. Ber.* **119**, 1868 (1986).
17. R. W. Begland and D. R. Hartter, *J. Org. Chem.* **37**, 4136 (1972).
18. D. Günther and D. Bosse, *Angew. Chem., Int. Ed. Engl.* **19**, 130 (1980).
19. Y. Yamada, T. Zama and I. Kumashiro, Japanese Patent 04 373 (1971); (*Chem. Abstr.* **74**, 125 693 (1971)).
20. J. F. O'Connell, J. Parquette, W. E. Yelle, W. Wang and H. Rapoport, *Synthesis*, 767 (1988).
21. N. Asai, German Patent 2,160,673 (1972); (*Chem. Abstr.* **77**, 114 409 (1972)).
22. Y. Ohtsuka, *J. Org. Chem.* **41**, 713 (1976).
23. D. W. Woodward, US Patent 2,534,331 (1950); (*Chem. Abstr.* **45**, 5191 (1951)).
24. L. De Vries, *J. Org. Chem.* **36**, 3442 (1971).
25. W. A. Sheppard and O. W. Webster, *J. Am. Chem. Soc.* **95**, 2695 (1973).
26. R. W. Begland, D. R. Hartter, F. N. Jones, D. J. Sam, W. A. Sheppard, O. W. Webster and F. J. Weigert, *J. Org. Chem.* **39**, 2341 (1974).
27. F. L. Merchan, J. Garin and T. Tejero, *Synthesis* 984 (1982).
28. H. Bredereck and G. Schmotzer, *Liebigs Ann. Chem.* **600**, 95 (1956).
29. M. P. Sammes and A. R. Katritzky, *Adv. Heterocycl. Chem.* **35**, 375 (1984).
30. J. Sundermeyer and H. W. Roesky, *Chem. Ber.* **124**, 1517 (1991).
31. S. S. Minovici, *Ber.* **29**, 2097 (1896); (*Br. Abstr.* **189**, 703 (1896)).
32. E. Brunn, E. Funke, H. Gotthardt and R. Huisgen, *Chem. Ber.* **104**, 1562 (1971).
33. R. S. Hosmane, *Liebigs Ann. Chem.* 831 (1984).
34. U. Pfeiffer, M. T. Riccaboni, R. Erba and M. Pinza, *Liebigs Ann. Chem.* 993 (1988).
35. K. Mai and G. Patil, *Synth. Commun.* **15**, 157 (1985).
36. J. Barrans, *Compt. Rend. Soc. Biol.* **258**, 6185 (1964); (*Chem. Abstr.* **61**, 7003 (1965)).
37. M. Mengelberg, *Chem. Ber.* **93**, 2230 (1960).
38. G. La Parola, *Gazz. Chim. Ital.* **75**, 216 (1945).
39. L. B. Volodarsky, A. N. Lisak and V. A. Koptyug, *Tetrahedron Lett.* 1565 (1965).
40. H. Lettau, *Z. Chem.* **10**, 211 (1970).
41. L. B. Volodarskii and A. Ya. Tikhonov, *Synthesis* 704 (1986).
42. L. N. Grigor'eva, S. A. Amitina and L. B. Volodarskii, *Chem. Heterocycl. Compd. USSR (Engl. Transl.)* **19**, 1104 (1983).
43. L. B. Volodarskii, *Chem. Heterocycl. Compd. USSR (Engl. Transl.)* **9**, 1175 (1973).
44. Yu. G. Putsykin and L. B. Volodarskii, *Izv. Sibirsk. Otd. Akad. Nauk SSSR, Ser. Khim. Nauk*, 86 (1969); (*Chem. Abstr.* **72**, 54 514 (1970)).
45. L. B. Volodarskii and A. N. Lysak, *Chem. Heterocycl. Compd. USSR (Engl. Transl.)* **3**, 713 (1967); (*Chem. Abstr.* **68**, 87 237 (1968)).
46. L. B. Volodarskii and A. N. Lysak, *Khim. Geterotsikl. Soedin., Sb. 1: Azotsoderzhashchie Geterotsikly* 109 (1967); (*Chem. Abstr.* **71**, 49 850 (1969)).
47. L. B. Volodarskii and E. I Vityaeva, *J. Org. Chem. USSR (Engl. Transl.)* **8**, 1936 (1972); (*Chem. Abstr.* **78**, 16 097 (1973)).

REFERENCES

48. I. K. Korobeinicheva, M. M. Mitasov, V. S. Kobrin and L. B. Volodarskii, *Izv. Sib. Otd. Akad. Nauk SSSR, Ser. Khim. Nauk* 96 (1976); (*Chem. Abstr.* **85**, 26 913 (1976)).
49. H. Gnichtel and K.-E. Schuster, *Chem. Ber.* **111**, 1171 (1978).
50. G. V. Nikitina and M. S. Pevzner, *Chem. Heterocycl. Compd. USSR (Engl. Transl.)* **29**, 127 (1993).
51. K. Hartke and A. Brutsche, *Synthesis* 1199 (1992).
52. P. Molina, C. Lopez-Leonardo, J. Llamas-Botia, C. Foces-Foces and A. L. Llamas-Saiz, *Synthesis* 449 (1995).
53. J. B. Wright, *Chem. Rev.* **48**, 397 (1951).
54. P. N. Preston, *Benzimidazoles and Congeneric Tricyclic Compounds* (ed. P. N. Preston). Interscience-Wiley, New York, 1981.
55. M. A. Phillips, *J. Chem. Soc.* 1143 (1931).
56. M. Moazzam, Z. H. Chohan, A. Tabassum, I. A. Chughtai and N. A. Haq, *J. Pure Appl. Sci.* **5**, 37 (1986).
57. A. Sykes and J. C. Tatlow, *J. Chem. Soc.* 4078 (1952).
58. L. Weinberger and A. R. Day, *J. Org. Chem.* **24**, 1451 (1959).
59. R. Bossio, S. Marcaccini, V. Parrini and R. Pepino, *Heterocycles* **23**, 2705 (1985).
60. D. W. Hein, R. J. Alheim and J. J. Leavitt, *J. Am. Chem. Soc.* **79**, 427 (1957).
61. T. Hisano, M. Ichikawa, K. Tsumoto and M. Tasaki, *Chem. Pharm. Bull.* **30**, 2996 (1982).
62. D. A. Rowlands, in *Synthetic Reagents* (ed. J. S. Pizey). Ellis Harwood, Chichester, 1985.
63. M. F. Brana, J. M. Castellano and M. J. R. Yunta, *J. Heterocycl. Chem.* **27**, 1177 (1990).
64. E. Elcade, I. Dinarés, L. Perez-Garcia and T. Roca, *Synthesis* 395 (1992).
65. L. N. Pushkina, S. A. Mazalov and I. Ya. Postovskii, *J. Gen. Chem. USSR (Engl. Transl.)* **32**, 2585 (1962).
66. B. Rigo, D. Valligny, S. Taisne and D. Couturier, *Synth. Commun.* **18**, 167 (1988).
67. L. Ya. Shteinberg, S. A. Kondratov, V. D. Boiko and S. M. Shein, *J. Org. Chem. USSR (Engl. Transl.)* **22**, 2215 (1986).
68. M. Rope, R. W. Isensee and L. Joseph, *J. Am. Chem. Soc.* **74**, 1095 (1952).
69. G. Crank and M. I. H. Makin, *J. Heterocycl. Chem.* **26**, 1163 (1989).
70. V. M. Berezovskii, L. S. Tul'chinskaya and N. G. Volikova, *J. Gen Chem. USSR (Engl. Transl.)* **30**, 3401 (1960).
71. K. L. Kirk and L. A. Cohen, *J. Org. Chem.* **34**, 384 (1969).
72. A. J. Freyer, C. K. Lowe-Ma, R. A. Nissan and W. S. Wilson, *Aust. J. Chem.* **45**, 525 (1992).
73. I. Sekikawa, *Bull. Chem. Soc. Jpn.* **31**, 252 (1958).
74. R. Belcher, A. Sykes and J. Tatlow, *J. Chem. Soc.* 4159 (1954).
75. R. Iemura, T. Kawashima, T. Fukuda, K. Ito and G. Tsukamoto, *J. Heterocycl. Chem.* **24**, 31 (1987).
76. G. Holan, J. J. Evans and M. Linton, *J. Chem. Soc., Perkin Trans. 1* 1200 (1977).
77. L. Mathias and C. G. Overberger, *Synth. Commun.* **5**, 461 (1975).
78. J. B. Hendrickson and M. S. Hussoin, *J. Org. Chem.* **52**, 4137 (1987).
79. G. Sandera, R. W. Isensee and L. Joseph, *J. Am. Chem. Soc.* **76**, 5173 (1954).
80. K. Matsumoto, A. Sera and T. Uchida, *Synthesis* 1 (1985).
81. A. F. Pozharskii, A. D. Garnovskii and A. M. Simonov, *Russ. Chem. Rev. (Engl. Transl.)* **35**, 122 (1966).
82. A. R. Katritzky, S. Rachwal and R. Ollmann, *J. Heterocycl. Chem.* **31**, 775 (1994).
83. P. C. Vyas, C. K. Oza and A. K. Goyal, *Chem. Ind. (London)* 287 (1980).
84. J. B. Hendrickson, D. A. Judelson and T. Chancellor, *Synthesis* 320 (1984).
85. R. J. Perry and B. D. Wilson, *J. Org. Chem.* **58**, 7016 (1993).
86. H. Lettre, W. Fritsch and J. Porath, *Chem. Ber.* **84**, 719 (1951).
87. H. Irving and O. Weber, *J. Chem. Soc.* 2296 (1959).

88. D. N. Gray, *J. Heterocycl. Chem.* **7**, 947 (1970).
89. N. Suzuki, T. Yamabayashi and Y. Izawa, *Bull. Chem. Soc. Jpn* **49**, 353 (1976).
90. Kh. A. Suerbaev and Ch. Sh. Kadyrov, *Chem. Heterocycl. Compd. USSR (Engl. Transl.)* **10**, 989 (1974).
91. A. R. Katritzky, B. Rachwal, S. Rachwal, P. J. Steel and K. A. Zaklika, *Heterocycles* **38**, 2415 (1994).
92. J. H. Musser, T. T. Hudec and K. Bailey, *Synth. Commun.* **14**, 947 (1984).
93. A. Hunger, J. Kebrle, A. Rossi and K. Hoffmann, *Helv. Chim. Acta* **43**, 800 (1960).
94. E. C. Lane, *J. Chem. Soc.* 2238 (1953).
95. S. H. Dandegaonker and G. R. Ravankar, *J. Karnatak Univ.* **6**, 25 (1961); (*Chem. Abstr.* **59**, 10 023 (1963)).
96. A. V. El'tsov, E. R. Zakhs and T. I. Frolova, *J. Org. Chem. USSR (Engl. Transl.)* **12**, 1097 (1976).
97. V. I. Kelarev and V. N. Koshlev, *Russ. Chem. Rev. (Engl. Transl.)* **64**, 317 (1995).
98. A. Hunger, J. Kebrle, A. Rossi and K. Hoffmann, *Helv. Chim. Acta* **43**, 800, 1727 (1960).
99. G. Holan, E. L. Samuel, B. C. Ennis and R. W. Hinde, *J. Chem. Soc. (C)* 20 (1967).
100. N. A. R. Nabulsi and R. D. Gandour, *J. Org. Chem.* **56**, 2260 (1991).
101. J. Grimshaw and J. Trocha-Grimshaw, *Tetrahedron Lett.* 2601 (1975).
102. H. B. Gillespie, F. Spano and S. Graaf, *J. Org. Chem.* **25**, 942 (1960).
103. V. I. Troitskaya, V. I. Rudyk, G. G. Yakobson and L. M. Yagupol'skii, *Chem. Heterocycl. Compd. USSR (Engl. Transl.)* **17**, 1232 (1981).
104. R. A. B. Copeland and A. R. Day, *J. Am. Chem. Soc.* **65**, 1072 (1943).
105. F. H. Pinkerton and S. F. Thames, *J. Heterocycl. Chem.* **9**, 67 (1972).
106. P. N. Preston, *Chem. Rev.* **74**, 279 (1974).
107. A. Hunger, J. Kebrle, A. Rossi and K. Hoffmann, *Helv. Chim. Acta* **43**, 1032 (1960).
108. F. Cramer, K. Pawelzik and H. J. Baldauf, *Chem. Ber.* **91**, 1049 (1958).
109. D. Jerchel, M. Kracht and K. Krucker, *Liebigs Ann. Chem.* **590**, 232 (1954).
110. D. Jerchel, H. Fischer and M. Kracht, *Liebigs Ann. Chem.* **575**, 162 (1952).
111. R. C. DeSelms, *J. Org. Chem.* **27**, 2163 (1962).
112. N. V. Subba Rao and C. V. Ratnam, *J. Ind. Chem. Soc.* **38**, 631 (1961).
113. N. Latif, N. Mishriky and F. M. Assad, *Recl. Trav. Chim. Pays-Bas Belg.* **102**, 73 (1983).
114. G. P. Ellis and R. T. Jones, *J. Chem. Soc., Perkin Trans. 1* 903 (1974).
115. B. Yadagiri and J. W. Lown, *Synth. Commun.* **20**, 955 (1990).
116. V. Milata and D. Ilavsky, *OPPI Briefs* **25**, 703 (1993).
117. E. P. Nesynov and P. S. Pel'kis, *J. Org. Chem. USSR (Engl. Transl.)* **3**, 831 (1967).
118. Yu. M. Yutilov and I. A. Svertilova, *Chem. Heterocycl. Compd. USSR (Engl. Transl.)* **12**, 1057 (1976).
119. A. V. El'tsov, V. S. Kuznetsov and M. B. Kolesova, *J. Org. Chem. USSR (Engl. Transl.)* **1**, 1126 (1965).
120. W. B. Wright, *J. Heterocycl. Chem.* **2**, 41 (1965).
121. G. F. Duffin and J. D. Kendall, *J. Chem. Soc.* 361 (1956).
122. T. L. Cairns, D. D. Coffman and W. W. Gilbert, *J. Am. Chem. Soc.* **79**, 4405 (1957).
123. M. S. Khajavi, M. Hajihadi and R. Naderi, *J. Chem. Res. (S)* 92 (1996).
124. S. Mataka, T. Irie, Y. Ikezaki and M. Tashiro, *Chem. Ber.* **126**, 1819 (1993).
125. T. Miyata, T. Mizumo, I. Nishiguchi, T. Hirashima, S. Nakatani and Y. Nakahara, *Chem. Express* **7**, 949 (1992).
126. M. S. Khajavi, M. Hajihadi and F. Nikpour, *J. Chem. Res. (S)* 94 (1996).
127. J. A. Van Allan and B. D. Deacon, *Org. Synth.* **30**, 56 (1950); *Coll. Vol.* **4**, 569 (1963).
128. H. A. Staab and G. Walther, *Liebigs Ann. Chem.* **657**, 98 (1962).
129. R. Dahiya and H. K. Pujari, *J. Fluorine Chem.* **42**, 245 (1989).
130. K. C. Joshi, R. A. Misra, R. Jain and K. Sharma, *J. Heterocycl. Chem.* **26**, 409 (1989).

REFERENCES

131. R. C. DeSelms, *Tetrahedron Lett.* 3001 (1970).
132. W. Tian and S. Grivas, *Synthesis* 1283 (1992).
133. W. Knobloch and H. Schaefer, *J. Prakt. Chem.* **17**, 187 (1962).
134. V. A. Kuznetsov, A. V. Garabadzhiu and O. F. Ginzburg, *J. Org. Chem. USSR (Engl. Transl.)* **23**, 576 (1987).
135. A. P. Thomas, C. P. Allott, K. H Gibson, J. S. Major, B. B. Masek, A. A. Oldham, A. H. Ratcliffe, D. A. Roberts, S. T. Russell and D. A. Thomason, *J. Med. Chem.* **35**, 877 (1992).
136. J. Kriedl and P. Turcsanyi, Hungarian Patent 3304 (1972); *Chem. Abstr.* **76**, 113 216 (1972).
137. J. M. Gardiner and C. R. Loyns, *Synth. Commun.* **25**, 819 (1995).
138. J. M. Gardiner, C. R. Loyns, C. H. Schwalbe, G. C. Barrett and P. R. Lowe, *Tetrahedron* **51**, 4101 (1995).
139. A. Rossi, A. Hunger, J. Kebrle and K. Hoffmann, *Helv. Chim. Acta* **43**, 1046, 1298 (1960).
140. K. Bougrin and M. Soufiaoui, *Tetrahedron Lett.* **36**, 3683 (1995).
141. Z. F. Solomko, V. S. Tkachenko, A. N. Kost, V. A. Budylin and V. L. Pikalov, *Chem. Heterocycl. Compd. USSR (Engl. Transl.)* **11**, 470 (1975).
142. V. I Sheremet, V. G. Dryuk, Z. F. Solomko and M. M. Kremlev, *Chem. Heterocycl. Compd. USSR (Engl. Transl.)* **17**, 941 (1981).
143. V. V. Rudchenko, V. A. Buevich, V. S. Grineva and V. V. Perekalin, *Chem. Heterocycl. Compd. USSR (Engl. Transl.)* **11**, 1341 (1975).
144. H. Schaefer and K. Gewald, *Z. Chem.* **16**, 272 (1976).
145. W. Schroth and F. Raabe, East German Patent 93,559 (1972); *Chem. Abstr.* **78**, 136 288 (1973).
146. V. A. Buevich, V. S. Grineva, V. V. Rudchenko and V. V. Perekalin, *J. Org. Chem. USSR (Engl. Transl.)* **11**, 2698 (1975).
147. N. Ishikawa and T. Muramatsu, *Nippon Kagaku Kaishi* 563 (1973); *Chem. Abstr.* **78**, 147 873 (1973).
148. M. Hammad and A. Emran, *Z. Naturforsch. B* **32**, 304 (1977).
149. D. B. Murphy, *J. Org. Chem.* **29**, 1613 (1964).
150. J. Garin, E. Melendez, F. L. Merchan, C. Tejel and L. Tejero, *Synthesis* 375 (1983).
151. E. Melendez, F. L. Merchan, P. Merino, J. Orduna and R. Urchegui, *J. Heterocycl. Chem.* **28**, 653 (1991).
152. J. Garin, V. Martinez, J. Mayoral, E. Melendez and F. Merchan, *Synthesis* 961 (1981).
153. G. W. Raiziss, L. W. Clemence and M. Freifelder, *J. Am. Chem. Soc.* **63**, 2739 (1941).
154. C. C. Price and R. H Reitsema, *J. Org. Chem.* **12**, 269 (1947).
155. J. V. Rodricks and H. Rapoport, *J. Org. Chem.* **36**, 46 (1971).
156. R. Gompper and R. Kunz, *Chem. Ber.* **99**, 2900 (1966).
157. F. L. Merchan, J. Garin and T. Tejero, *Synthesis* 984 (1982).
158. W. W. Hartman and E. E. Dreger, *Org. Synth. Coll. Vol.* **2** 150 (1943).
159. J. H. Wikel, C. J. Paget, D. C. DeLong, J. D. Nelson, C. Y. E. Wu, J. W. Paschal, A. Dinner, R. J. Templeton, M. O. Chaney, N. D. Jones and J. W. Chamberlin, *J. Med. Chem.* **23**, 368 (1980).
160. L. Joseph, *J. Med. Chem.* **6**, 601 (1963).
161. B. Vercek, B. Ogorevc, B. Stanovnik and M. Tisler, *Monatsh. Chem.* **114**, 789 (1983).
162. R. Rastogi, S. Sharma and R. N. Iyer, *Eur. J. Med. Chem.* **14**, 489 (1979).
163. B. Adcock, A. Lawson and D. H. Miles, *J. Chem. Soc.* 5120 (1961).
164. S. Weiss, H. Michaud, H. Prietzel and H. Krommer, *Angew. Chem., Int. Ed. Engl.* **12**, 841 (1973).
165. R. J. Gyurik and V. J. Theodorides, US Patent 3,915,986 (1975); *Chem. Abstr.* **84**, 31 074 (1976).

166. R. Aries, French Patent 2,063,815 (1971); *Chem. Abstr.* **76**, 126 984 (1958).
167. R. M. Acheson, G. Taylor and M. L. Tomlinson, *J. Chem. Soc.* 3750 (1958).
168. A. Keller, *Ber. Dtsch. Chem. Ges.* **24**, 2498 (1891).
169. A. Hunger, J. Kebrle, A. Rossi and K. Hoffmann, *Helv. Chim. Acta* **44**, 1273 (1961).
170. J. B. Fell and G. M. Coppola, *Synth. Commun.* **25**, 43 (1995).
171. F. L. Merchan, J. Garin, V. Martinez and E. Melendez, *Synthesis* 482 (1982).
172. C. P. Whittle, *Aust. J. Chem.* **33**, 1545 (1980).
173. R. I. Ho and A. R. Day, *J. Org. Chem.* **38**, 3084 (1973).
174. A. V. Zieger and M. M. Joullie, *J. Org. Chem.* **42**, 542 (1977).
175. A. V. Zieger and M. M. Joullie, *Synth. Commun.* **6**, 457 (1976).
176. Z. Budesinsky, J. Sluka, J. Novac and J. Danek, *Collect. Czech. Chem. Commun.* **40**, 1089 (1975).
177. D. Takiguchi, Japanese Patent 75 12,087 (1975); *Chem. Abstr.* **83**, 43 322 (1975).
178. S. Ram, D. S. Wise, L. T. Wotring, J. W. McCall and L. B. Townsend, *J. Med. Chem.* **35**, 539 (1992).
179. K. Takahashi, T. Suzuki, Y. Suzuki, H. Takeda, T. Zaima and K. Mitsuhashi, *Nippon Kagaku Kaishi* 1595 (1974); *Chem. Abstr.* **81**, 169 478 (1974).
180. J. Vanden Eynde, A. Mayence, A. Maquestiau and E. Anders, *Bull. Soc. Chim. Belg.* **102**, 357 (1993).
181. N. S. Koslov and M. N. Stepanova, *Dokl. Akad. Nauk Beloruss. SSR* **13**, 541 (1969); *Chem. Abstr.* **71**, 124 337 (1969).
182. N. S. Koslov and M. N. Tovshtein, *Vesti Akad. Navuk. Belaruss SSR, Ser. Khim. Navuk.* 72 (1968); *Chem. Abstr.* **70**, 115 063 (1969).
183. T. Kondo, S. Yang, K. T. Huh, M. Kobayashi, S. Kotachi and Y. Watanabe, *Chem. Lett.* 1275 (1991).
184. D. W. Russell, *J. Med. Chem.* **10**, 984 (1967).
185. M. Z. Nazer, M. J. Haddadin, J. P. Petridou and C. H. Issidorides, *Heterocycles* **6**, 541 (1977).
186. R. Marshall, D. J. Sears and D. M. Smith, *J. Chem. Soc (C)* 2144 (1970).
187. E. Yu. Belyaev, V. P. Kumarev, L. E. Kondrat'eva and E. I. Shakhova, *Chem. Heterocycl. Compd. (Engl. Transl.)* **6**, 1576 (1970).
188. F. Paetzold, H. J. Niclas and E. Gruendemann, *J. Prakt. Chem.* **332**, 345 (1990).

3.2 FORMATION OF 1,2 AND 1,5 BONDS

Methods which fit this classification have not been used for the synthesis of benzimidazoles.

The major strategies leading to uncondensed imidazoles revolve around reactions of amines, ammonia or hydrazines with suitable acylamides, N-alkenylamides, 2-azabutadienes and N-cyanoalkyl- or N-ω-acylimidates. Alkyl N-cyanoalkylimidates (**1**) are converted by primary amines into 1-substituted 5-aminoimidazoles (**2**) (Scheme 3.2.1) [1, 2]. With hydrazines, 1,5-diaminoimidazoles are formed [3–5]. The reaction is quite general for

3.2. FORMATION OF 1,2 AND 1,5 BONDS

a wide variety of imidates with primary alkyl- and arylamines, substituted (but not 1,2-disubstituted, hydrazines, and semicarbazides. The R^4 substituent can be alkyl, aryl, acyl or amido [6, 7].

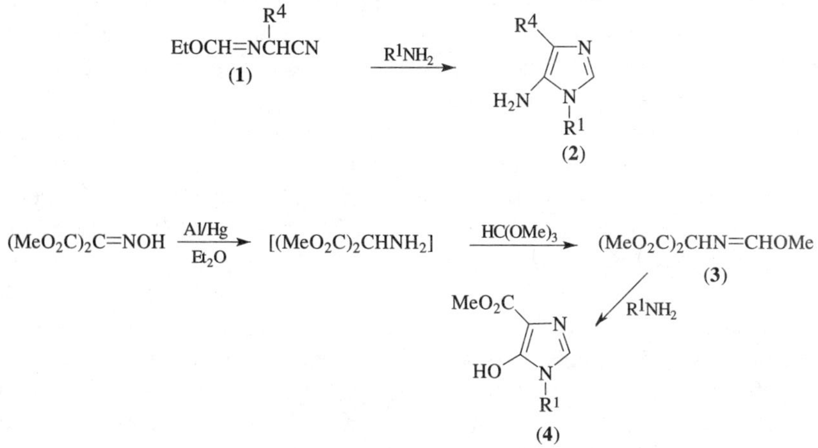

$$EtOCH=NCHCN(R^4) \quad (1) \xrightarrow{R^1NH_2} \quad (2)$$

$$(MeO_2C)_2C=NOH \xrightarrow[Et_2O]{Al/Hg} [(MeO_2C)_2CHNH_2] \xrightarrow{HC(OMe)_3} (MeO_2C)_2CHN=CHOMe \quad (3)$$

$$\xrightarrow{R^1NH_2} (4)$$

Scheme 3.2.1

Ethyl 5-amino-1-benzylimidazole-4-carboxylate **(2)** ($R^1 = CH_2Ph$, $R^4 = CO_2Et$) *[1]*

A solution of ethyl 2-amino-2-cyanoacetate (6.25 g, 49 mmol) in acetonitrile (50 ml) and triethyl orthoformate (7.6 g, 51 mmol) is refluxed (35 min), then evaporated to give an oil. A solution of this oil in acetonitrile (50 ml) with added benzylamine (5.35 g, 50 mmol) is refluxed (35 min), then evaporated to dryness, and the residue is dissolved in chloroform (100 ml) and washed in turn with 2 M sodium hydroxide (25 ml) and water (25 ml) before drying over Na_2SO_4. Removal of the solvent gives a gum which solidifies on trituration with ethyl acetate. The product (4.56 g, 37%) separates from ethyl acetate as needles, m.p. 160°C (dec.). Similarly prepared are the 1-(-methoxybenzyl) (40%) and 1-t-butyl (40%) analogues.

The imidates are readily made from alkyl 2-amino-2-cyanoesters refluxed with an ortho ester, while the 2-amino-2-cyano esters are available by aluminium amalgam reduction of the oximino precursors [2]. Although yields are usually only in the range 35–60%, the process has useful application to the synthesis of 5-aminoimidazole nucleosides (using amino sugars as the amine reagents) [8, 9]. A more recent example converts the analogous methyl *N*-(dicarbomethoxymethyl)methanimidate **(3)** into

1-substituted "5-hydroxy" imidazoles (**4**) (Scheme 3.2.1) [10]. The unstable imidate (**3**) is made in about 50% yield by slow, dropwise addition of dimethyl aminomalonate to a large excess of refluxing trimethyl orthoformate. It is conveniently stored as the stable -toluenesulfonic acid salt, and regenerated as needed by treatment with aqueous sodium hydroxide. When treated with primary amines or hydrazines at room temperature, (**3**) is converted into (**4**) in 65–90% yields [10]. This is possibly the best method of synthesis of 2-imidazolin-5-ones or 5-hydroxyimidazoles.

Similar reactions take place between α-amidoketones and amines in the presence of dehydrating agents. Thus, a range of anilines react with ω-benzamidoacetophenone heated in the presence of phosphorus trichloride to give 1-aryl-2,5-diphenylimidazoles in 25–65% yields [11]. In spite of the moderate yields the method has had considerable application and success in some specific instances. Suzuki and co-workers used it to make 2-substituted 4-(2'-thienyl)- and 2,5-disubstituted 4-(2'-thienyl)imidazoles in moderate to good yields (43–86%). The required ω-acylamides (**5**) are made from the α-acylamines under Schotten–Baumann conditions (65–99% yields) (Scheme 3.2.2) [12]. Subsequent reflux with ammonium acetate in glacial acetic acid achieves cyclization [13]. This essentially uses Davidson's earlier approach to imidazole ring formation [14]. The required α-acylamine starting compounds are readily made by treating methyl α-isocyanoacetate (commercially available) with an acid chloride in the presence of triethylamine or potassium t-butoxide. This generates an oxazole ester, which ring opens when treated with mineral acid [15]. Such α-acylamines are also widely used in cyclizations with thiocyanate (see Section 3.3). α-Amidoketones can also be made in 55–70% yields from ketone-derived nitrones reacting with an N-methylcarboximidoyl chloride [16], while direct alkylation in the presence of sodium hydride in DMF of the carbon anions derived from the α-aminoketones ($R^5 = H$) provides an alternative route to a variety of ω-acylamides (**5**) [13].

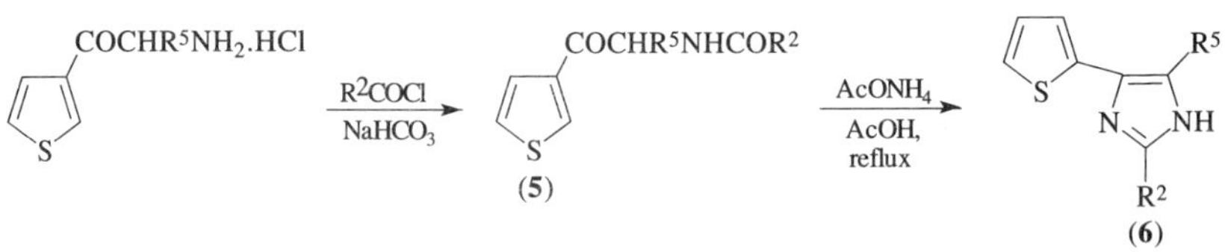

Scheme 3.2.2

General method for synthesis of ω-acylamides (**5**) [12]

To a suspension of aminomethyl 3-thienylketone hydrochloride (95 g, 0.54 mol) in ethyl acetate (900 ml) and water (500 ml) is added $NaHCO_3$

(118 g, 1.4 mol) at 0°C. The appropriate acid chloride (0.55 mol) is then added dropwise with vigorous stirring, which is continued at ambient temperature overnight. The precipitate is collected. The organic layer is washed with water, dried ($MgSO_4$) and concentrated. The solid residue is recrystallized from ethyl acetate and added to the earlier precipitate.

N-Methylbenzamide acetone [16]

A solution of *N*-2-propylidenemethanamine *N*-oxide [17] (1.24 g, 10 mmol) in anhydrous THF (30 ml) is cooled to 0°C and stirred under N_2 while triethylamine (3.03 g, 30 mmol) is introduced by syringe. A solution of *N*-methylbenzenecarboximidoyl chloride [18] (2.02 g, 13 mmol) in THF (5 ml) is then added dropwise. The resulting suspension is then allowed to warm to room temperature before stirring (24 h). The reaction is quenched into a solution of acetate buffer and stirred (30 min) at 25°C. The product is extracted into dichloromethane (3 × 30 ml), and the combined organic extracts are dried ($MgSO_4$) and rotary evaporated. Purification of the product by flash chromatography (5:1 hexane–ethyl acetate) gives a colourless oil (1.15 g, 60%).

General method for synthesis of 2,5-disubstituted 4-(2′-thienyl)imidazoles (**6**) [13]

Ammonium acetate (20 g, 0.26 mol) is added gradually to a solution of the *N*-acylamino ketone (0.01 mol) in glacial acetic acid (2 ml) at 130–140°C over a 2 h period. The reaction mixture is then cooled in an ice bath before addition of ethyl acetate (100 ml), diethyl ether (50 ml) and water (100 ml). The organic layer is separated, washed with water, dried ($MgSO_4$) and evaporated under reduced pressure. The residue is purified by column chromatography using chloroform as the eluent. Recrystallization from a mixture of chloroform and hexane gives the product, e.g. (**6**) (R^2, R^5, yield given): iPr, H, 81%; Ph, H, 54%; -FC_6H_4, H, 63%; -BrC_6H_4, H, 54%; -MeC_6H_4, H, 53%; -ClC_6H_4, Ph, 75%.

A number of related strategies utilize suitable "2-azabutadienes" in reaction with a nitrogen source (Scheme 3.2.3). Examples of suitable substrates include 4-amino-2-azabutadienes (**7**) and 1-isocyano-1-tosylalkenes (**8**). These reactions have been the subject of a review article [19]. In the former compound (**7**) the amino function is displaced in the cyclization. For example, 1-aminoimidazoles can be made by the reactions of hydrazines with suitable compounds (**7**) [20]. Yields are not particularly good, but the method may have application to the synthesis of some specific 1-aminoimidazoles. There are other examples of acid-catalysed cyclizations of similar 2-azabutadienes with electron donor and acceptor substituents, but yields are seldom high (Scheme 3.2.3) [21].

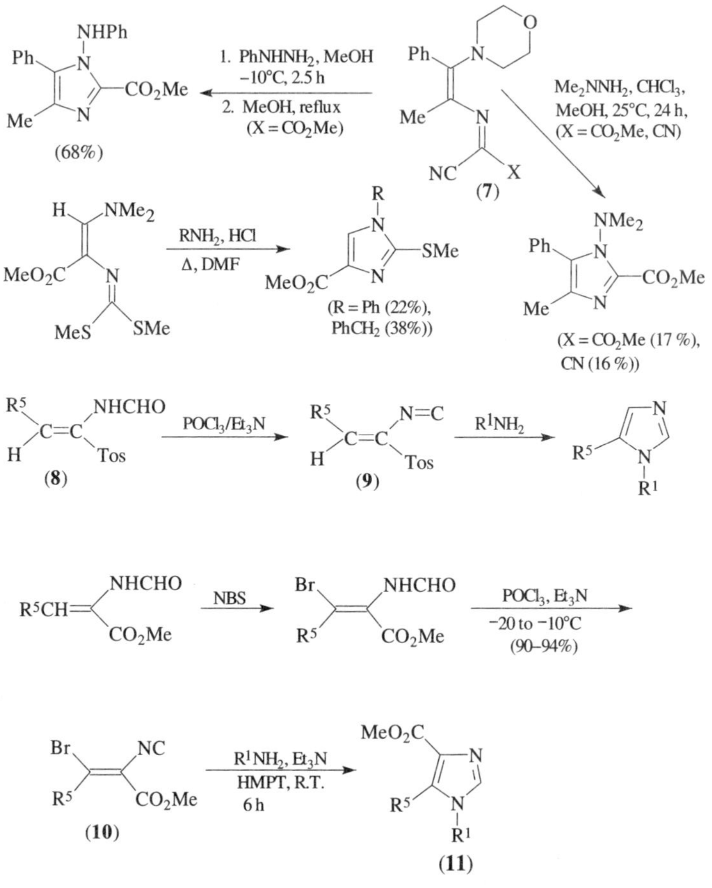

Scheme 3.2.3

The initial problem in such processes has been to make the 2-azabutadienes (**7**), but they are available from the reaction of suitable azomethines with dimethylformamide diethylacetal in yields which range between 35 and 87%. Azomethines with suitable displaceable substituents (amino, methylthio) can be converted into imidazoles. They can be made in high yields from amino acid esters [21].

When TOSMIC reacts with an aldehyde or ketone the products formed are N-(1-tosyl-1-alkenyl)formamides (**8**) (yields are high, 50–85%), which can be dehydrated in 54–77% yields to give 1-isocyano-1-tosyl-1-alkenes (**9**)

3.2. FORMATION OF 1,2 AND 1,5 BONDS

(Scheme 3.2.3) . These react smoothly with ammonia or primary aliphatic amines to give either 4-substituted or 1,5-disubstituted imidazoles [22]. Two equivalents of the amine are needed (the second is to promote the -elimination of -toluenesulfonic acid), and the reactions are usually complete in less than half an hour at room temperature unless the R^1 substituent is bulky. The process is not suitable for making 1-arylimidazoles, but offers a useful method of achieving 1,5 regiochemistry for 1-alkylimidazoles.

1-Isocyano-2-phenyl-1-tosylethene (**9**) ($^5 =$) [22]

To a solution of the formamide (**8**) ($R^5 =$ Ph) (7.53 g, 25 mmol) in dry DMF (25 ml) under N_2 at $-5°C$ is added all at once triethylamine (17.5 ml, 125 mmol), followed by the slow addition (in $\sim$5 min at -10 to $-5°C$) of $POCl_3$ (2.5 ml, 27.3 mmol) in DME (2.5 ml). After stirring (1 h) at $0°C$, the mixture is added to ice-water (500 ml) and immediately extracted with CH_2Cl_2, and the organic extracts are dried ($MgSO_4$), concentrated and treated with methanol (30 ml) at $-20°C$ to give the product (3.67 g, 54%), m.p. $80°C$ (dec.). Recrystallization from methanol gives an analytically pure product with the same melting point.

1-Cyclohexyl-5-phenylimidazole [22]

Cyclohexylamine (1.10 g, 11 mmol) is added to a solution of (**9**) ($R^5 =$ Ph) (1.42 g, 5.0 mmol) in methanol (25 ml). The slightly exothermic reaction (temperature increases to $\sim$30°C) is complete in 30 min. After dilution with water the product is extracted into CH_2Cl_2, and the solvent is dried ($MgSO_4$) and concentrated. Vacuum sublimation (twice, 0.1 mmHg, bath temperature 80–140°C) gives an oil (1.1 g, 97%) which slowly solidifies, m.p. $\sim$50°C. Similarly prepared are 1-t-butyl-5-phenyl- (82%), 1-methyl-5-phenyl- (87%), 1-methyl-5-(-nitrophenyl)- (88%) and 1-methyl-5-t-butylimidazoles (46%).

The above process has been adapted to high-yielding synthesis of some C-5-linked imidazole nucleosides [23].

Methyl 3-bromo-2-isocyanoacrylates (**10**), which can also be made by dehydration of the *N*-alkenylformamides, react with a variety of primary amines in the presence of triethylamine, to give moderate to good yields of 1,5-disubstituted imidazole-4-carboxylates (**11**) (Scheme 3.2.3) [24, 25].

Methyl (E)- and (Z)-3-Bromo-2-isocyanocinnamate (**10**) ($^5 =$) [24]

Phosphoryl chloride (5.1 g, 33 mmol) is added dropwise to a mixture of methyl ( )- and ( )-3-bromo-2-formylaminocinnamate [26] (8.52 g, 30 mmol) and triethylamine (8.41 g, 83 mmol) in dichloromethane (30 ml) at −10 to −20°C with stirring. After stirring at room temperature (2 h), the mixture is poured into 20% aqueous potassium carbonate (30 ml). The organic layer is separated, washed with water, dried (MgSO$_4$) and concentrated *in vacuo* to give an oil. Chromatography on a silica gel column using chloroform as the eluent gives a mixture of the *cis* and *trans* products as a colourless oil (7.2 g, 90%).

Methyl 1-phenethyl-5-phenylimidazole-4-carboxylate (**11**) ($R^1 =$ (CH$_2$)$_2$, $R^5 =$)

Phenylethylamine (0.53 g, 4.4 mmol) is added dropwise to a solution of (**10**) (R^5 = Ph) (1.06 g, 4 mmol) and triethylamine (0.62 ml, 4.4 mmol) in HMPT (4 ml) under ice cooling and stirring. Stirring is continued at room temperature (6 h). The mixture is then poured into a mixture of diethyl ether and saturated sodium bicarbonate, and the ether layer is separated, dried (MgSO$_4$) and rotary evaporated. Chromatography of the residue on a silica gel column eluted with ethyl acetate–hexane (1:1) gives the product as a colourless oil (0.96 g, 78%). Similarly prepared are the following compounds (**11**) (R^1, R^5, yield listed): PhCH$_2$, Ph, 80%; PhCH$_2$, Et$_2$CH, 61%; 3-picolyl, Ph, 71%; Me, Ph, 74%; -MeOC$_6$H$_4$, Et$_2$CH, 54%; Ph, Ph, 59%.

It is possible to make 4-acylimidazoles from 1,3-dicarbonyl compounds nitrated in their enolic forms (Scheme 3.2.4). The resultant nitroalkenes form *N*-alkenylformamides when reduced, and subsequent cyclization with formamide in formic acid gives the products, albeit in only moderate yields [27].

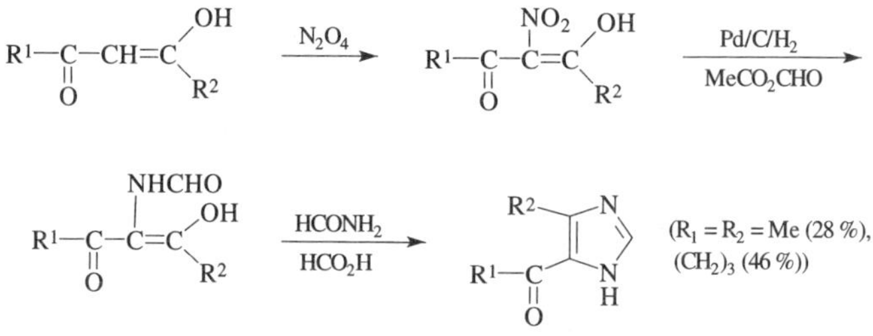

Scheme 3.2.4

When oxoketene acetals (**12**) are heated at 200°C in a sealed tube with nitrosoaromatics in acetic anhydride, imidazoles (**13**) are formed in moderate to good yields (Scheme 3.2.5) [28, 29]. Best results use excess of the nitrosoaromatic. Presumably the reaction is initiated by attack of the nitroso nitrogen at the carbon α to the carbonyl function. The α-oxoketene- ,N-acetals (**12**) are easily derived from a variety of active methylene ketones, and represent a novel class of functionalized enaminones or vinylogous amides [30, 31].

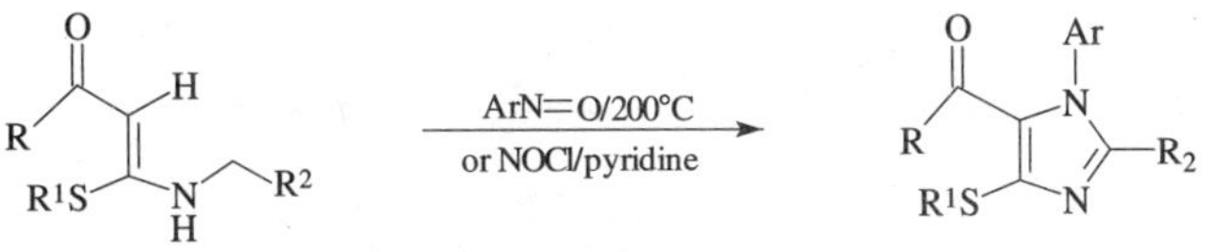

Scheme 3.2.5

General procedure for synthesis of imidazoles (**13**) [28]

A solution of (**12**) (0.01 mol) and nitrosobenzene (0.03 mol) in acetic anhydride (25 ml) is heated in a sealed tube (200°C, 1 h). The acetic anhydride is removed under reduced pressure, the residue is diluted with water (50 ml) and extracted with chloroform (3 × 250 ml). The combined organic extracts are dried (Na_2SO_4) and evaporated to give the various compounds (**13**) which can be purified by chromatography on silica gel using benzene-hexane (1:4) as the eluent. Examples of (**13**) (Ar, R, R^1, R^2, yield listed): Ph, Ph, Me, Ph, 85%; Ph, -MeC$_6$H$_4$, Me, Ph, 83%; Ph, -MeOC$_6$H$_4$, Me, Ph, 84%; Ph, -ClC$_6$H$_4$, Me, Ph, 85%; Ph, Ph, Et, Ph, 80%; Ph, Ph, CH$_2$Ph, Ph, 82%; Ph, Me, Me, Ph, 78%; Ph, Ph, Me, H, 75%; Ph, -MeC$_6$H$_4$, Me, H, 72%; Ph, -ClC$_6$H$_4$, Me, H, 75%; Ph, Ph, Me, Me, 80%; Ph, -MeC$_6$H$_4$, Me, Me, 77%; Ph, -ClC$_6$H$_4$, Me, Me, 78%; -MeC$_6$H$_4$, Ph, Me, Ph, 84%; -MeC$_6$H$_4$, Ph, Me, H, 70%; -MeC$_6$H$_4$, Ph, Me, Me, 72%.

Similar results are obtained when the oxoketene acetals are heated with a slight excess of nitrosyl chloride in dry diethyl ether–pyridine. Again yields range between 55 and 87% [29].

REFERENCES

1. T. Brown, G. Shaw and G. J. Durant, *J. Chem. Soc., Perkin Trans. 1* 2310 (1980).
2. D. H. Robinson and G. Shaw, *J. Chem. Soc., Perkin Trans. 1* 1715 (1972).
3. R. N. Naylor, G. Shaw, D. V. Wilson and D. N. Butler, *J. Chem. Soc.* 4845 (1961).
4. G. Shaw, R. N. Warrener, D. N. Butler and R. K. Ralph, *J. Chem. Soc.* 1648 (1959).
5. C. L. Leese and G. M. Timmis, *J. Chem. Soc.* 3816 (1961).
6. A. F. Pozharskii, A. D. Garnovskii and A. M. Simonov, *Russ. Chem. Rev. (Engl. Transl.)* **35**, 122 (1966).
7. M. R. Grimmett, *Adv. Heterocycl. Chem.* **27**, 241 (1980).
8. G. Mackenzie and G. Shaw, *J. Chem. Res. (S)* 184 (1977).
9. C. G. Beddows and D. V. Wilson, *J. Chem. Soc., Perkin Trans. 1* 1773 (1972).

10. R. S. Hosmane and B. B. Lim, *Tetrahedron Lett.* **26**, 1915 (1985).
11. O. N. Popilin and G. Tishchenko, *Chem. Heterocycl. Compd. USSR (Engl. Transl.)* **8**, 1142 (1972).
12. T. Moriya, S. Takabe, S. Maeda, K. Matsumoto, K. Takashima, T. Mori and S. Takeyama, *J. Med. Chem.* **29**, 333 (1986).
13. M. Suzuki, S. Maeda, K. Matsumoto, T. Ishizuka and Y. Iwasawa, *Chem. Pharm. Bull.* **34**, 3111 (1986).
14. D. Davidson, M. Weiss and M. Jelling, *J. Org. Chem.* **2**, 319 (1937).
15. S. Maeda, M. Suzuki, T. Iwasaki, K. Matsumoto and Y. Iwasawa, *Chem. Pharm Bull.* **32**, 2536 (1984).
16. I. Lantos and W.-Y. Zhang, *Tetrahedron Lett.* **35**, 5977 (1994).
17. O. Exner, *Collect. Czech. Chem. Commun.* **16**, 258 (1951).
18. I. Lantos, W.-Y. Zhang, X. Shui and D. S. Eggleston, *J. Org. Chem.* **58**, 7092 (1993).
19. J. Barluenga and M. Thomas, *Adv. Heterocycl. Chem.* **57**, 1 (1993).
20. D. Legroux, J.-P. Schoeni, C. Pont and J.-P. Fleury, *Helv. Chim. Acta* **70**, 187 (1987).
21. R. Gompper and V. Heinemann, *Angew. Chem., Int. Ed. Engl.* **20**, 296 (1981).
22. A. M. van Leusen, F. J. Schaart and D. Van Leusen, *Recl. Trav. Chim. Pays-Bas Belg.* **98**, 258 (1979).
23. D. E. Bergstrom and P. Zhang, *Tetrahedron Lett.* **32**, 6485 (1991).
24. K. Hiramatsu, K. Nunami, K. Hayashi and K. Matsumoto, *Synthesis* 781 (1990).
25. K. Nunami, M. Yamada, T. Fukui and K. Matsumoto, *J. Org. Chem.* **59**, 7635 (1994).
26. K. Nunami, K. Hiramatsu, K. Hayashi and K. Matsumoto, *Tetrahedron* **44**, 5467 (1988).
27. E.-P. Krebs and E. Bondi, *Helv. Chim. Acta* **62**, 497 (1979).
28. A. Rahman, R. T. Chakrasali, H. Ila and H. Junjappa, *Ind. J. Chem., Sect. B* **24**, 463 (1985).
29. A. Rahman, H. Ila and H. Junjappa, *J. Chem. Soc., Chem. Commun.* 430 (1984).
30. V. Aggarwal, H. Ila and H. Junjappa, *Synthesis* 147 (1983).
31. R. Gompper and H. Schaefer, *Chem. Ber.* **100**, 591 (1967).

—4—
Ring Syntheses Involving Formation of Two Bonds: [3 + 2] Fragments

4.1 FORMATION OF THE 1,2 AND 3,4 (OR 1,5 AND 2,3) BONDS

Only approaches to the uncondensed imidazole ring are available using this strategy.

The earliest method of this type was the old Marckwald synthesis [1] in which a suitable α-aminocarbonyl compound is cyclized with cyanate, thiocyanate or isothiocyanate. More recent modifications have employed the acetals of the α-amino aldehyde or ketone or an α-amino acid ester. The two-carbon fragment can also be provided by cyanamide, a thioxamate, a carbodiimide or an imidic ester. When cyanates, thiocyanates or isothiocyanates are used, the imidazolin-2-ones or -2-thiones (**1**) are formed initially, but they can be converted into 2-unsubstituted imidazoles quite readily by oxidative or dehydrogenative means (Scheme 4.1.1). The chief limitations of the method are the difficulty of making some α-aminocarbonyls and the very limited range of 2 substituents which are possible in the eventual imidazole products. The method is nonetheless valuable and widely used, and typically condenses the hydrochloride of an α-amino aldehyde or ketone (or the acetals or ketals), or an α-amino-β-ketoester with the salt of a cyanic or thiocyanic acid. Usually the aminocarbonyl hydrochloride is warmed in aqueous solution with one equivalent of sodium or potassium cyanate or thiocyanate. An alkyl or aryl isocyanate or isothiocyanate will give an *N*-substituted imidazole product (**2**), as will a substituted aminocarbonyl compound (Scheme 4.1.1) [2–4].

The most convenient method of making α-aminoaldehydes is by reduction of α-amino acid esters using sodium amalgam [5, 6] by what has come to be known as the Akabori method. Alternatively, an α-halogenocarbonyl compound can be converted via the Gabriel synthesis into the aminocarbonyl

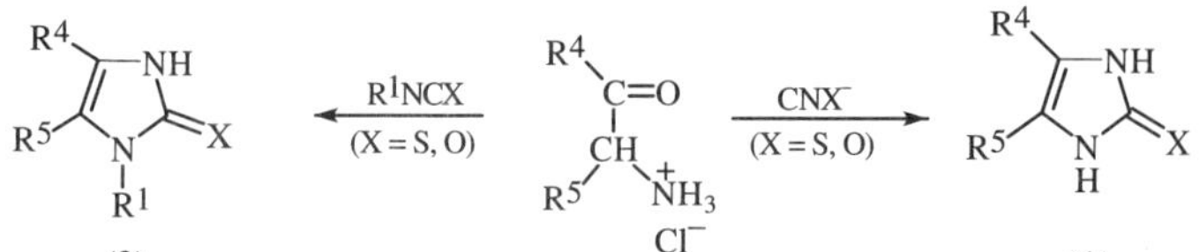

Scheme 4.1.1

analogue [7]. An appealing alternative approach to the preparation of α-bromoketones treats a vinyl bromide with NBS (1.1–1.5 eq.) in aqueous acetonitrile (1:4 H$_2$O:MeCN) containing a catalytic amount of hydrogen bromide. The bromoketones are obtained in 60–85% yields [8]. A recent easy and fast conversion of N^2-[(t-butoxy)carbonyl] (Boc) L-amino acids into the corresponding α-aminoaldehydes is based on the reduction of mixed anhydrides with LiAl(tBuO)$_3$H. The t-butoxy-protected amino acids are commercially available (Fluka), and yields of Boc-aminoaldehydes are reported in the range 76–85% [9]. Nitrosation of a ketone in the α position followed by reduction is yet another alternative [10]. When ethyl azidoformate is thermolysed in an enol trimethylsilyl ether, followed by treatment with silica gel, the products are N-ethoxycarbonyl-α-aminoketones. Such reactions are carried out in sealed tubes at 110°C (~15 h) with ethyl azidoformate and the substrate in a 1:10 volume ratio. Excess substrate is distilled off at the end of the reaction, and chromatography of the crude products gives 35–65% yields of the α-aminoketones [11].

Direct aminolysis of an α-bromoketone also gives the α-aminoketone provided that conditions are carefully controlled [12].

General method [12]

Under an argon atmosphere, a pressure-equalizing dropping funnel charged with the bromomethyl ketone (10.0 g) in diethyl ether (20 ml) is attached to a 300 ml RB flask containing a magnetic stirrer bar and a solution of primary amine (3 or 2 eq.) in ether (70 ml). The solution is stirred while cooling to −78°C in a dry-ice–acetone bath. The solution of the bromoketone is then added dropwise over 15 min, and the mixture is stirred for several hours until precipitation of the hydrobromide salts appears complete. The product is then shaken in a separating funnel with a small amount of 15% aqueous NaOH until the white solids dissolve. The ether layer is washed with water and then with brine, separated, dried (MgSO$_4$), filtered and concentrated to give the crude α-aminoketone as a pale yellow oil or solid. The individual compounds can be isolated by vacuum distillation or flash chromatography in 80–90% yields. Usually it is not necessary to purify the products before conversion into imidazoles.

4.1. FORMATION OF THE 1,2 AND 3,4 (OR 1,5 AND 2,3) BONDS

Aminoketones can also be made from amino acids by successive acylation–hydrolysis steps.

Acetamidoacetone [13]

A mixture of glycine (75.0 g, 1.0 mol), pyridine (485 ml, 6 mol) and acetic anhydride (1.1 l, 11.67 mol) is heated under reflux with stirring (6 h). Excess pyridine and acetic anhydride are removed under reduced pressure, and the residue is distilled using a Claisen head to give a pale yellow oil, b_1 120–125°C (80–90 g, 70–78%).

Aminoacetone hydrochloride [13]

A mixture of concentrated hydrochloric acid (175 ml) and water (175 ml) is added to the acetamidoacetone (52 g, 0.45 mol). The mixture is boiled under reflux in a nitrogen atmosphere (6 h) and concentrated by flash evaporation below 60°C using a condensation trap for solvent cooled in a dry-ice–acetone bath. A dark, oily residue (40–45 g) is obtained. This is very hygroscopic but can be dried over P_2O_5 in a vacuum desiccator. Purification is achieved by recrystallization from absolute ethanol, adding ether until cloudy. This reagent can be stored as the semicarbazone, from which aminoacetone can be generated as required *in situ*.

α-C-Acylamino acid esters can be made quite readily in good yields by acid hydrolysis of α-acyl-α-isocyanoacetate analogues (or oxazole-4-carboxylates), which in turn are prepared from α-isocyanoacetates with acyl halides or acid anhydrides in the presence of metallic or organic bases. Hydrolysis (6 M HCl, 90–95°C, 4 h) of either the oxazole derivatives or the α-C-acylamino esters gives good yields (50–65%) of the α-aminoketones as hydrochloride salts [14].

α-Benzoylaniline methyl ester hydrochloride [14]

To a mixture of DBU (6.0 g, 0.04 mol) and DMF (40 ml) is added a mixture of methyl α-isocyanopropanoate (3.39 g, 0.03 mol) and DMF (10 ml) at ambient temperature with stirring over a period of 10 min. After further stirring (3 h), benzoic anhydride (6.78 g, 0.03 mol) in DMF (10 ml) is added dropwise with vigorous stirring during 30 min at 30°C. After further stirring (3 h), water (100 ml) is added with cooling, and the resulting solution is extracted with ethyl acetate (100 ml). The extracts are washed in turn with water (2 × 50 ml) and 10% aqueous Na_2CO_3 (2 × 30 ml), dried ($MgSO_4$), and the solvent is removed. Distillation of the residual oil under reduced pressure gives methyl α-benzoyl-α-isocyanopropanoate (4.2 g, 65%), $b_{0.15}$ 95–97°C. The isocyano product (2.17 g, 0.01 mol) is then dissolved in 2 M HCl (30 ml) and heated at 40°C (1 h), the solvent and excess acid are removed under reduced pressure below 30°C, and the crystals obtained are washed with ethyl acetate. Recrystallization from ethyl acetate–methanol gives the above product (2.4 g, quantitative), m.p. 154–155°C.

Typical hydrolysis procedure to α-aminoketone hydrochloride

α-Butyrylglycine ester hydrochloride (0.5 g, 2.4 mmol) dissolved in 6 M HCl (20 ml) is heated (4 h) at 90–95°C. The solvent is removed under reduced pressure, and the resulting crystalline product is washed with ethyl acetate, and then recrystallized from ethyl acetate–methanol as colourless leaflets (0.29 g, 90%), m.p. 172–174°C. A variety of analogues have also been made in yields of 75–95%.

Reactions between α-aminocarbonyl compounds (or their acetals) and cyanates, thiocyanates and isothiocyanates are accomplished in the presence of acid, usually at a pH around 4. Table 4.1.1 lists some examples. Acetals are usually much more stable than α-aminoaldehydes, and they become the substrates of choice in many instances, being hydrolysed *in situ* [21].

Alkylaminoethanal dimethylacetals [23]

A mixture of 2-chloroethanal dimethylacetal (12.45 g, 0.1 mol) and alkylamine (0.4 mol) is heated under reflux (24 h), cooled, poured into a stirred mixture of KOH (25 g), water (50 ml) and chloroform (100 ml). The organic layer is separated, and the aqueous layer is extracted three times with chloroform. The combined chloroform extracts are dried (K_2CO_3), concentrated and distilled.

One can prepare a wide range of isothiocyanates conveniently from 1,1'-thiocarbonyldiimidazole and the appropriate amine, e.g. phenylisothiocyanate (78%) using aniline, cyclohexylisothiocyanate (72%) using cyclohexylamine or n-butylisothiocyanate (81%) using n-butylamine [24], thereby allowing synthesis of *N*-substituted imidazoles (see Table 4.2.1). An improved procedure refluxes the α-aminoketone hydrochloride with the isothiocyanate in toluene containing one equivalent of triethylamine. The thiourea derivative which is formed eliminates water to give the imidazole product. Continuous removal of the water with a Dean–Stark apparatus improves yields, while the triethylamine serves as an HCl scavenger [4].

Methyl 2-mercapto-4-phenylimidazole-5-carboxylate [15]

A mixture of methyl α-benzoylglycinate hydrochloride (6.0 g, 0.02 mol) and potassium thiocyanate (2.3 g, 0.02 mol) in water (6 ml) is heated at 60–90°C (4 h). The mixture is cooled in ice, and the precipitate is filtered, washed with

4.1. FORMATION OF THE 1,2 AND 3,4 (OR 1,5 AND 2,3) BONDS

TABLE 4.1.1
Imidazolin-2-thiones (**1**), (**2**); (X=S) made from α-aminocarbonyls (or their acetals) with thiocyanates or isothiocyanates

R^1	R^4	R^5	Yield (%)	Ref.
H	H	Pr	50	10
H	H	iPr	50	10
H	H	iBu	45	10
H	H	n-C_6H_{13}	60	10
H	H	CH_2Ph	85	21
H	H	2-FC_6H_4	64	21
H	H	2-ClC_6H_4	79	21
H	H	4-MeC_6H_4	95	21
H	H	4-$MeOC_6H_4$	98	21
H	H	CH_2CH_2Ph	78	21
H	H	2'-Thienyl	81	21
H	H	2'-Pyridyl	61	21
H	H	3'-Pyridyl	29	16
H	H	CO_2Me	50	17
H	Ph	CO_2Me	92	15
H	4-FC_6H_4	CO_2Me	86	15
H	2-ClC_6H_4	CO_2Me	94	15
H	CH_2Ph	CO_2Me	83	15
H	4-MeC_6H_4	CO_2Me	96	15
Me	H	Me	91	18
Me	H	CO_2Me	44	19, 20
CH_2Ph	H	CO_2Me	97	17
Pr	H	Me	84	5
Ph	H	CO_2Me	86	17
n-C_5H_{11}	H	H	36	23
n-C_7H_{15}	H	H	29	23
CH_2Ph	H	CH_2COCH_2Ph	31	5
CH_2COMe	H	Me	35	5
Pr	CO_2Et	CO_2Et	59	22
iPr	CO_2Et	CO_2Et	55	22
Bu	CO_2Et	CO_2Et	47	22
Et	CO_2Et	CO_2Et	57	22
Me	H	tBu	31	4
Me	H	iPr	56	4
Me	Me	H	40	18
CH_2Ph	H	tBu	30	4
n-C_6H_{11}	H	iPr	18	4
$CH_2-CH=CH_2$	H	tBu	43	4
$(CH_2)_2SMe$	H	tBu	28	4
$(CH_2)_2Ac$	H	tBu	57	4

cold water and recrystallized from methanol, to give the product as colourless prisms (4.3 g, 92%), m.p. 186–187°C.

5-t-Butyl-1-methoxyethylimidazolin-2-thione [4]

A stirred solution of 1-amino-3,3-dimethyl-2-butanone hydrochloride [10] (5.0 g, 37 mmol), 2-methoxyethyl isothiocyanate (4.3 g, 37 mmol) and triethylamine (3.7 g, 37 mmol) in toluene (60 ml) is refluxed (24 h) while collecting water in a Dean–Stark trap. The solvent is rotary evaporated to leave a semi-solid residue which is partitioned between dichloromethane and water. The dried (MgSO$_4$) organic extract is chromatographed on a column containing Florisil (50 g), and the eluted product is recrystallized from ethyl acetate to give colourless crystals (4.5 g, 57%), m.p. 93–96°C.

Imidazole 4(5)-carboxylic esters and their 1-alkyl-5-ester derivatives can be made from glycine and its *N*-aryl derivatives (Scheme 4.1.2) [17]. The 1,5 isomer is the unique product in this case. The appropriate glycine ester is initially formylated, then treated with an alkyl formate in the presence of sodium alkoxide, to give the sodium enolate salt of the *N*-formylglycine ester. Care must be taken at this stage to avoid transesterification, e.g. treating the ethyl ester with methyl formate in sodium methoxide converts it into the methyl ester [25]. We have also found that addition of a catalytic quantity of methanol to this reaction mixture hastens the rate of Claisen condensation. Subsequent cyclocondensation with potassium thiocyanate and either oxidative or reductive desulfurization leads to the imidazole ester. The reaction works well for *N*-formyl or *N*-acetyl glycine esters but not for the *N*-benzoyl analogues. The cyclocondensation process rapidly cleaves the *N*-acyl group [17].

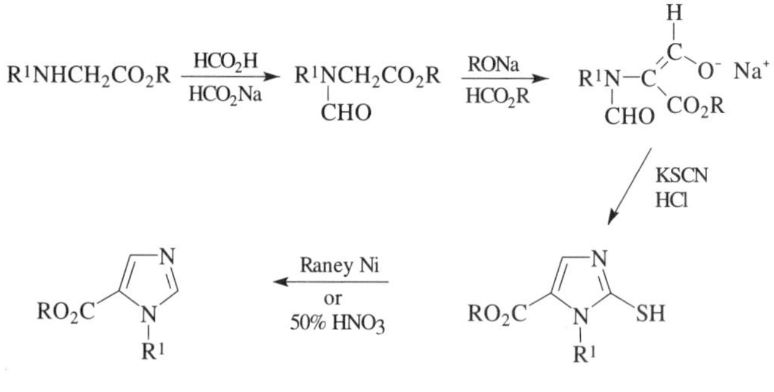

Scheme 4.1.2

Methyl imidazole-4-carboxylate [25]

To a warmed solution of glycine ethyl ester hydrochloride (75.0 g, 0.54 mol) in 90% formic acid (150 ml) is added a hot solution of sodium formate (45.0 g, 0.66 mol) in 90% formic acid (100 ml). After standing (1 h), the precipitated NaCl is filtered off, and acetic anhydride (250 ml) is added in portions. When the vigorous reaction has subsided, the mixture is heated at 100°C (30 min), then evaporated under reduced pressure to remove the acetic and formic acids. The residue is dissolved in acetone (500 ml), and the solution is filtered from additional NaCl, and then evaporated. Distillation gives N-formylglycine ethyl ester (59.5 g, 85%), b_2 127°C. To this ester (57 g, 0.43 mol) dissolved in dry benzene (200 ml) is added dry methyl formate (78.0 g, 1.30 mol) and a catalytic quantity of methanol. The mixture is cooled in an ice bath, and, with continuous stirring, freshly prepared sodium methoxide (27.0 g, 0.50 mol) suspended in dry benzene (50 ml) is added at such a rate that the temperature does not exceed 15°C. After stirring for a further 2 h the mixture is chilled overnight, and the crude enolate salt is filtered and washed several times with dry benzene. The crude salt is then dissolved in the minimum volume of water, cooled in an ice bath, and then concentrated HCl (85 ml, 1.0 mol) is added slowly. Potassium thiocyanate (49.0 g, 0.50 mol) is then added and the solution is warmed (2 h) at 50–70°C. After chilling overnight the crystalline methyl 2-mercaptoimidazole-4-carboxylate is filtered, washed with a small quantity of water, air dried, then recrystallized from water to give the pure product (12.9 g, 19% based on the N-formyl ester), m.p. 189°C (dec.).

To the above mercaptoester (3.0 g, 19 mmol) in ethanol (20 ml) is added a wet suspension of freshly prepared Raney nickel [26] in ethanol (∼5 g). The mixture is heated under reflux and monitored by TLC on silica (ethyl acetate). The desulfurization is complete in about 1 h. The suspension is then treated with activated charcoal, filtered (Celite) and washed with a little ethanol, and then the filtrate is rotary evaporated. Recrystallization of the residue from aqueous ethanol gives methyl imidazole-4-carboxylate (1.27 g, 53%), m.p. 149°C.

α-Aminocarbonyl compounds react rather less readily with cyanates to form 2-"hydroxy" imidazoles. Yields are usually poor to moderate only [17, 27], and it seems more productive to make these compounds by addition of isocyanates to N-unsubstituted 4-oxazolin-2-ones. The 2-oxo-4-oxazoline-3-carboxamides which form are cleaved at elevated temperatures in the presence of strong acid. Subsequent ring closure leads to good yields of substituted imidazolin-2-ones (44–92%) [28].

1-Ethyl-5-methyl-4-phenylimidazolin-2-one [27]

Triethylamine (3.0 g, 30 mmol) is added dropwise during 30 min to a solution of α-amino-α-phenylacetone hydrochloride (1.86 g, 10 mmol) and ethyl isocyanate (1.0 g, 14.1 mmol) in acetone (100 ml) at 0°C with stirring. After 1 h the solution is filtered, the acetone is evaporated, and the residue recrystallized from aqueous ethanol to give the product (1.0 g, 50%), m.p. 215-217°C. Similarly prepared in 61% yield is 1,4-dimethyl-5-phenylimidazolin-2-one.

When cyanamide replaces the cyanate, thiocyanate or isothiocyanate in the general method a useful route to 2-aminoimidazoles is opened up [29-31]. The intermediate guanidine which is formed in these reactions can often be isolated, and so the process could also be classified as a 1,5 bond formation (see Section 2.2).

2-Aminoimidazole [31]

Aminoacetaldehyde diethyl acetal (4 g, 30 mmol) is heated (1 h) at 100°C with cyanamide (2.5 g, 60 mmol) dissolved in water containing a few drops of acetic acid. After concentration under reduced pressure the resulting syrup is triturated with dry ether, and the gummy residue is treated with acetone (30 ml) to give N-(2,2-diethoxyethyl)guanidine acetate (1.8 g, 28%) as colourless crystals. The acetate (1 g, 4.7 mmol) is then warmed in concentrated HCl (3 ml) for a few minutes in a water bath. After cooling, water is added and the solution is evaporated to dryness *in vacuo*. The process is repeated. The crystalline hydrochloride obtained in quantitative yield is recrystallized from ethanol-ether as hygroscopic plates, m.p. 155°C.

Ethyl 2-amino-1-methylimidazole-5-carboxylate hydrochloride [29]

A solution of N-methyl-β,β-diethoxyalanine ethyl ester (10 g, 45 mmol) in 10% aqueous HCl (137 ml) is heated at 60°C (3 h). After cooling to −30°C and adjusting the pH to 4.5 with 10% aqueous NaOH, cyanamide (5.2 g, 120 mmol) is added, and the mixture is again stirred at 60°C (2 h) while maintaining the pH at 4.5 by addition of dilute hydrochloric acid. The resulting solution is decolorized by boiling with charcoal, filtered and rotary evaporated to dryness before trituration with diethyl ether to remove unreacted cyanamide. The remaining solid is then extracted several times with anhydrous ethanol containing 5% dry hydrogen chloride, the extracts are concentrated, and the syrupy product is separated by addition of ether. Crystallization from isopropanol gives the pure ester hydrochloride (5.8 g, 62%) as a hygroscopic solid, m.p. 200-205°C.

4.1. FORMATION OF THE 1,2 AND 3,4 (OR 1,5 AND 2,3) BONDS

Other N–C species which have been successfully condensed with α-aminocarbonyl compounds include thiooxamates [13, 19, 32], imidic esters [33–38], formamide [12] and N-substituted carbonimidodithioates [39]. Thiooxamates are readily made by treatment of ethyl cyanoformate with hydrogen sulfide in pyridine. They are easily alkylated using trialkyloxonium fluoroborates or other powerful alkylating agents before reaction with the aminoketone, to provide a useful route to 4-alkylimidazole-2-carboxylates (**3**) in greater than 70% yields (Scheme 4.1.3). Imidic esters (**4**) react similarly, e.g. (**4**) ($R^2 = CD_3$) as its hydrochloride has been converted by aminoacetaldehyde diethyl acetal into 2-trideuteriomethylimidazole in 66% yield [36], while (**4**) ($R^2 = 2,6$-dimethoxybenzoyl) reacted in turn with the same acetal and p-toluenesulfonic acid to give 2-(2′,6′-dimethoxy)benzoylimidazole (79%) (Scheme 4.1.3) [37]. Such imidic esters are readily made by treatment of a nitrile with ethanol in the presence of acid [34]. When N-substituted α-aminoketones are heated with formamide the product is a 1,4-disubstituted imidazole (**5**). Although reported yields are only moderate, this would still appear to be the method of choice for preparing imidazoles with 1,4 regiochemistry. The nature of the 4 substituent is only limited by the availability of bromomethyl ketones (which react with a primary amine in ether at $-78°C$ to give the substituted α-aminoketone), but the recent synthesis of α-haloketones from haloalkenes increases the range of such substrates (see above). The major problem seems to be some restriction to the range of nitrogen substituents, since hindered anilines or amines of low nucleophilicity hamper the eventual reaction with formamide [12]. Reaction of N-alkyl- or N-arylcarbonimidodithioates with aminoacetaldehyde acetals in refluxing acetic acid gives 1-substituted 2-methylthioimidazoles (**6**) in high yields (80–90%) (Scheme 4.1.3). Quite long reaction times are needed (10–16 h), and it is not clear if α-aminoketones will take part in the reaction. The carbonimidodithioates have been known for some time, and they can be made from a wide range of alkyl- and arylamines [39, 40].

2-Methylimidazole-d_3 [36]

To ethyl imidoacetate-d_3 hydrochloride (20 g) in dry CH_2Cl_2 is added, with ice cooling, aminoacetaldehyde diethyl acetal (21.2 g). After standing overnight at 25°C, the solvent is removed, the residue is treated at 0°C (3 h) with concentrated sulfuric acid (40 ml), and then neutralized with 50% aqueous NaOH, again with cooling in ice. The resulting solution is saturated with sodium chloride and extracted repeatedly with chloroform, and the organic extracts are dried (Na_2SO_4) and concentrated. Crystallization from benzene gives the product (0.52 g, 66%), m.p. 246–247°C.

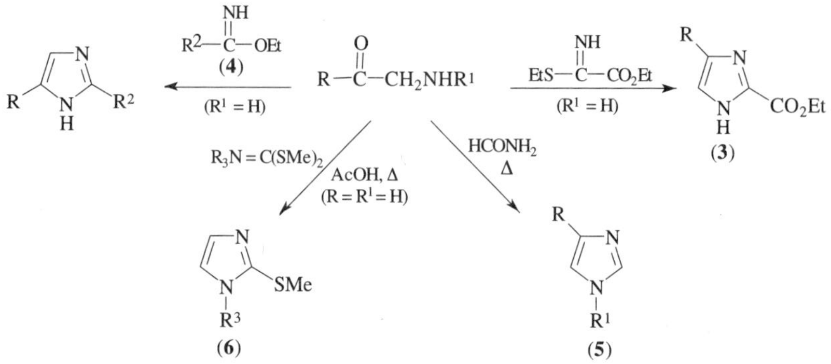

Scheme 4.1.3

2-(p-Fluorophenyl)-4-methylimidazole [34]

2-Aminopropanal diethylacetal (5.0 g, 34 mmol) and ethyl 4-fluorobenzimidate (5.7 g, 34 mmol) combined with glacial acetic acid (4 ml) are heated at 100°C (2 h) before addition of 5 M HCl (14 ml) and further heating at 100°C (30 min). The solution is then diluted with water (150 ml) and extracted with ether (3 × 150 ml). The ethereal solution is then dried (Na_2SO_4) and evaporated, to give a residue (3.5 g), which on recrystallization from acetonitrile gives the pure product (2.4 g, 39%), m.p. 185–187°C.

General procedure for 1,4-disubstituted imidazoles using formamide [12]

A 300 ml two-necked flask with an attached air condenser is charged with formamide (35 ml) and heated to 180°C under argon with stirring. The aminoketone is added, from a pressure-equalizing dropping funnel, dropwise during 1 h. Solid aminoketones are added in portions over 1 h. The mixture is then allowed to react for an additional 2–8 h at 180°C. After cooling, the dark reaction mixture is treated with an equal volume of water and 15% aqueous NaOH (20 ml) before extracting with toluene (2 × 200 ml). The combined toluene extracts are washed with water and brine, dried (Na_2SO_4), filtered and rotary evaporated to give a yellow-brown oil which is purified by flash chromatography using ethyl acetate–methanol (9:1) or ethyl acetate as the eluent. Kugelrohr distillation and recrystallization from hexanes can also be used to give the hygroscopic products (**5**) (R, R^1, yield given): Me, tBu, 21%; iPr, Et, 27%; iPr, iPr 47%; iPr, tBu, 34%; tBu, Et, 30%; tBu, iPr, 39%; tBu, tBu, 28%; tBu, $PhCH_2$, 82%; tBu, p-tolyl, 20%. Some tetraalkyl-1,2-dihydropyrazines may also be formed by self-condensation of the α-aminoketones, but they are readily identifiable by their complex proton NMR spectra.

4.1. FORMATION OF THE 1,2 AND 3,4 (OR 1,5 AND 2,3) BONDS

General procedure for 1-alkyl- and 1-aryl-2-methylthioimidazoles (**6**) [39]

A solution of the dimethyl *N*-aryl- or *N*-alkyl-carbonimidodithioate (10 mmol) and aminoacetaldehyde diethyl acetal (2.0 g, 15 mmol) in glacial acetic acid (10 ml) is boiled (10–15 h). The acetic acid is removed under reduced pressure, the residue is dissolved in chloroform (50 ml), washed with water (3 × 30 ml), dried (Na_2SO_4) and evaporated, to give the crude product. Purification is achieved by column chromatography on silica gel using ethyl acetate–hexane (1:4) and recrystallization from dichloromethane. Typical products (**6**) (R^3, reaction time, yield given): Ph, 10 h, 80%; *p*-MeC_6H_4, 10 h, 89%; *p*-$MeOC_6H_4$, 11 h, 85%; *p*-ClC_6H_4, 10 h, 81%; *p*-BrC_6H_4, 10 h, 90%; *o*-MeC_6H_4, 15 h, 82%; *o*-ClC_6H_4, 16 h, 83%; Me, 11 h, 80%; Et, 12 h, 81%; CH_2Ph, 15 h, 85%.

Reaction of an imidate with aminomalonodinitrile gives a 1-substituted 5-aminoimidazole-4-nitrile in low to moderate yields (10–60%) (Scheme 4.1.4). The reactions are carried out at room temperature in an acetate buffer, and there are similarities to reactions of DAMN (see Sections 2.1.1, 2.2.1 and 3.1.1) [35], and to the synthesis of 5-amino-4-cyanoimidazole from aminomalonodinitrile and formamidine acetate [41]. The reaction with an ethyl formimidate may initially involve replacement of the ethoxy group by amino followed by ring closure. Yields are reported to be low to moderate, with ethyl *N*-heteroarylformimidates reacting more readily than the *N*-aryl analogues. Indeed, the phenyl derivative will not react at all.

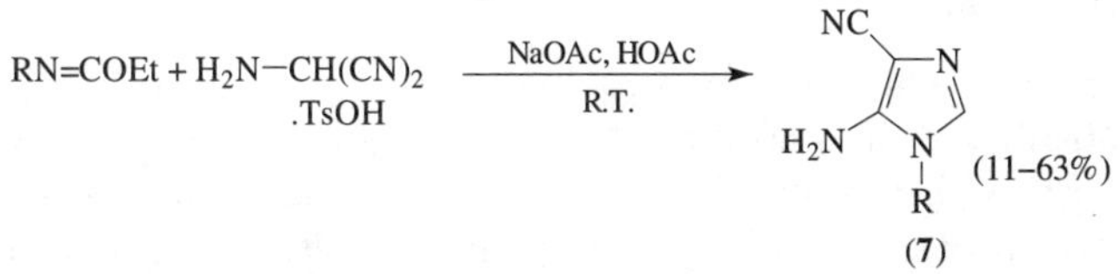

RN=COEt + $H_2N-CH(CN)_2$ · TsOH $\xrightarrow{\text{NaOAc, HOAc}}_{\text{R.T.}}$ [imidazole structure] (11–63%)

(**7**)

Scheme 4.1.4

General method for 1-substituted 5-aminoimidazole-4-nitriles (**7**) [35]

To a stirred solution containing aminomalonodinitrile *p*-toluenesulfonate (Aldrich or Fluka) (5.1 g, 20 mmol), sodium acetate (1.6 g, 20 mmol) and acetic acid (30 ml), the ethyl formimidate (20 mmol) is added portionwise over 20 min. The solution is stirred at room temperature (16 h), before addition of ice–water (200 ml). The precipitate is filtered, washed with water (100 ml) and recrystallized from ethyl acetate–hexane. Typical products (**7**) (R, yield): 2-pyridyl, 28%; 3-pyridyl, 30%; 2-pyrimidinyl, 30%; *p*-ClC_6H_4, 11%; *o*-ClC_6H_4, 33%; 2,4-$Cl_2C_6H_3$, 21%; 5-Cl-2-pyridyl, 29%.

A one-step synthesis of 4,5-disubstituted imidazole-2-carboxamides reacts a 1,2-diketone monophenylhydrazone with aminomalonamide (Scheme 4.1.5). Yields lie in the range 42–77%. Reactions are carried out in 1-methyl-2-pyrrolidinone at 100°C, and the method can be adapted to the preparation of 2-cyanoimidazoles by subsequent dehydration of the amides with phosphoryl chloride [42].

A novel imidazole synthesis utilizes the salt N-isopropylacetonitrilium tetrachloroferrate (**8**) in condensation with an amino acid ester. The salt is readily made from acetonitrile, ferric chloride and isopropyl chloride. For example, with norvaline methyl ester a 40% yield of 5-hydroxy-1-isopropyl-2-methyl-4-propylimidazole results. The reaction is believed to proceed via an intermediate amidine. Similarly, 1-isopropyl-2,5-dimethylimidazole is formed when (**8**) reacts with propargylamine (Scheme 4.1.5) [43, 44].

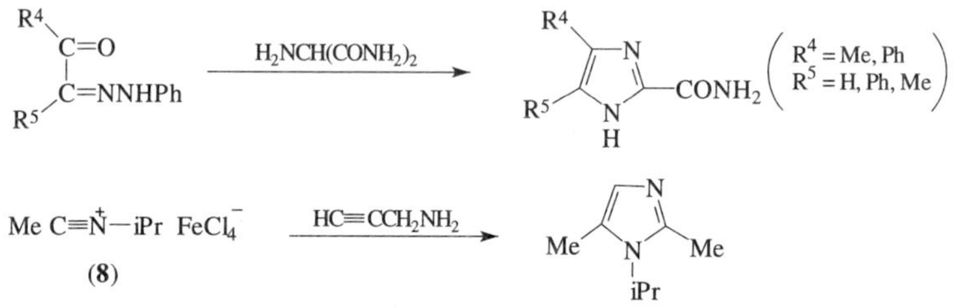

Scheme 4.1.5

1-Isopropyl-2,5-dimethylimidazole [43]

To a suspension containing ferric chloride (12.5 g, 77 mmol) in isopropyl chloride (50 ml) at 0°C is added, dropwise with stirring, acetonitrile (3.16 g, 76 mmol) dissolved in a little isopropyl chloride. The mixture turns red and then yellow. After 3 h at 0°C the precipitated salt can be filtered off in around 80% yield. To the crude salt (**8**) in dichloromethane (40 ml) is now added, dropwise with stirring and ice cooling, a solution of propargylamine (5 g, 90 mmol) in dichloromethane (10 ml). The mixture is kept refrigerated overnight, then treated with 30% aqueous sodium hydroxide and ether extracted. The combined ether extracts are dried (Na_2SO_4), and then removed before vacuum distillation of the imidazole product in 40% yield, $b_{0.5}$ 60°C.

Another reaction which proceeds through an intermediate amidine is that in which 4-formylimidazoles are made from 5-aminopyrimidine and an N-substituted imidoyl chloride in the presence of phosphoryl chloride. Once the intermediate amidine has formed, the pyrimidine ring opens up. Yields, however, are rather low (~25%) [45].

4.1. FORMATION OF THE 1,2 AND 3,4 (OR 1,5 AND 2,3) BONDS

1,2,5-Trisubstituted 4-mercaptoimidazoles can be made by the exothermic reaction of N-unsubstituted α-oxothionamides (**9**) with aldimines (Scheme 4.1.6). Yields vary between 9 and 81%, but many lie in the 50–60% region [46, 47]. Alkylation of the sulfur atom of ethyl 2-thioxamate (**9**) ($R^5 = OEt$) followed by treatment with aminoacetone gives ethyl 4-methylimidazole-2-carboxylate (**10**) in a reaction which provides a useful, general synthetic approach to imidazole-2-carboxylates (Scheme 4.1.6) [32].

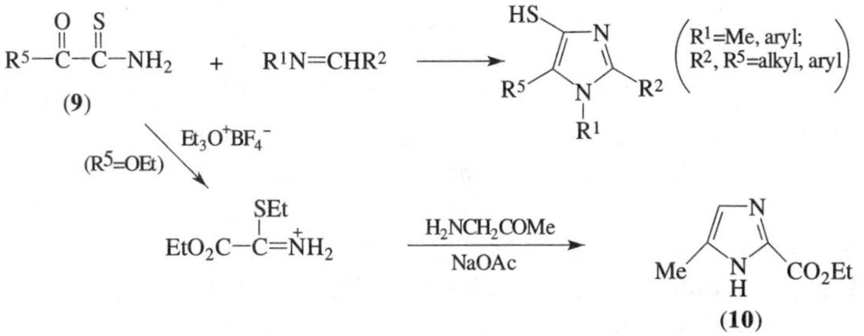

Scheme 4.1.6

α-Oximinoketones will react with aldimines, aldehydes and ammonia or amines, or aldoximes, to form imidazole N-oxides or 1-hydroxyimidazole 3-oxides [48–52]. A number of these cyclizations have been discussed elsewhere (see Sections 2.1 and 3.2, and Chapter 5), where they are classified as 1,2, 1,2 and 1,5, and 1,2, 3,4 and 1,5 bond formations.

1-Hydroxy-2,4,5-trimethylimidazole 3-oxide [49]

A mixture of acetaldoxime (20.0 g, 0.34 mol) and butanedione monoxime (34.2 g, 0.34 mol) is warmed to give a homogeneous solution, then allowed to stand at room temperature (2 days), refluxed (1 h), cooled, and diluted with diethyl ether (500 ml). The product precipitates as a tan solid (30.4 g, 63%), m.p. 194°C (dec.). Similarly prepared in 65% yield is 2-ethyl-1-hydroxy-4,5-dimethylimidazole 3-oxide.

1-Benzyl-4,5-dimethylimidazole 3-oxide [48]

N-Methylenebenzylamine (11.9 g, 0.09 mol) is added dropwise to a solution of biacetyl monoxime (10.1 g, 0.12 mol) in glacial acetic acid, and the solution is allowed to stand (12 h). After saturation with dry HCl and pouring into diethyl ether the semi-solid which separates is washed with ether, taken up in methanol, and reprecipitated as a fine, white solid (12.7 g, 53%) by addition of ether. The hydrochloride is made basic with ammonia (d. 0.88) and extracted with chloroform. The organic extracts are dried (MgSO$_4$), the solvent is removed, and the solid product recrystallized from acetone, to give the oxide (10.2 g, 51%), m.p. 196-198°C. Similarly prepared are 1-benzyl-5-methylimidazole 3-oxide (35%) and 1-benzyl-5-hydroxymethyl-4-methylimidazole 3-oxide (35%).

Such hydroxyimidazoles or oxides are capable of complete reduction to the unoxygenated imidazoles, while the 1-hydroxyimidazole 3-oxides can also be partically reduced with, for example, $NaBH_4$ [53], or completely deoxygenated with Raney nickel [52, 54]. Although most cyclizations of α-ketooximes lead to N-oxygenated imidazoles, there are exceptions. When an α-oximino-β-dicarbonyl compound is refluxed with benzylamine in a suitable solvent (e.g. DMSO, acetonitrile, toluene), 4-acylimidazoles (**11**) are formed in moderate to good yields. The reaction is readily adapted to the synthesis of imidazole-4-carboxylates and -amides (Scheme 4.1.7) [55]. Similarly, substitution of allylamine for benzylamine in the process gives a useful, one-pot synthesis of 2-vinylimidazoles in 40-65% yields [56]. Previous routes to such compounds have included dehydration of 2-(β-hydroxyethyl)imidazoles [55] (or dehydrobromination of the analogous bromoethyl compounds), pyrolysis of 2-(5-norbornen-2-yl) imidazoles [57] or thermal rearrangement of 1-vinylimidazoles [58].

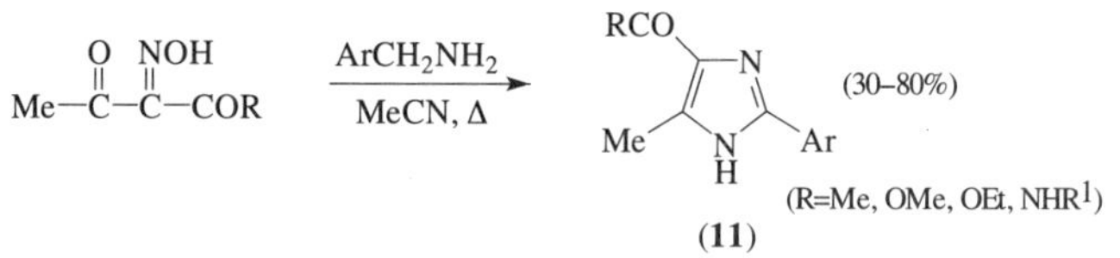

Scheme 4.1.7

General method for synthesis of 4-acylimidazoles (**11**) [59]

A 0.5 M solution of the dicarbonylmonoxime (5 mmol) in DMSO or acetonitrile (10 ml) is refluxed with the appropriate benzylamine (5.5 mmol) for 2-4 h. Products separate on cooling or concentration, and can be recrystallized from solvents such as toluene, acetonitrile or ethanol, e.g. (**9**) (R, Ar, reaction time yield listed): Me, Ph, 2 h, 68%; Me, p-MeC$_6$H$_4$, 2 h, 60%; Me, p-ClC$_6$H$_4$, 2.5 h, 66%; Me, p-NO$_2$C$_6$H$_4$, 2 h, 72%; OMe, Ph, 3 h, 32%; OEt, Ph, 3 h, 31%; NHCH$_2$Ph, p-NO$_2$C$_6$H$_4$, 4 h, 52%.

4-Acetyl-5-methyl-2-vinylimidazole [56]

Allylamine (0.415 ml, 5.05 mmol) is added to a solution of 3-hydroximino-2,4-pentanedione (0.645 g, 5 mmol) in anhydrous acetonitrile (5 ml). The violet solution is first stored at room temperature (12 h) and then refluxed (2 h). After concentration, the product is purified by flash chromatography (Merck silica gel 60, 230–400 mesh ASTM) using ethyl acetate as eluent. The product is recrystallized from diethyl ether–light petroleum (b. 40–70°C) as white crystals (0.75 g, 65%), m.p. 142–144°C.

Less common as a synthetic method is the rearrangement under mild conditions of some hydrazinium salts. In contrast to the severe conditions necessary (200°C, solid KOH) to rearrange 1,1,1-trialkylhydrazinium salts [60], 1,1-dimethyl-1-phenacylhydrazinium bromide is transformed merely by refluxing in pentanol or pyridine into 2-benzoyl-4-phenylimidazole (65–73%). This is a Stevens-type rearrangement [61, 62].

REFERENCES

1. W. Marckwald, *Ber. Dtsch. Chem. Ges.* **25**, 2354 (1892).
2. M. R. Grimmett, in *Comprehensive Heterocyclic Chemistry* (ed. A. R. Katritzky and C. W. Rees). Pergamon Press, Oxford, 1984, Vol. 5 (ed. K. T. Potts), p. 457.
3. M. R. Grimmett, *Adv. Heterocycl. Chem.* **12**, 103 (1970).
4. J. J. Doney and H. W. Atland, *J. Heterocycl. Chem.* **16**, 1057 (1979).
5. A. Lawson and H. V. Morley, *J. Chem. Soc.* 1695 (1955).
6. A. J. Lawson and H. V. Morley, *J. Chem. Soc.* 566 (1957).
7. H. Schubert, B. Ruehberg and G. Friedrich, *J. Prakt. Chem.* **32**, 249 (1966).
8. H. E. Morton and M. R. Leanna, *Tetrahedron Lett.* **34**, 4481 (1993).
9. P. Zlatoidsky, *Helv. Chim. Acta* **77**, 575 (1994).
10. M. Jackman, M. Klenk, B. Fishburn, B. F. Tullar and S. Archer, *J. Am. Chem. Soc.* **70**, 2884 (1948).
11. S. Lociuro, L. Pellacani and P. A. Tardella, *Tetrahedron Lett.* **24**, 593 (1983).
12. T. N. Sorrell and W. E. Allen, *J. Org. Chem.* **59**, 1589 (1994).
13. J. D. Hepworth, *Org. Synth.* **45**, 1 (1965).
14. M. Suzuki, T. Iwasaki, M. Miyoshi, K. Okumura and K. Matsumoto, *J. Org. Chem.* **38**, 3571 (1973).
15. S. Maeda, M. Suzuki, T. Iwasaki, K. Matsumoto and Y. Iwasawa, *Chem. Pharm. Bull.* **32**, 2536 (1984).
16. G. R. Clemo, T. Holmes and G. C. Leitch, *J. Chem. Soc.* 753 (1938).
17. R. G. Jones, *J. Am. Chem. Soc.* **71**, 644 (1949).
18. R. Burtles, F. L. Pyman and J. Roylance, *J. Chem. Soc.* 581 (1925).
19. H. Yamanaka, M. Mizugaki, T. Sakamoto, M. Sagi, Y. Nakagawa, H. Takayama, M. Ishibashi and H. Miyazaki, *Chem. Pharm. Bull.* **31**, 4549 (1983).
20. J. F. O'Connell, J. Parquette, W. E. Yelle, W. Wang and H. Rapoport, *Synthesis* 767 (1988).
21. J. M. Bobbitt and A. J. Bourque, *Heterocycles* **25**, 601 (1987).
22. H. Schubert and W. D. Rudorf, *Z. Chem.* **11**, 175 (1971).
23. E. Baeuerlein and H. Trasch, *Liebigs Ann. Chem.* 1818 (1979).

24. H. A. Staab and G. Walther, *Liebigs Ann. Chem.* **657**, 104 (1962).
25. P. Benjes and M. R. Grimmett, unpublished (P. Benjes, Ph.D. thesis, University of Otago, New Zealand, 1994).
26. M. Fieser and L. F. Fieser, *Reagents for Organic Synthesis*. Wiley, New York, 1967, Vol. 1, p. 729.
27. G. Holzmann, B. Krieg, H. Lautenschläger and P. Konieczny, *J. Heterocycl. Chem.* **16**, 983 (1979).
28. B. Krieg and H. Lautenschlaeger, *Liebigs Ann. Chem.* 208 (1976).
29. B. Cavalleri, R. Ballotta and G. C. Lancini, *J. Heterocycl. Chem.* **9**, 979 (1972).
30. A. Kreutzberger and R. Schuecker, *Arch. Pharm. (Weinheim, Ger.)* **306**, 169 (1973).
31. A. Lawson, *J. Chem. Soc.* 307 (1956).
32. J. E. Oliver and P. E. Sonnet, *J. Org. Chem.* **38**, 1437 (1973).
33. D. P. Matthews, J. P. Whitten and J. R. McCarthy, *Synthesis* 336 (1986).
34. J. S. Walsh, R. Wang, E. Bagan, C. C. Wang, P. Wislocki and G. T. Miwa, *J. Med. Chem.* **30**, 150 (1987).
35. I. Frank and M. Zeller, *Synth. Commun.* **20**, 2519 (1990).
36. M. Miyano and J. N. Smith, *J. Heterocycl. Chem.* **19**, 659 (1982).
37. J. P. Dirlam, R. B. James and E. V. Shoop, *J. Org. Chem.* **47**, 2196 (1982).
38. A. Lawson, *J. Chem. Soc.* 4225 (1957).
39. D. Pooranchand, H. Ila and H. Junjappa, *Synthesis* 1136 (1987).
40. A. Liden and J. Sandström, *Tetrahedron* **27**, 2893 (1971).
41. J. P. Ferris and L. E. Orgel, *J. Am. Chem. Soc.* **88**, 3829 (1966).
42. B. Singh, *Heterocycles* **34**, 2373 (1992).
43. R. Fuks, *Tetrahedron* **29**, 2147 (1973).
44. R. Fuks, *Tetrahedron* **29**, 2153 (1973).
45. C. Gönczi, Z. Swistun and H. C. van der Plas, *J. Org. Chem.* **46**, 608 (1981).
46. H. Offermanns, P. Krings and F. Asinger, *Tetrahedron Lett.* **15**, 1809 (1968).
47. F. Asinger, A. Sans, H. Offermanns, P. Krings and H. Andree, *Liebigs Ann. Chem.* **744**, 51 (1971).
48. I. J. Ferguson and K. Schofield, *J. Chem. Soc., Perkin Trans. 1* 275 (1975).
49. J. B. Wright, *J. Org. Chem.* **29**, 1620 (1964).
50. L. B. Volodarskii, *Chem. Heterocycl. Compd. (Engl. Transl.)* **9**, 1175 (1973).
51. B. Krieg and W. Wohlleben, *Chem. Ber.* **108**, 3900 (1975).
52. K. Bodendorf and H. Towliati, *Arch. Pharm.* **298**, 293 (1965); *Chem. Abstr.* **63**, 5629 (1965).
53. K. Akagane and G. G. Allan, *Chem. Ind. (London)* 38 (1974).
54. H. Towliati, *Chem. Ber.* **103**, 3952 (1970).
55. J. K. Lawson, *J. Am. Chem. Soc.* **75**, 3398 (1953).
56. A. C. Veronese, G. Vecchiati, S. Sferra and P. Orlandini, *Synthesis* 300 (1985).
57. A. S. Rothenberg, D. L. Dauplaise and H. P. Panzer, *Angew. Chem., Int. Ed. Engl.* **22**, 560 (1983).
58. C. G. Begg, M. R. Grimmett and P. D. Wethey, *Aust. J. Chem.* **26**, 2435 (1973).
59. A. C. Veronese, G. Cavicchioni, G. Servadio and G. Vecchiati, *J. Heterocycl. Chem.* **17**, 1723 (1980).
60. K. König and B. Zeeh, *Chem. Ber.* **103**, 2052 (1970).
61. M. Koga, A. P. Stamegna, D. J. Burke and J.-P. Anselme, *J. Chem. Ed.* **54**, 111 (1977).
62. L. Lessinger, M. K. Killoran, J.-P. Anselme and T. N. Margulis, *Tetrahedron Lett.* 3333 (1977).

4.2 FORMATION OF 1,2 AND 4,5 (OR 2,3 AND 4,5) BONDS

Among the most useful routes to imidazoles (especially those with 1,5 regiochemistry) are those which make use of the reagent TOSMIC (toluene-*p*-sulfonylmethyl isocyanide (**1**), available from Aldrich) (Scheme 4.2.1; see also Section 3.2 and Chapter 6) [1]. TOSMIC provides the NCN fragment in a number of cycloadditions, which give imidazoles, although it is not readily applicable to the synthesis of imidazoles with C-2 substituents. A variety of CN species (aldimines, imidoyl chlorides, nitriles, isothiocyanates and imino ethers) can take part in these reactions. Furthermore, a two-step synthesis of imidazoles from TOSMIC with an aldehyde proceeds via a 4-tosyloxazoline, which then reacts with ammonia or a primary amine. The process could be classified under a number of headings, e.g. from other heterocycles, or ultimately 1,2 and 1,5 bond formation. The aldehyde, however, becomes the C-4 of imidazole when it combines with TOSMIC to form the 4,5 bond. Hence there are similarities with 1,2 and 4,5 bond formation methods, and the process will be described in this section. Modification of TOSMIC by alkylation increases the variety of substituted imidazoles which can be made by these methods.

From Scheme 4.2.1 it can be seen that there is considerable potential for making 4-tosylimidazoles, 4(5)-mercaptoimidazoles, and imidazoles with 1,5 (and 1,4,5) regiochemistry. Such regiochemistry is usually difficult to achieve by *N*-alkylation or *N*-arylation of a preformed imidazole ring.

The original mild synthesis reported by van Leusen involved cycloaddition of an aldimine to a tosylmethylisocyanide in basic medium. The products were either 1,5-disubstituted imidazoles or (from a *C*-substituted TOSMIC) 1,4,5-trisubstituted imidazoles [2]. A variety of bases can be used to deprotonate the TOSMIC. Diarylaldimines with electron-donating substituents are converted into imidazoles (**2**) in poor to moderate yields when treated with sodium hydride in dimethoxyethane for 1–3 h at −20°C followed by refluxing in methanol for half an hour with potassium carbonate. The intermediates formed in these reactions are unstable 4-tosyl-2-imidazolines, which can be isolated. If the diarylaldimines have electron-withdrawing substituents it is necessary to treat the reagents in methanol at room temperature with potassium carbonate for about 16 h to achieve good yields. The use of t-butylamine in methanol or

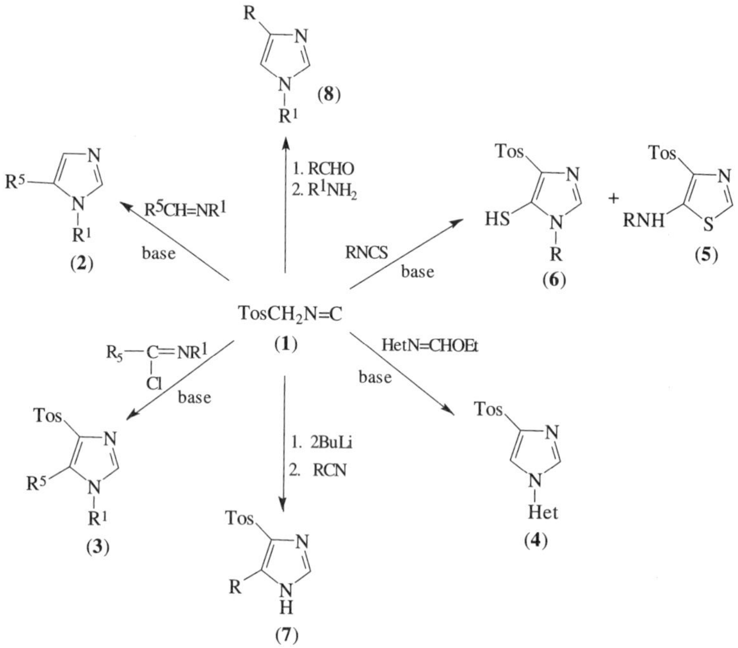

Scheme 4.2.1

dimethoxyethane at room temperature leads to high yields with dialkylimines, but *C*-alkyl- and *N*-arylimines do not react.

5-(p-Nitrophenyl)-1-phenylimidazole [2]

A solution of TOSMIC (0.975 g, 5.0 mmol) and *N*-*p*-nitrobenzylideneaniline (0.678 g, 3.0 mmol) in DME (10 ml) and methanol (20 ml) is stirred (16 h) with potassium carbonate (0.828 g, 6.0 mmol) at room temperature. After removal of the solvent at reduced pressure, the residue is digested with aqueous sodium chloride (25 ml), to give a yellow precipitate which is dissolved in dichloromethane (25 ml). The solvent is rotary evaporated, and the solid residue washed with diethyl ether to give the above product (0.650 g, 82%), m.p. 162–165°C.

1-t-Butyl-5-methylimidazole [2]

A mixture of TOSMIC (0.98 g, 5.0 mmol), *N*-ethylidene-t-butylamine (0.60 g, 6.0 mmol) and t-butylamine (0.44 g, 6.0 mmol) in methanol (10 ml) is stirred at room temperature (20 h). After removal of the solvent, the residue is distilled to give a colourless liquid product (0.65 g, 94%), $b_{0.05}$ 105°C, m.p. 42–44°C

4.2. FORMATION OF 1,2 AND 4,5 (OR 2,3 AND 4,5) BONDS

(hygroscopic). Similarly prepared are other 1,5-disubstituted imidazoles (**2**) (R^1, R^5, yield listed): p-$NO_2C_6H_4$, Ph, 70%; p-$NO_2C_6H_4$, p-$NO_2C_6H_4$, 87%; p-ClC_6H_4, p-ClC_6H_4, 43%; Ph, Ph, 56%; Ph, p-MeC_6H_4, 19%; Et, Me, 70%; Me, iPr, 96%; iPr, Me, 94%; iPr, iPr, 75%; Me, Ph, 10%; Me, p-MeC_6H_4, 37%; Me, p-$NO_2C_6H_4$, 14% [2].

If a *C*-substituted derivative of TOSMIC is used in the reaction, the product is a 1,4,5-trisubstituted imidazole. Such substituted isocyanides can be made either by dehydration of the *N*-(α-tosylalkyl)formamide, or by alkylation of an appropriate isocyanide followed by reaction with tosyl fluoride. It is also possible to alkylate TOSMIC directly using phase transfer conditions [3]. Such alkylations work best with primary alkyl halides (75-95%), with the isopropyl (40%), allyl (75%) and benzyl (80%) derivatives proving quite accessible.

α-Tosylbenzyl isocyanide [2]

n-Butyllithium (20% in hexane solution, 100 ml, 0.22 mol) is added dropwise to benzyl isocyanide (11.7 g, 0.10 mol) over 0.5 h in THF (180 ml) at −65°C. After stirring (5 min), a solution of tosyl fluoride (17.4 g, 0.10 mol) in THF (60 ml) is added dropwise, maintaining the temperature at −60°C (about 0.5 h). After stirring (5 min) without cooling, the mixture is poured into water (1.2 l), and the water layer is neutralized with aqueous HCl. After extraction with benzene (2 × 200 ml), the combined extracts are dried ($MgSO_4$), and concentrated, to give a pale yellow solid which is washed with carbon tetrachloride, to give the product (22.2 g, 82%), m.p. 128-130°C (dec.).

1-t-Butyl-5-methyl-4-phenylimidazole [2]

To a stirred solution of α-tosylbenzyl isocyanide (0.542 g, 2.0 mmol) and *N*-ethylidene-t-butylamine (0.300 g, 3.0 mmol) in DME (10 ml) is added (during 10 min) a solution of t-butylamine (0.212 g, 3.0 mmol) in DME (5 ml). After 0.5 h the mixture is filtered, the solvent removed, and the residue distilled to give a pale yellow oil (0.380 g, 89%), $b_{0.02}$ 140°C. The pure compound has an m.p. of 74-75°C. Similarly prepared are a number of other 1,4,5-trisubstituted imidazoles (1-, 4- and 5-substituents, yield quoted): iPr, Ph, Me, 89%; Me, Ph, Ph, 90%; p-$NO_2C_6H_4$, Ph, Ph, 82%; p-$NO_2C_6H_4$, Me, Ph, 75%.

Imidoyl chlorides are readily available reagents [4] which can be converted by the sodium derivative of TOSMIC into 4-tosylimidazoles (**3**) (Scheme 4.2.1) [2, 5]. This cycloaddition is achieved by loss of HCl rather than *p*-toluenesulfonic acid. Under aprotic conditions the primary adducts are imidazolines, but these aromatize readily. Good yields are obtained with C-aryl imidoyl chlorides but the C-alkyl derivatives do not react.

1,5-Diphenyl-4-tosylimidazole **(3)** *($R^1 = R^5 = Ph$)* [2]

A solution of TOSMIC (0.390 g, 2.0 mmol) and N-phenylbenzimidoyl chloride (0.426 g, 2.0 mmol) in DME (distilled from LiAlH$_4$, 5 ml) is added over 15 min to a suspension of sodium hydride (0.05 g, 2.0 mmol) in dry DME (5 ml) at room temperature under N$_2$. The reaction mixture is stirred (45 min), then poured slowly into water. The precipitate is collected and recrystallized from benzene–petroleum ether (b. 40–60°C, 1:1) as a white solid (0.45 g, 60%), m.p. 186–187°C. Similarly prepared are other 4-tosylimidazoles (1- and 5-substituents, yields given): Ph, p-NO$_2$C$_6$H$_4$, 85%; p-NO$_2$C$_6$H$_4$, Ph, 81%; cyclohexyl, Ph, 80%; cyclohexyl, p-NO$_2$C$_6$H$_4$, 75%.

Heterocyclic imino ethers are readily made in 40–80% yields by heating a heterocyclic amine with triethyl orthoformate in the presence of p-toluenesulfonic acid. When these ethers are treated with the monoanion of TOSMIC they are converted into 1-heteroaryl substituted 4-tosylimidazoles (**4**) in high yields (75–90%) (Scheme 4.2.1) [6].

1-(4'-Pyridinyl)-4-tosylimidazole [6]

A mixture of TOSMIC (19.5 g, 0.1 mol) and 4-(ethoxymethylene)aminopyridine (15.0 g, 0.1 mol) in dry DME (100 ml) is added dropwise under N$_2$ to a stirred slurry of sodium hydride (2.4 g, 0.1 mol) in dry DME (40 ml) during 20 min at room temperature. The reaction mixture is stirred (2 h), then poured into water (1 l) to precipitate the product (26.9 g, 90%), m.p. 198–199°C.

When TOSMIC reacts with isothiocyanates in the presence of base, the cycloaddition can give rise either to a thiazole or an imidazole, depending on the reaction conditions. Some reaction methods give rise to mixtures of the two heterocycles, but separation is simple since the thiazoles (**5**) can be directly extracted from alkaline aqueous solution, whilst the imidazoles (**6**) need to be acidified first. Furthermore, it is known that reactions of alkyl isothiocyanates with the monoanion of TOSMIC are more likely to lead to the 5-amino-2-tosylthiazoles, whereas dilithiated TOSMIC leads mainly to the 4-tosylimidazole-5-thiol products. With aryl isothiocyanates the reactions are less clear cut and seem to offer poorer synthetic potential. Nevertheless, the discovery that thiazoles (**5**) rearrange irreversibly to imidazoles when treated with more than one equivalent of butyllithium can be taken advantage of. Presumably an electrocyclic ring opening and ring closure are responsible,

4.2. FORMATION OF 1,2 AND 4,5 (OR 2,3 AND 4,5) BONDS

and the thiazoles result from kinetically controlled reactions, while imidazole formation is under thermodynamic control. The imidazole products are best prepared then by the use of two equivalents of butyllithium in THF at −65°C [7].

1-Methyl-4-tosylimidazole-5-thiol [7]

To a stirred solution of TOSMIC (1.95 g, 10 mmol) in dry THF (40 ml) kept at −70 to −60°C is added n-butyllithium (13 ml, 1.6 N in hexane, 21 mmol). After stirring (10 min) at −70°C, a solution of methyl isothiocyanate (0.8 g, 11 mmol) in THF (2 ml) is added rapidly, the cooling medium is removed, and the mixture is allowed to warm up to −10°C (in ~10 min). Water (50 ml) is then added, and the aqueous alkaline solution is washed with dichloromethane (2 × 50 ml) before the addition of 2 M HCl to adjust the pH to 6.5, followed by further extraction with dichloromethane (3 × 50 ml). The combined organic extracts are dried (Na_2SO_4) and concentrated, and the residue is stirred with methanol–diethyl ether (1:2, 6 ml) to give the product (1.40 g, 54%). Recrystallization from methanol gives the pure imidazole product (**6**) (R = Me), m.p. 159–161°C. Similarly prepared are (**6**) (R, yield given): Et, 20%; cyclohexyl, 40%.

The dianion of TOSMIC will react readily by a $[4\pi + 2\pi]$ cycloaddition with nitriles to give *N*-unsubstituted 4-substituted imidazoles (**7**) (Scheme 4.2.1). These reactions also occur with other C−N multiple bonds with much more facility than with the monoanion, while other isocyanides susceptible to α-metallation can also take part [8, 9]. As mentioned above, the dilithiated derivative of TOSMIC is much more reactive than the monolithio derivative. It is also considerably more stable, e.g. the half-life of the monoanion at 20°C under nitrogen in THF–hexane is about 3 h; under the same conditions the dianion is still 80% recoverable after 24 h. It is possible, therefore, to prepare 4-phenyl-5-tosylimidazole quite rapidly in 33% yield from the dianion and benzonitrile. The same product is formed only reluctantly from the monoanion. The dianion will also react even with azaaromatics (e.g. isoquinoline) with weakly electrophilic C=N bonds [9].

Such reactions can be extended to other isocyanides which are prone to α-metallation (Scheme 4.2.2). Thus, *p*-tolylthiomethyl isocyanide, which forms a more nucleophilic anion than TOSMIC, reacts with a variety of nitriles to give imidazole products [8]. Nitriles are added at −75°C to the monoanions formed from a variety of thiomethyl isocyanides (butyl lithium in THF–hexane at −75°C). The process is completed by allowing the reaction mixtures to come to 0°C, followed by treatment with water (method A). Alternatively, a THF solution of equimolar proportions of the isocyanide and the nitrile

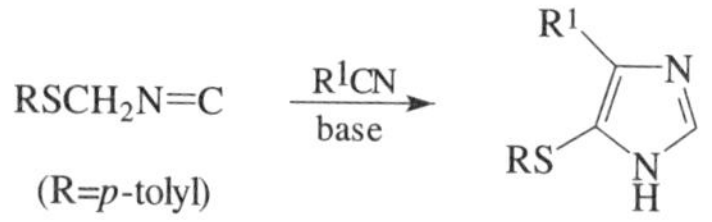

4. RING SYNTHESES INVOLVING FORMATION OF TWO BONDS

$$RSCH_2N=C \xrightarrow[\text{base}]{R^1CN} \quad (9)$$

(R=p-tolyl)

Scheme 4.2.2

is simply added at room temperature to a stirred suspension of potassium t-butoxide in THF. Addition of water after 15 min leads to high yields of the 4,5-disubstituted imidazoles (method B). The former method is of particular advantage with sterically hindered nitriles. The alkylthio or arylthio groups can be removed later using Raney nickel if required [8]. Table 4.2.1 lists some representative examples.

TOSMIC can be converted into an N-tosylmethylimidic ester or thioester (**10**) which will react with an aldimine to form a 1,2,5-trisubstituted imidazole (Scheme 4.2.3). These esters (**10**) can be made from N-tosylmethylacetamide (from the Mannich condensation of p-toluenesulfonic acid, formaldehyde and acetamide [10]), which is smoothly converted by P_4S_{10} in DME into the thioamide which forms the S-methylated imidate when treated with methyl fluorosulfonate in dichloromethane. Yields of the N-tosylmethylimidic thioesters are good (65–93%); they are fairly stable crystalline solids which are best stored under nitrogen at $-20°C$. In reaction with an aldimine in the presence of sodium hydride or potassium t-butoxide (in DME–DMSO or

TABLE 4.2.1
Imidazoles (**9**) prepared from [3 + 2] cycladdition of p-tolylmethyl isocyanides and nitriles [8]

R^1	Yield (%)	
	Method A	Method B
Ph	90	97
p-MeO-C_6H_4	94	94
p-Me-C_6H_4	86	93
p-NO$_2$-C_6H_4	—	0
2-Pyridyl	73	93
Me	86	85
Et	64	84
iPr	—	85
tBu	82	49
MeOCH$_2$	—	91
MeSCH$_2$	68	75
PhCH$_2$	0	—

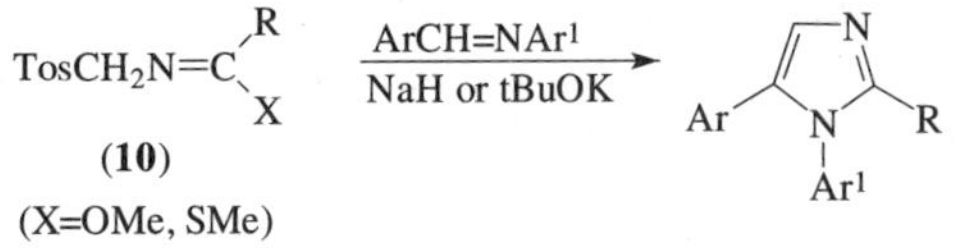

4.2. FORMATION OF 1,2 AND 4,5 (OR 2,3 AND 4,5) BONDS

$$TosCH_2N=C\underset{X}{\overset{R}{\diagup}} \quad \xrightarrow{ArCH=NAr^1}_{NaH \text{ or } tBuOK} \quad \underset{Ar^1}{\overset{}{Ar\diagdown\underset{N}{N}\diagup R}}$$

(10)
(X=OMe, SMe)

Scheme 4.2.3

t-butanol–DME) the trisubstituted imidazoles are formed in poor to moderate yields, e.g. 1,2,5-triphenylimidazole (23%) and 1,5-di(*p*-chlorophenyl)-2-methylthioimidazole (64%) [11].

A two-step synthesis of 1,4-disubstituted imidazoles (**8**) from TOSMIC (**1**) plus an aldehyde, followed by reaction with ammonia or a primary amine, proceeds via a 4-tosyloxazoline (**11**). The reaction sequence could be classified as 1,2 and 1,5 bond formation, 1,5 bond formation, or transformation of another heterocycle. There are, however, analogies to the aldimine reactions, and so the process is detailed at this stage. Certainly the synthesis is carried out in two steps often with isolation of the oxazoline (see also Chapter 6). Heating (**11**) with a saturated solution of methanolic ammonia gives a 4-substituted imidazole; with methanolic methylamine a 1,4-disubstituted product is isolated as a single regioisomer (Scheme 4.2.4). Some of the oxazolines cannot be isolated as they are unstable oils which have to be heated immediately with the amino compound [12]. Related is the synthesis of 2-carbamoyl-4-(2′-deoxy-β-D-ribofuranosyl)imidazole [13].

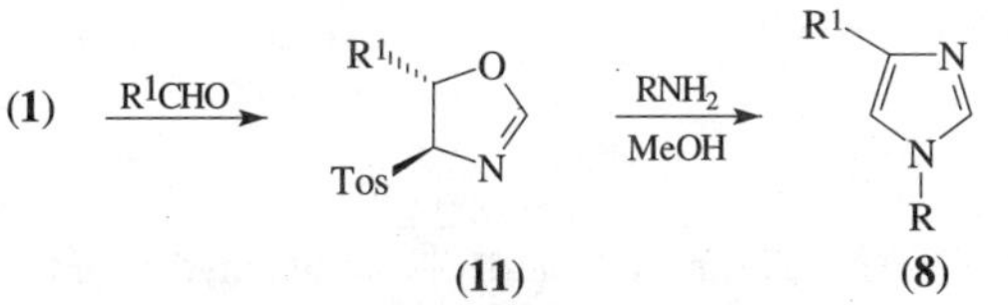

$$(1) \xrightarrow{R^1CHO} \underset{Tos}{\overset{R^1}{\diagdown}}\!\!\!\diagdown\!\!\!\diagup^O_N \xrightarrow{RNH_2}_{MeOH} \underset{\underset{R}{|}}{\overset{R^1}{\diagdown}}\!\!\!\diagdown\!\!\!\diagup^N_N$$

(11)   (8)

Scheme 4.2.4

General preparation of a 4-tosyloxazoline (**11**) [12]

To a stirred suspension of TOSMIC (5.0 mmol) and the aldehyde (5.1 mmol) in dry ethanol (15 ml) is added finely powdered sodium cyanide (0.5 mmol). Within minutes a slightly exothermic reaction gives a clear solution from which the oxazoline separates (~15 min). Stirring is continued for an additional 10 min, or until TLC (dichloromethane–diethyl ether, 95:5) shows that all of the TOSMIC has disappeared. The mixture is filtered, the crystals are washed with ether–hexane (1:1, 15 ml) and dried. A second crop is obtained by concentration of the mother liquors and trituration with ether–hexane (1:5).

If diastereoisomeric mixtures form, crystallization may be prevented. In such cases the solution is concentrated *in vacuo* to give an oil of sufficient purity to allow subsequent reaction. Yields lie in the range 84–93%.

Conversion of a 4-tosyloxazoline into a 1,4-disubstituted imidazole **(8)**

(i) Conversion into **(8)** *(R = H)*   In a resealable pressure tube a solution of **(11)** (1 mmol) in a saturated solution of ammonia in dry methanol (8 ml) is heated at 90–110°C (15–20 h). TLC (dichloromethane–methanol, 9:1) can be used to monitor the reaction. Concentration followed by flash chromatography (using the above solvent mixture) of the residue gives the 4-substituted imidazole **(8)**. (See Table 4.2.2.)

(ii) Conversion into **(8)** *(R = alkyl/aryl)*   In a resealable pressure tube a solution of **(11)** (1 mmol) and the amine (4 mmol) in benzene (5 ml, for methylamine) or xylene (5 ml, for n-butyl- or benzylamine) is heated at 110–120°C (for benzene) or 135–140°C (for xylene) for 15–20 h. The mixture is worked up as above to give **(8)** (R = alkyl) as the sole regioisomer. The *N*-alkylamide by-product is readily removed by chromatography. (See Table 4.2.2.)

The process can also be adapted to the synthesis of 4-ethyl-5-substituted imidazoles if α-(*p*-tosyl)propyl isocyanide is used with an aldehyde in the presence of potassium t-butoxide.

Analogous to the TOSMIC reactions is the cyclocondensation of an isothiourea with the enolate of ethyl isocyanoacetate **(12)** to give an alkyl 5-aminoimidazole-4-carboxylate **(13)**. This regioselective synthesis provides

TABLE 4.2.2
Imidazoles **(8)** made from TOSMIC with aldehydes and amino compounds [12]

R	R^1	Yield (%)
H	Et	80
H	tBu	78
H	CH$_2$CH$_2$Ph	73
H	CH$_2$OCH$_2$Ph	72
H	Ph	61
H	*p*-MeO–C$_6$H$_4$	75
Me	tBu	63
Me	Ph	48
Me	*p*-MeO–C$_6$H$_4$	47
Bu	Et	57
PhCH$_2$	CH$_2$CH$_2$Ph	55

4.2. FORMATION OF 1,2 AND 4,5 (OR 2,3 AND 4,5) BONDS

an alternative to the multistep cyclization of α-formamidoamidines [14] (see Section 2.1.1) or the condensation of a primary amine with an α-(ethoxymethyleneamino) cyanoacetate [15] (see Section 3.2). The method shown in Scheme 4.2.5 is a short route which uses mild conditions and gives reasonable control of regiochemistry when unsymmetrical isothioureas are used, e.g. (R^1, R^2, yield, **(13a)**:**(13b)** ratio given): Ph, Ph, 75%, — ; $PhCH_2$, Ph, 66%, 10:1; H, $PhCH_2$, 30%, ≤ 1:20; $PhCH_2$, PhCO, 70%, 1:4 [16].

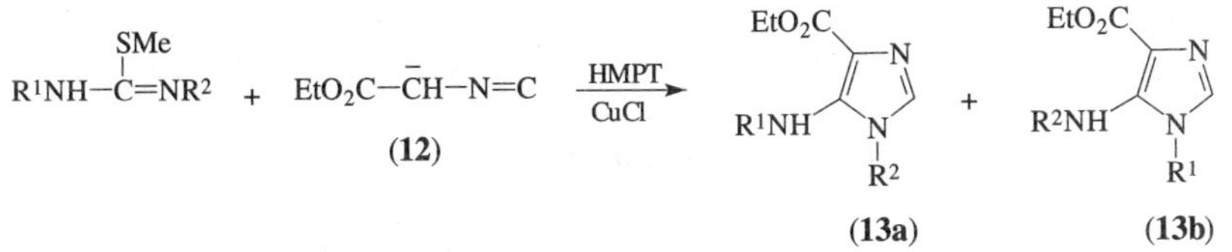

Scheme 4.2.5

Ethyl 1-phenyl-5-phenylaminoimidazole-4-carboxylate [16]

A solution of **(12)** prepared in hexamethylphosphoric triamide (HPMT, 0.5 ml) from ethyl isocyanoacetate (1.7 mmol) and potassium hydride (1.7 mmol), is added to a mixture of *S*-methyl-*N*,*N*'-diphenylisothiourea (0.19 g, 0.8 mmol) and copper(I) chloride (0.03 g, 0.3 mmol) in HMPT (0.2 ml) at 0°C. After stirring at 25°C (4 h), the reaction is quenched with aqueous ammonium chloride, and extracted three times with ethyl acetate. The combined organic extracts are dried ($MgSO_4$) and evaporated. Column chromatography on silica gel (diethyl ether–ethyl acetate, 1:3) gives **(13a)** (R^1, R^2 = Ph) (0.19 g, 75%), m.p. 157–159°C.

Treatment of an imidoyl chloride with triethylamine forms a nitrile ylide, which undergoes cycloaddition with a variety of reagents in rather low yields [17]. The 1,3-dipolar cycloadditions of mesoionic oxazolones with electron-deficient nitriles also fall under this general classification [18]. Such oxazolones are known as "munchnones", and are masked 1,3 dipoles. Such munchnones can be prepared *in situ* (they are highly reactive) from the corresponding *N*-acyl-α-amino acids and *N*,*N*-dicyclohexylcarbodiimide, before treating them with an appropriate imine **(14)** (or nitrile). These reactions are carried out under nitrogen in toluene or DMF solutions for about 12 h at temperatures which range between 25 and 60°C depending on the reactivity of the imine (Scheme 4.2.6). The imidazole product **(15)** is obtained by column chromatography after removal of the dicyclohexylurea by filtration and evaporation of the solvent. The method provides a versatile entry to 1,4,5-tri- and 1,2,4,5-tetrasubstituted imidazoles in moderate yields. The control of regiochemistry is a function of the fact that a bond is always formed between C-2 of the dipole and the imine nitrogen atom [19, 20]. Because the phenylsulfonyl group is a good leaving group, it promotes the tendency to form an aromatic product.

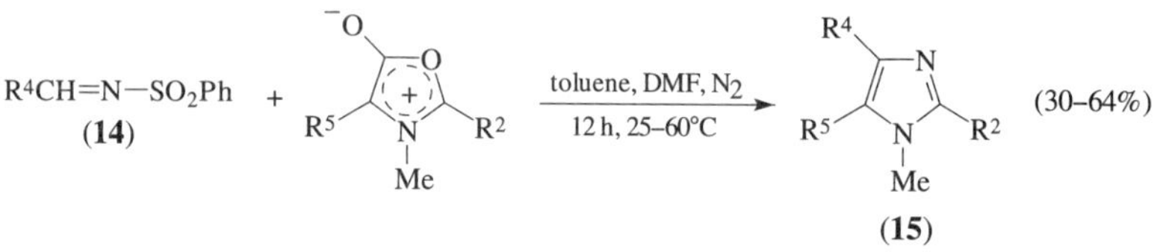

$$R^4CH=N-SO_2Ph + \underset{Me}{\underset{|}{R^5 \diagdown N \diagup R^2}} \text{(O-O)}^+ \xrightarrow[12\,h,\,25-60°C]{\text{toluene, DMF, N}_2} \underset{Me}{\underset{|}{R^5 \diagdown N \diagup R^2}}^{R^4 \diagdown N} \quad (30-64\%)$$
(14)   (15)

Scheme 4.2.6

1-Methyl-2,5-diphenyl-4-styrylimidazole (**15**) *(R^2, R^5 = Ph; R^4 = PhCH = CH)* [20]

A suspension of N-benzoyl-N-methylphenylglycine (1.0 g, 3.7 mmol) in DMF (15 ml) is treated with DCCI (0.84 g, 4.1 mmol) in DMF (10 ml) and then with (**14**) (R^4 = PhCH = CH) (1.0 g, 3.7 mmol) in DMF (10 ml). The mixture is stirred under N_2 at room temperature (3 h). The dicyclohexylurea is filtered off and washed with solvent, and the filtrates are rotary evaporated, the residue is taken up in dichloromethane (30 ml), washed in turn with water (150 ml), aqueous $NaHCO_3$ (2 × 15 ml), and water (100 ml) before drying (Na_2SO_4). Evaporation to dryness gives the product, which is purified by flash chromatography on silica gel (toluene–ethyl acetate, 19:1) (0.62 g, 50%). Suitable recrystallization solvents for such compounds are dipropyl ether and isopropanol.

Similarly prepared are (**15**) (R^2, R^4, R^5, yield given) [19, 20]: Ph, Ph, Ph, 64%; Ph, Ph, Me, 29%; Ph, Me, Ph, 40%; H, Ph, Ph, 20%; Ph, p-$NO_2C_6H_4$, Ph, 65%; Ph, p-$NO_2C_6H_4$, Me, 50%; Me, p-$NO_2C_6H_4$, Ph, 30%; Ph, p-$MeOC_6H_4$, Ph, 45%; Ph, p-$MeOC_6H_4$, Me, 55%; Me, p-$MeOC_6H_4$, Ph, 42%; Ph, p-$NO_2C_6H_4$CH=CH, Ph, 30%; Ph, p-$MeOC_6H_4$CH=CH, 35%.

The 4-styryl products are of value in that they offer an alternative route to imidazole-4-carbaldehydes when they are oxidized by osmium tetroxide and sodium periodate in about 70% yield [20].

The most recent synthetically useful modification of TOSMIC reactions involves the cycloaddition of N-trimethylsilylimines with lithiotosylmethyl isocyanides. This allows the preparation of 4-mono- and 4,5-disubstituted imidazoles from readily accessible aldehydes and organolithium compounds in a one-pot reaction. Although yields are only moderate, there are advantages which accrue from the easy availability of the starting

4.2. FORMATION OF 1,2 AND 4,5 (OR 2,3 AND 4,5) BONDS

materials [21]. The process has been recommended as a mild alternative to the rather harsh reaction conditions of Bredereck's "formamide synthesis" (see Chapter 5). Two alternative synthetic procedures generate the trimethylsilylimines (**16**): method A — an alkyl- or aryllithium is treated with N,N-bis(trimethylsilyl)formamide; method B — an aldehyde is treated with lithium bis(trimethylsilyl)amide. It is not necessary to isolate the silylamine (**16**); it can be trapped by the anion of the TOSMIC derivative to give (**17**) (Scheme 4.2.7) [21].

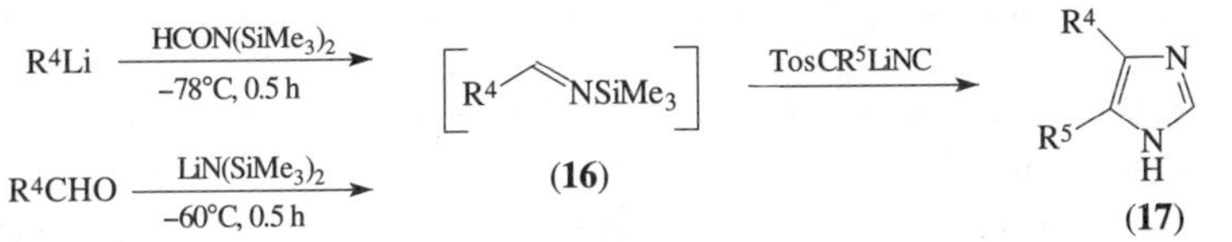

$$R^4Li \xrightarrow[-78°C, 0.5h]{HCON(SiMe_3)_2}$$

$$R^4CHO \xrightarrow[-60°C, 0.5h]{LiN(SiMe_3)_2} \left[R^4 \diagup NSiMe_3 \right] \xrightarrow{TosCR^5LiNC} \underset{R^5}{\overset{R^4}{\diagdown}} \underset{H}{\overset{N}{\diagup}}$$

(**16**)   (**17**)

Scheme 4.2.7

5-Benzyl-4-butylimidazole (method A) [21]

To a cold ($-78°C$) solution of N,N-bis(trimethylsilyl)formamide (1.63 ml, 7.5 mmol) in anhydrous THF (10 ml) is slowly added a solution of n-butyllithium in hexane (4.7 ml, 1.5 N, 7.5 mmol), and the mixture is stirred at $-78°C$ (30 min) before adding via cannula a solution of the anion of tosylbenzylmethylisocyanide. This is prepared by addition of a solution of lithium bis(trimethylsilyl)amide (7.15 ml, 1.0 N, 7.16 mmol) to a cold ($-55°C$) solution of tosylbenzylmethylisocyanide (2.035 g, 7.16 mmol) in dry THF (5 ml) followed by stirring (30 min) at -50 to $-60°C$. The resultant solution is stirred (30 min) at $-78°C$, allowed to warm to $0°C$ (2 h), then stirred at room temperature (16 h). Concentration of the reaction mixture, dilution of the residue with water (30 ml), adjustment of the pH to 10-11 with 1 M HCl, saturation with sodium chloride, extraction of the aqueous solution with ethyl acetate-methylene chloride (4:1), drying of the extracts (Na_2SO_4 and K_2CO_3), concentration, and purification by flash chromatography on silica gel gives the product (**17**) ($R^4 = CH_2Ph$, $R^5 = nBu$) (1.01 g, 66%).

General procedure (method B) [21]

To a solution of the aldehyde (1.5 mmol) in anhydrous THF (2 ml) at $-60°C$ is added dropwise a solution of lithium bis(trimethylsilyl)amide. The resulting solution is warmed to $-30°C$ (20 min), and then a solution of the anion of tosylmethylisocyanide (1.4 mmol) is added. After stirring (30 min) at $-78°C$, the solution is allowed to warm up to $0°C$ (2 h), then stirred at room temperature (16 h). The reaction mixture is then worked up as above (method A).

TABLE 4.2.3
Imidazoles (17) [21]

Starting material	R^4	R^5	Method	Yield (%)
MeLi	Me	H	A	55
MeCHO	Me	H	B	25
BuLi	Bu	H	A	51
PhLi	Ph	H	A	23
PhCHO	Ph	H	B	24
BuLi	Bu	Me	A	51
BuLi	Bu	$PhCH_2$	A	66

Table 4.2.3 lists some representative examples of 4,5-disubstituted imidazoles prepared using those methods.

An addition to the earlier cycloadditions is the self condensation of an *N*-methyl-*C*-aryl nitrone (**18**) in the presence of potassium cyanide (Scheme 4.2.8) [22, 23]. The reactions are carried out in mild conditions (room temperature overnight in aqueous ethanol), but although the yields are good there are no obvious advantages over other methods which are available for making 4,5-diarylimidazoles. The nitrones (**18**) are, however, readily available from *N*-aryl- or *N*-alkylhydroxylamines and aryl aldehydes [24], and so there may be occasions when the approach could be useful.

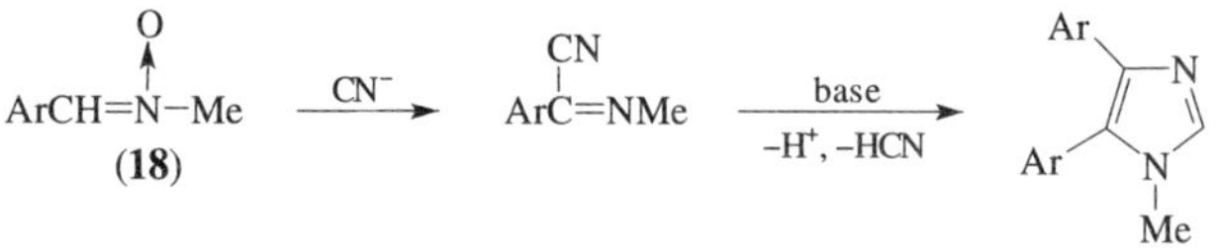

Scheme 4.2.8

N,α-Diphenylnitrone [24]

A solution of *N*-phenylhydroxylamine (27.3 g, 0.25 mol) in ethanol (50 ml) is prepared by swirling and brief warming to 40–60°C (on prolonged heating the hydroxylamine begins to decompose). To the clear solution is added benzaldehyde (26.5 g, 0.25 mol), when an exothermic reaction occurs. After storing overnight at ambient temperature in the dark in a stoppered flask, the colourless needles of the nitrone are filtered, washed with ethanol (20 ml) and recrystallized from ethanol (35–39 g, 71–79%), m.p. 113–114°C.

4,5-Diaryl-1-methylimidazole [23]

A solution of *N*-(substituted benzylidene)methylamine *N*-oxide (**18**) (0.01 mol) in a mixture of ethanol (50 ml) and water (20 ml) is shaken with KCN

4.2. FORMATION OF 1,2 AND 4,5 (OR 2,3 AND 4,5) BONDS

(0.01 mol) until solution is complete. The mixture is allowed to stand at room temperature (2-4 days), when the product is filtered, washed with water, dried, and recrystallized from ethanol or toluene. Products (Ar, yield given): Ph, 65%; p-MeC$_6$H$_4$, 51%; p-ClC$_6$H$_4$, 72%; 2-naphthyl, 37%; 4-pyridyl, 57%.

The previous cycloaddition reaction discussed is believed to proceed through an aldimine anion (**19**). Such delocalized anions can also be generated by treatment of suitable aldimines with a strong base. Subsequent cyclocondensation with a nitrile produces imidazoles [25-28]. The "2-azaallyl lithium" compounds (**19**) are made by treatment of an azomethine with lithium diisopropylamide in THF-hexane (~5:1) (Scheme 4.2.9) [29]. To stirred solutions of (**19**) one adds an equimolar amount of a nitrile in THF at $-60°$C. Products are obtained after hydrolysis with water (see also Section 2.3). If the original Schiff base is disubstituted on carbon, the product can only be a 3-imidazoline, but anions (**19**) eliminate lithium hydride to give aromatic products (**20**) in 37-52% yields (Scheme 4.2.9). It is, however, not possible to make delocalized anions (**19**) with R^1 = alkyl, and aliphatic nitriles react only very reluctantly. Examples of (**20**) (Ar, R^2, R^5, yield listed) include: Ph, Ph, Ph, 52%; Ph, Ph, m-MeC$_6$H$_4$, 50%; Ph, Ph, p-MeC$_6$H$_4$, 52%; Ph, Ph, 3-pyridyl, 47%; Ph, Ph, nPr, 1% [25]. Closely related is the synthesis of tetrasubstituted imidazoles (**22**) by regioselective deprotonation of (**21**) and subsequent reaction with an aryl nitrile. Even better yields and reactivity are observed when one equivalent of potassium t-butoxide is added to the preformed monolithio anion of (**21**) (Scheme 4.2.9) [30].

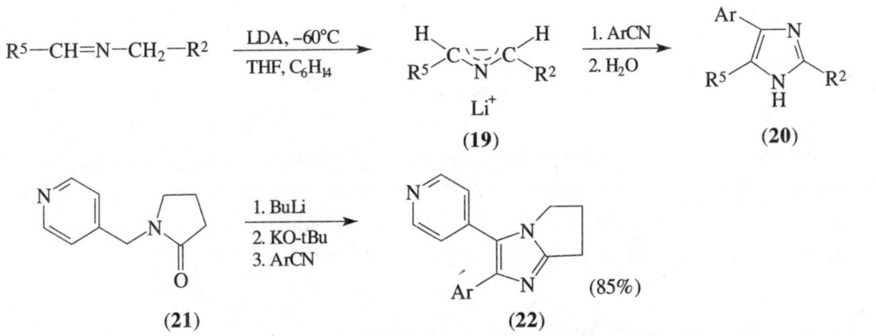

Scheme 4.2.9

A further [3 + 2]-cycloaddition process takes place when imines react with 2-azaallenyl radical cations (**23**) derived from azirines by photolysis (Scheme 4.2.10) [31, 32]. Yields of (**24**) are variable, e.g. 2,4,5-triphenyl-1-propyl- (87%), 4-butyl-1,5-dipropyl- (40%), 4-phenyl-1,5-dipropyl- (35%),

4,5-diphenyl-1-propyl- (12%), 2,4-diphenyl-1,5-dipropyl- (25%) and 4-butyl-5-phenyl-1-propylimidazoles (3%) [31]. This approach suffers from the twin problems that both the azirines and imines must be prepared before the cyclization is attempted, and there is no guarantee that good yields will be obtained (see also Chapter 6).

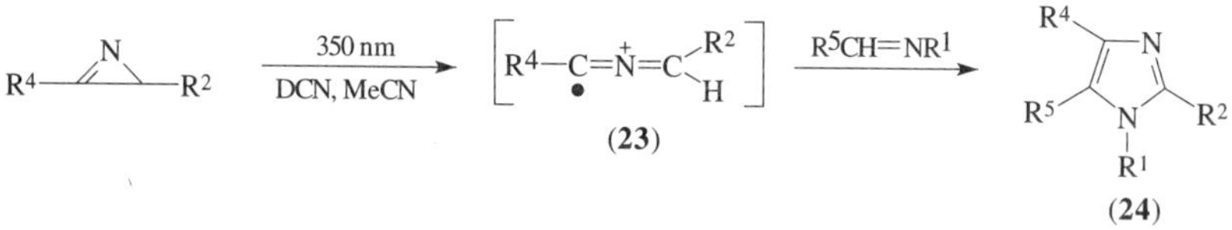

Scheme 4.2.10

Anionic cycloaddition of methyl isocyanoacetate to diethoxyacetonitrile gives methyl 5-diethoxymethylimidazole-4-carboxylate in high yield. The reaction is quite general for nitriles which are sufficiently activated by $-I$ or $-M$ effects, thus aliphatic nitriles do not react. The reactions are also strongly dependent on the base and solvent employed. Whereas n-butyllithium in THF at $-70°C$ is unsuccessful in effecting the above transformation, potassium hydride in diglyme works very well, giving an 82% yield (Scheme 4.2.11) [33].

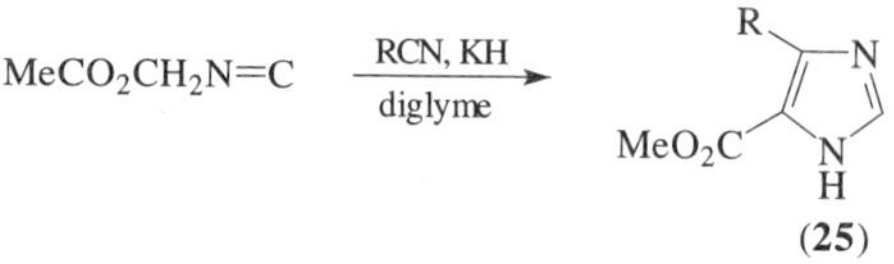

Scheme 4.2.11

Using a process similar to other 2-azaallenyl cyclizations, a number of novel histamine analogues have been made by applying the ring-chain transfer concept to the reaction of semicyclic 2-aza-3-methylthio-3-propeniminium iodides with methylamines possessing an acidic CH group [34].

Methyl 5-diethoxymethylimidazole-4-carboxylate [33]

To a stirred suspension of potassium hydride (0.55 g, 1.38 mmol) in freshly distilled diglyme (2.0 ml), diethoxyacetonitrile [35] (0.129 g, 1.0 mmol) and methyl isocyanoacetate [36] (0.099 g, 1.0 mmol) in diglyme (1.5 ml) are added slowly with ice cooling under an argon atmosphere. The solution is heated at 70–80°C (5 h), then cooled, quenched with saturated aqueous ammonium chloride, and extracted with dichloromethane (3 × 10 ml) and ethyl acetate

(10 ml). The combined organic extracts are dried (Na_2SO_4), concentrated, and subjected to preparative TLC on silica gel (developed with ethyl acetate) to give (**25**) (R = $CH(OEt)_2$) (82%, m.p. 148–150°C) after recrystallization from ethanol. Similarly prepared are (**25**) (R, yield given): $CH(OMe)_2$, 67%; Ph, 13%; 2-pyridyl, 85%; 3-pyridyl, 24%; 4-pyridyl, 81%; 2-furyl, 40%.

When 4-nitrobenzyl isocyanide reacts with an arylsulfenyl chloride it is converted into an N-(4-nitrobenzyl)-S-arylisothiocarbamoyl chloride. In the presence of triethylamine a solution containing the 1,3-dipolar species (**26**) is obtained. If (**26**) is generated in the presence of ethyl cyanoformate, cyclo-addition gives an imidazole (Scheme 4.2.12) [37]. Reactions are performed at 60°C in the presence of a large excess of the dipolarophile, and yields are reported in the range 38–50%.

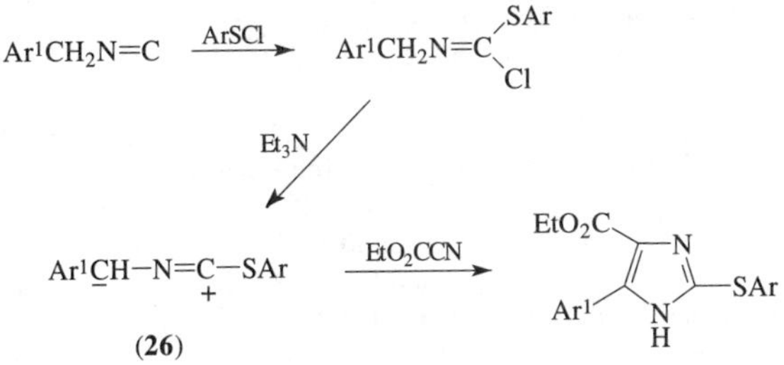

Scheme 4.2.12

When cyanuric chloride reacts with DMF in t-butyl methyl ether it forms Gold's salt, [3-(dimethylamino)-2-azaprop-3-enylidene]dimethylammonium chloride, in almost quantitative yield. Methylate-promoted condensation of this salt with the methyl ester of sarcosine gives methyl 1-methylimidazole-5-carboxylate in up to 75% yield. The synthetic utility of this method is dependent on interception of the eliminated dimethylamine by addition of dimethyl oxalate and the bubbling of nitrogen through the solution [38].

REFERENCES

1. A. M. van Leusen, G. J. M. Boerma, R. B. Helmholdt, H. Siderius and J. Strating, *Tetrahedron Lett.* 2367 (1972).
2. A. M. van Leusen, J. Wildeman and O. H. Oldenziel, *J. Org. Chem.* **42**, 1153 (1977).
3. A. M. van Leusen, R. J. Bouma and O. Possel, *Tetrahedron Lett.* 3487 (1975).
4. F. Cramer and U. Baer, *Chem. Ber.* **93**, 1231 (1960).
5. A. M. van Leusen and O. H. Oldenziel, *Tetrahedron Lett.* 2373 (1972).
6. E. C. Taylor, J. L. La Mattina and C.-P. Tseng, *J. Org. Chem.* **47**, 2043 (1982).

7. S. P. J. M. van Nispen, J. H. Bregman, D. G. van Engen, A. M. van Leusen, H. Saikachi, T. Kitagawa and H. Sasaki, *Recl. Trav. Chim. Pays-Bas Belg.* **101**, 28 (1982).
8. A. M. van Leusen and J. Schut, *Tetrahedron Lett.* 285 (1976).
9. S. P. J. M. van Nispen, C. Mensink and A. M. van Leusen, *Tetrahedron Lett.* **21**, 3723 (1980).
10. T. Olijnsma, J. B. F. N. Engberts and J. Strating, *Recl. Trav. Chim. Pays-Bas, Belg.* **86**, 463 (1967).
11. H. Houwing, J. Wildeman and A. M. van Leusen, *Tetrahedron Lett.* 143 (1976).
12. D. A. Horne, K. Yakushijin and G. Büchi, *Heterocycles* **39**, 139 (1994).
13. D. E. Bergstrom and P. Zhang, *Tetrahedron Lett.* **32**, 6485 (1991).
14. E. Shaw, *J. Org. Chem.* **30**, 3371 (1965).
15. G. Shaw, R. N. Warrener, D. N. Butler and R. K. Ralph, *J. Chem. Soc.* 1648 (1959).
16. J. T. Hunt and P. A. Bartlett, *Synthesis* 741 (1978).
17. K. Bünge, R. Huisgen, R. Raab and H. J. Sturm, *Chem. Ber.* **105**, 1307 (1972).
18. E. Brunn, E. Funke, H. Gotthardt and R. Huisgen, *Chem. Ber.* **104**, 1562 (1971).
19. R. Consonni, P. Dalla Croce, R. Ferraccioli and C. La Rosa, *J. Chem. Res. (S)* 188 (1991).
20. P. Dalla Croce, R. Ferraccioli, C. La Rosa and T. Pilati, *J. Chem. Soc., Perkin Trans. 2* 1511 (1993).
21. N.-Y. Shih, *Tetrahedron Lett.* **34**, 595 (1993).
22. N. G. Clark and E. Cawkill, *Tetrahedron Lett.* 2717 (1975).
23. R. P. Soni, *Aust. J. Chem.* **35**, 1493 (1982).
24. I. Bruning, R. Grashey, H. Hauck, R. Huisgen and H. Seidl, *Org. Synth.* **46**, 127 (1966).
25. T. Kauffmann, A. Busch, K. Habersaat and E. Köppelmann, *Angew. Chem., Int. Ed. Engl.* **12**, 569 (1973).
26. T. Kauffmann, *Angew. Chem., Int. Ed. Engl.* **13**, 627 (1974).
27. T. Kauffmann, A. Busch, K. Habersaat and E. Köppelmann, *Chem. Ber.* **116**, 492 (1983).
28. V. V. Kuznetsov and N. S. Prostakov, *Chem. Heterocycl. Compd. (Engl. Transl.)* **26**, 1 (1990).
29. Th. Kauffmann, H. Berg and E. Köppelmann, *Angew. Chem., Int. Ed. Engl.* **9**, 380 (1970).
30. J. F. Hayes, M. B. Mitchell and G. Procter, *Tetrahedron Lett.* **35**, 273 (1994).
31. F. Müller and J. Mattay, *Angew. Chem., Int. Ed. Engl.* **30**, 1336 (1991).
32. F. Müller and J. Mattay, *Chem. Ber.* **126**, 543 (1993).
33. T. Murakami, M. Otsuka and M. Ohno, *Tetrahedron Lett.* **23**, 4729 (1982).
34. M. Pätzel and J. Liebscher, *J. Org. Chem.* **57**, 1831 (1992).
35. K. Utimoto, Y. Wakabayashi, Y. Shishiyama, M. Inoue and H. Nozaki, *Tetrahedron Lett.* **22**, 4279 (1981).
36. G. D. Hartman and L. M. Weinstock, *Org. Synth.* **59**, 183 (1979).
37. R. Bossio, S. Marcaccini, R. Paoli, R. Pepino and C. Polo, *Heterocycles* **31**, 1855 (1990).
38. R. Kirchlechner, M. Casutt, U. Heywang and M. W. Schwarz, *Synthesis* 247 (1994).

4.3 FORMATION OF 1,5 AND 3,4 BONDS

The most common NCN synthons used are amidines, guanidines, ureas and thioureas. The two-carbon units are usually suitably functionalized carbonyl compounds.

4.3. FORMATION OF 1,5 AND 3,4 BONDS

When amidines or guanidines cyclize in the presence of α-functionalized carbonyl reagents they form imidazoles with a variety of 1-, 2-, 4- and 5-substituents [1, 2]. Formamidine will react with α-hydroxy- [8] or α-halogenoketones [9], to give mixtures of oxazoles and imidazoles. Usually, the formamidine is liberated from its hydrochloride by treatment with sodium butoxide in butanol. When the two-carbon synthon is an α-hydroxyketone the aliphatic members mainly form imidazoles (35–70%), while benzoins give oxazoles (67–80%) preferentially [8]. The competing pathways are shown in Scheme 4.3.1.

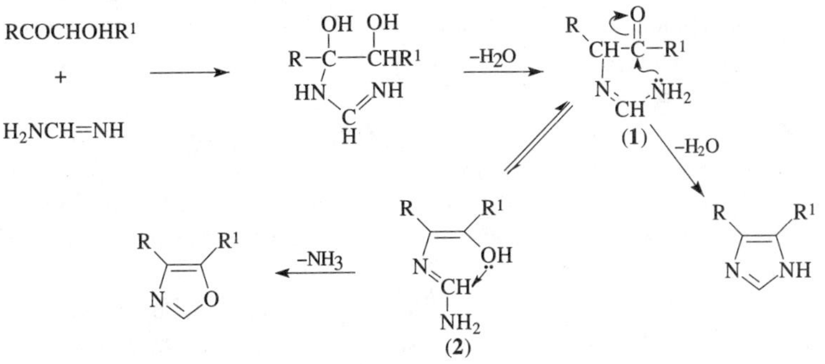

Scheme 4.3.1

Unlike formamidine, acetamidine and benzamidine react with both aromatic and aliphatic α-hydroxyketones to give imidazoles exclusively. It has been suggested that aryl groups favour the enolic form (2) of the tautomeric mixture, resulting in the formation of oxazoles as major products. Aliphatic groups favour the keto form (1), from which imidazoles are derived. That amidines more complex than formamidine favour imidazole formation may be a consequence of steric hindrance to reaction of the enolic hydroxy groups with the amidine carbon in (2). The general reaction has been used to prepare such compounds as 4,5-dipropylimidazole (25% yield from tris(formylamino)-methane and 5-hydroxyoctan-4-one), and a variety of 2-imidazolones and 2-aminoimidazoles [8]. The fact that oxazoles can be converted into imidazoles with some ease extends the applicability of this reaction.

There are standard methods available for synthesis of amidines [10] and guanidines. In particular, reaction of an acylated thiourea with an amine, followed by removal of the acyl group(s) from the acylguanidine intermediate provides a mild and efficient synthesis [11]. When dihydroxyacetone is treated with formamidine acetate in liquid ammonia, imidazole-4-methanol is isolated in 65–70% yield as its hydrochloride (3) (Scheme 4.3.2). This convenient synthesis is much less tedious than the old approach based on the reaction of fructose with ammonia [16].

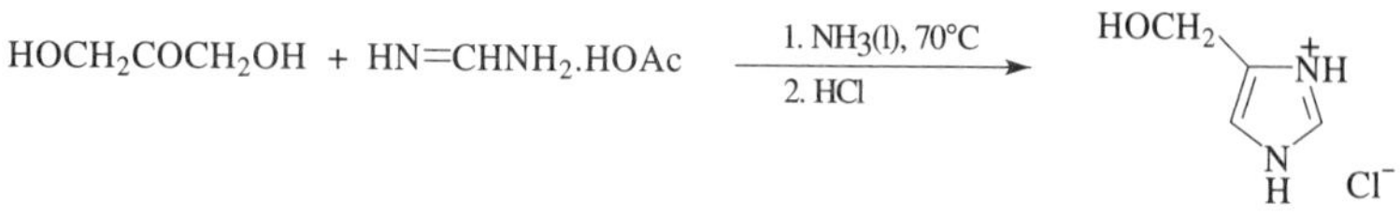

HOCH$_2$COCH$_2$OH + HN=CHNH$_2$·HOAc $\xrightarrow[\text{2. HCl}]{\text{1. NH}_3\text{(l), 70°C}}$ HOCH$_2$-[imidazole ring]·Cl$^-$

(3)

Scheme 4.3.2

Imidazole-4-methanol hydrochloride (3) [16]

In a 600 ml stainless steel bomb protected from moisture and cooled in a dry-ice–isopropanol bath, liquid ammonia (150 ml) is condensed before the addition of formamidine acetate (31.2 g, 0.3 mol). Dihydroxyacetone (27 g, 0.3 mol) is then added, and the bomb is sealed and heated to 70°C (pressure reaches 420 psi (29.0 bar)) with stirring (6 h). The bomb is then cooled and opened, and the ammonia is evaporated at room temperature under a stream of nitrogen. The resulting red-brown oil is dissolved in isopropanol (200 ml), cooled to 0°C in an ice bath, and then with vigorous stirring bubbled with dry hydrogen chloride gas until the solution is acidic (pH ≤ 2). The voluminous precipitate of inorganic salts is filtered and washed with hot isopropanol (200 ml), and diethyl ether (50 ml) is added with swirling to the filtrate, causing a tarry black material to settle out. The supernatant liquid is decanted, and the resultant cloudy solution is refrigerated at −5°C (48 h) to give a hygroscopic light tan precipitate which is filtered and dried in a vacuum desiccator to give (3) (28.2 g, 70%). This product is very hygroscopic; it may be recrystallized from isopropanol–ether (63:35) to give less hygroscopic colourless needles, m.p. 106.5–109°C (^{1}H NMR (DMSO-d_6) δ = 4.70 (s, 2H); 7.36 (s, 1H, H-5), 8.87 (s, 1H, H-2)). It may be more convenient to convert the initial tan precipitate directly into the non-hygroscopic *N*-trityl derivative [17], or basify it and isolate the free base form. Evaporation of the alkaline solution to dryness, extraction of the syrupy residue with anhydrous ethanol, filtration, and concentration *in vacuo* is followed by chromatography on a short silica gel column eluted with methanol. Recrystallization of the crystalline product from DMK gives 4-hydroxymethylimidazole as off-white needles, m.p. 92°C [18].

The general approach has been modified to provide a route to 2-aminoimidazoles in up to 90% yields by initially reacting a benzoin at 100°C in aqueous ethanolic sodium hydroxide with diguanylhydrazine. Air oxidation of the products gives 2,2′-azoimidazoles, which can be catalytically reduced to 2-aminoimidazoles [12]. This process should be compared with that which couples an imidazole with an aryldiazonium salt before hydrogenation [13, 14]. If benzaldehyde guanylhydrazone is treated with a 4-alkyl- or 4-aryl-ω-haloacetophenone, 2-amino-1-benzylideneamino-4-arylimidazoles are obtained. The free 1,2-diamino-4-arylimidazoles can be isolated by treating a mixture of 0.02 mol of the 1-benzylideneaminoimidazole with 20 ml of hydrazine hydrate in diethylene glycol (50 ml) heated to its boiling point, and stirred and refluxed until homogeneous (about 3 h) [15].

It is more common to use α-halogenoketones in such cyclization reactions with amidines and guanidines [5, 19–21], although sometimes pyrimidine by-products are also formed, e.g. when benzamidine reacts with 3-bromobenzo-4-pyrones both imidazoles and pyrimidines are formed with the former predominating in non-polar solvents. Yields in this particular example, however, are low [5]. Better examples are found in the condensation of substituted benzamidines with a phenacyl bromide; 2,5-diaryl-4-methylimidazoles are isolated in moderate to good yields [19]. Another instance is the synthesis of L-homohistidine in 71% overall yield from reaction of formamidine acetate with an α-chloroketone in liquid ammonia [21]. Formamidine acetate is also the cyclizing agent of choice in the synthesis of a series of 4,4'-alkylenebisimidazoles. It was found that these latter compounds are best isolated as the 1-trityl derivatives, which are easy to purify by flash chromatography on silica gel. The tritylation process involves treating the N-unsubstituted products with trityl chloride–triethylamine–DMF, and is reported to give, as sole regioisomers, the 1,4 products in 20–36% yields [20]. When α-haloacyl chlorides react with benzamidine in the presence of triethylamine, the products are mesoionic imidazolium 4-oxides [4].

4-Methyl-2,5-diphenylimidazole [19]

A solution of benzamidine hydrochloride dihydrate (48 g, 0.25 mol) in water (100 ml) is added to a solution of α-bromopropiophenone (53 g, 0.25 mol) in chloroform (250 ml) to form a two-phase mixture. While stirring vigorously at room temperature, a solution of KOH (28 g, 0.5 mol) in water (100 ml) is added dropwise, then heated to boiling and refluxed (3–4 h). The chloroform phase is then separated from the hot aqueous phase and cooled. The crystalline material which separates is washed with benzene, then with diethyl ether to give the above product (which may separate as an oil, but which solidifies gradually on addition of benzene). The yield is in the range 26–32 g (45–55%), m.p. 214–215°C. Similarly prepared are 4-methyl-2,5-bis-(m-nitrophenyl)- (39%), 4-methyl-2,5-bis-(m-bromophenyl)- (40%), and 4-methyl-2,5-bis-(p-nitrophenyl)imidazoles (64%).

1,2-Dicarbonyls are also capable of taking part in these reactions with amidines, but most of the references apply to guanidines or ureas acting as the NCN synthons (see below).

Appropriately functionalized alkenes (enamines, α-haloalkenes, silylenol ethers, acylvinyl phosphonium salts) are frequently employed as the CC fragment in reaction with amidines, guanidines and ureas. An N-chloro-N'-alkylamidine will react with the enamine derived from an aldehyde to give in the first instance a 2-imidazoline (**4**). Such compounds deaminate readily to produce the fully aromatic imidazoles (**5**), especially when heated with pyridinium or triethylammonium chlorides or 50% sulfuric acid (Scheme 4.3.3).

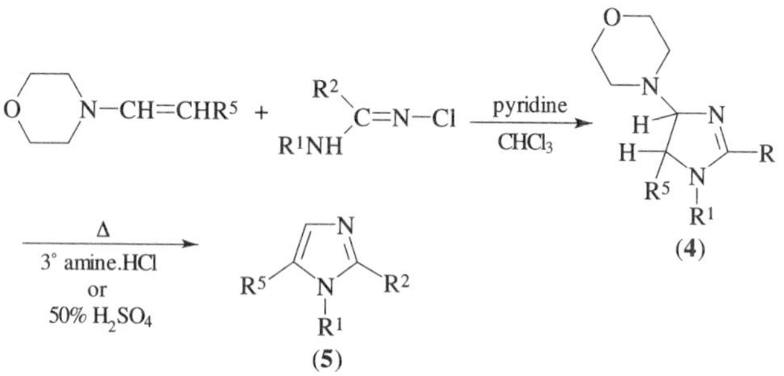

Scheme 4.3.3

They may eliminate so readily that the intermediate imidazoline cannot be isolated, but if the enamine is disubstituted, the 5,5-disubstituted imidazolines which are formed are so stable that it is necessary to heat them with 50% aqueous sulfuric acid, or in anhydrous medium with an equimolar amount of either triethylammonium or pyridinium chlorides, to effect aromatization. Enamines derived from ketones or hindered aldehydes do not lead to imidazoles, and there are other difficulties (some enamines are difficult to prepare, only aryleneamines give good yields). Nevertheless, the reported results demonstrate the potential of the method [22–24].

N-Chloroamidines (general procedure) [23]

To a solution of the amidine (0.1 mol) in dry dichloromethane (100 ml) is added N-chlorosuccinimide (0.105 mol). The mixture is stirred at room temperature (2 h) and washed twice with water, the organic layer is dried (Na_2SO_4), and the solvent removed, to give high yields of the chloroamidines.

1,2,5-Trisubstituted imidazoles (**5**) *(general procedure)* [23]

To a solution of enamine (0.01 mol) in dry pyridine (to remove HCl; 0.88 g, 0.011 mol) the N-chloroamidine (0.01 mol) is added. The mixture is heated under reflux until no more chloroamidine can be detected (TLC), at which time it is cooled to room temperature and washed with saturated sodium bicarbonate solution. The organic layer is dried (Na_2SO_4), and the solvent is removed under reduced pressure. The residue (5) is purified by column chromatography on silica gel using benzene–THF (85:15). Imidazole products will often crystallize when set aside, and can be further purified by recrystallization. Prepared by this method are (5) (R^1, R^2, R^5, yield, recrystallizing solvent given): Ph, Ph, Ph, 65%, EtOH; p-$NO_2C_6H_4$, Ph, Ph, 60%, iPrOH; p-$MeOC_6H_4$, Ph, Ph, 40%, EtOH; p-BrC_6H_4, Ph, Ph, 55%, EtOH; p-FC_6H_4, Ph, Ph, 70%, benzene; Ph, p-BrC_6H_4, Ph, 45%, benzene; Ph, Me, Ph, 25%, light petroleum; p-FC_6H_4, Ph, Et, 35%, diisopropyl ether; $PhCH_2$, Ph, Ph, 40%, diisopropyl ether.

Aromatization of 5,5-disubstituted 2-imidazolines [24]

Method A. The imidazoline (2 mmol) is heated at 130°C in 50% aqueous H_2SO_4 (10 ml) for 24 h. The cooled solution is diluted with water (50 ml), and the resultant precipitate is filtered and then crystallized from ethanol.

Method B. The imidazoline (2 mmol) is refluxed (15 h) in 1,1,2-trichloroethane (20 ml) with triethylamine hydrochloride (2 mmol). After cooling to room temperature the solvent is rotary evaporated, and the crude residue is washed with water and extracted twice with chloroform. The organic extracts are dried (Na_2SO_4) and evaporated. The residue is crystallized from ethanol.

Prepared in these ways are 5-methyl-1,2,4-triphenyl- (85% (method A); 65% (method B)), 5-methyl-2,4-diphenyl-1-p-tolyl- (70% (method A); 90% (method B)), 5-methyl-2,4-diphenyl-1-p-fluorophenyl- (60% (method B)) and 1,2,4,5-tetraphenylimidazoles (90% (method A); 85% (method B)).

The main deficiencies with the preceding approaches are that only arylene-amines give good yields (and there are other routes to 5-arylimidazoles), and since unsubstituted vinylamines ($R^5 = H$; Scheme 4.3.3) are not available this means that 4- and 5-unsubstituted imidazoles are not accessible. An alternative route involving silyl enol ethers (6) has been reported to overcome these deficiencies (Scheme 4.3.4). Silyl enol ethers can be made either by treating a ketone with chlorotrimethylsilane and triethylamine in DMF solution, or by sequential reactions of the ketone with LDA and chlorotrimethylsilane in 1,2-dimethoxyethane. This normally gives a mixture in which the less highly substituted enol ether is the major product (enolate formation is kinetically controlled) [25]. When (6) is heated with an N-chloroamidine for 12–24 h in chloroform solution in the presence of an equimolar amount of dry pyridine 1,2-disubstituted (5) ($R^5 = H$) or 1,2,5-trisubstituted (5) imidazoles are obtained. This means that the regiospecific synthesis of 5-alkylimidazoles (**5**) (R^5 = alkyl) also becomes possible using this approach. Examples include 1,2-diphenyl- (55%), 5-methyl-1,2-diphenyl- (65%), 5-ethyl-1,2-diphenyl- (70%) and 5-methyl-2-phenyl-1-(*p*-tolyl)imidazoles (65%) [26].

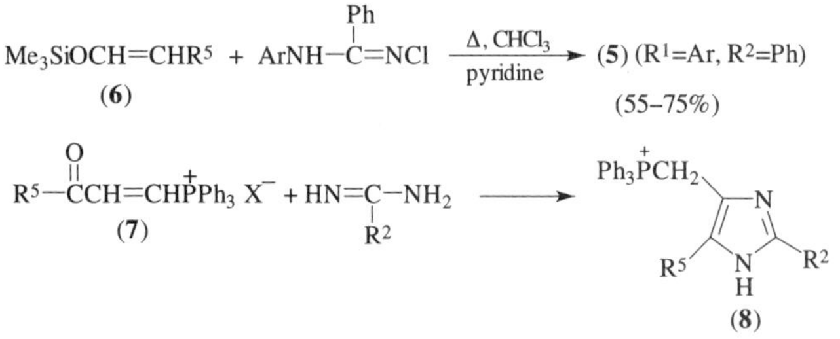

$$Me_3SiOCH=CHR^5 + ArNH-\underset{\underset{(6)}{}}{\overset{\overset{Ph}{|}}{C}}=NCl \xrightarrow[\text{pyridine}]{\Delta, CHCl_3} (\mathbf{5})\ (R^1=Ar, R^2=Ph)$$
$$(55-75\%)$$

$$R^5-\overset{\overset{O}{\|}}{C}CH=CH\overset{+}{P}Ph_3\ X^- + HN=\underset{\underset{R^2}{|}}{C}-NH_2 \longrightarrow$$
(**7**)

(**8**) imidazole structure with $Ph_3\overset{+}{P}CH_2$, R^5, R^2 substituents

Scheme 4.3.4

In a similar process in which acylvinyl phosphonium salts (**7**) act as the "masked dicarbonyl", amidines react with these reagents to give imidazoles (**8**) in high yields. The nucleophilic amino group of the amidine here attacks at the carbon adjacent to the carbonyl group in what is the reverse of the normal Michael addition [27]. If the amidine is present as its sulfenic acid derivative ($R_2 = SO_2H$), heating the mixture in DMSO gives a 2-unsubstituted imidazole (**8**) ($R_2 = H$), when SO_2 is eliminated (Scheme 4.3.4). The triphenylphosphonium derivatives (**8**) are also susceptible to nucleophilic displacement by reagents such as alkoxide, thereby providing access to a wide range of 4 substituents on the imidazole nucleus [28]. (See also below, reaction with *S*-methylisothiourea.)

Alternative activated enones (Michael acceptors) include 3,4-disubstituted 3-buten-2-ones (**9**) (X = Cl, OMe, SMe) [29, 30], and β-formylacrylates (**11**) (Scheme 4.3.5) [31]. The enolic analogue (**9**) (X = OH), which is readily prepared from the sodium derivative of acetoacetaldehyde, should theoretically be capable of cyclization with bifunctional nucleophiles such as ureas or amidine salts to give formyl or acetyl 5- or 6-membered rings. Somewhat surprisingly it was found that the neutral vinylogous ester (**9**) (X = OEt), prepared *in situ* from (**9**) (X = OH) in quantitative yield, reacts with amidines to give imidazoles rather than pyrimidines. Obviously the nature of the product is dependent upon the substituents on the enone; a good leaving group in the 3-position combined with a 4-substituent which is not easily eliminated preferentially gives the aromatic imidazole. Yields of (**10**) ($R^2 = Me$) are 10, 62 and 28% for X = Cl, OMe and SMe, respectively, but there are also pyrimidines formed in 35, 2 and 21% yields [30]. A more easily accessible alternative to (**9**) is 3-chloro-4,4-dimethoxy-2-butanone, which fulfils the above requirements and which can be made by exploiting Friedel–Crafts

4.3. FORMATION OF 1,5 AND 3,4 BONDS

chemistry. Its conversion into (**10**) ($R^2 = Me$) is only in 46% yield, however, because of the poor partitioning of the product between water and organic solvents. More lipophilic analogues can be isolated more readily [30].

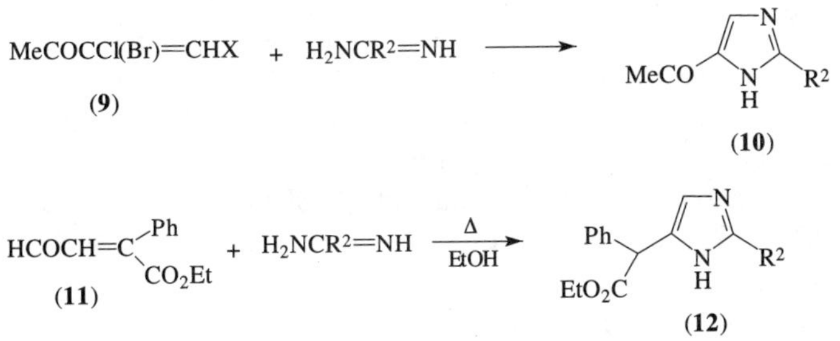

MeCOCCl(Br)=CHX + H_2NCR^2=NH ⟶ (**10**) MeCO-imidazole-R^2

(**9**)

HCOCH=C(Ph)(CO$_2$Et) + H_2NCR^2=NH $\xrightarrow{\Delta, \text{EtOH}}$ Ph, EtO$_2$C-imidazole-R^2

(**11**) → (**12**)

Scheme 4.3.5

3-Bromo-4-ethoxy-3-buten-2-one (**9**) *(X = OEt)* [29]

To a mixture of absolute ethanol (800 ml) and dry toluene (175 ml) is added 2-bromoacetoacetaldehyde (41.2 g, 0.25 mol). This mixture is refluxed (2.5 h) while solvent is gradually removed, then after cooling, the remaining solution is concentrated *in vacuo* to an oil (yield quoted as virtually quantitative by NMR).

4-Acetyl-2-methylimidazole (**10**) *($R^2 = Me$)* [29]

The above vinylogous ester (46.7 g, 0.24 mol) is combined with acetamidine acetate (28.6 g, 0.24 mol) and anhydrous sodium acetate (19.8 g, 0.24 mol) in 1,4-dioxane (1 l), and heated at reflux (60 h). After cooling, the solid is removed by filtration, and the dioxane solution is concentrated *in vacuo* to an oily solid which is dissolved in water (200 ml), and adjusted to pH 10 with concentrated sodium bicarbonate solution. The solution is decolorized with activated charcoal and concentrated under reduced pressure to give a solid. This is slurried in chloroform, and the solid is removed by filtration. The organic solution is concentrated to produce a heavy oil. Trituration with ethyl acetate gives (**10**) ($R^2 = Me$) (12.44 g, 42%) as a yellow solid, m.p. 127–128°C. Similarly prepared are 4-acetyl-2-hydroxymethyl- (12%) and 4-acetyl-2-(4'-pyridyl)imidazoles (39%).

3,4-Dichloro-3-buten-2-one [30]

Aluminium chloride (734 g, 5.5 mol) is added to a mixture of acetyl chloride (392 g, 5.0 mol) and 1,2-dichloroethene (1.92 l, 25 mol), keeping the temperature around 25°C. The mixture is then refluxed (16 h), cooled and poured into ice, and the organic layer is separated. The aqueous layer is extracted with dichloromethane (3 × 500 ml), and the combined organic extracts are filtered (Celite), dried (Na_2SO_4), and concentrated *in vacuo* to a black oil which is vigorously stirred with sodium carbonate (745 g, 6.0 mol) in water (2.5 l) for 1.5 h. The solids are filtered off and washed with dichloromethane. The organic layer of the filtrate is separated and the aqueous layer extracted with dichloromethane (2 × 200 ml). The combined dichloromethane extracts are dried ($MgSO_4$), filtered, concentrated *in vacuo* to a black oil, and distilled to give an almost colourless, mildly lachrymatory liquid (550.3 g, 79%), b_8 40–46°C.

3-Chloro-4,4-dimethoxy-2-butanone [30]

A solution of sodium methoxide (305 g, 5.64 mol) in methanol (3.78 l) is added to the above enone (523 g, 3.76 mol) in a slow stream, keeping the temperature at 0°C. After 30 min at 0°C, acetic acid (107 ml, 1.88 mol) is added, methanol is rotary evaporated off, and the residue is dissolved in isopropyl ether (1.5 l). After washing with water (1 l) and saturated sodium bicarbonate solution (250 ml), the extract is dried ($MgSO_4$), filtered and concentrated *in vacuo*. The residue is distilled to give the product as a colourless liquid (488.4 g, 78%), b_8 66–75°C.

4-Acetyl-2-phenylimidazole (**10**) *($R^2 = Ph$)* [30]

A mixture of 3-chloro-4,4-dimethoxy-2-butanone (3.33 g, 20 mmol), benzamidine hydrochloride (4.70 g, 30 mmol) and sodium acetate (4.10 g, 50 mmol) is refluxed (42 h) in dioxane (100 ml). After cooling, the salts are removed by filtration and the filtrate concentrated to an oil under reduced pressure. This oil is taken up in ethyl acetate (100 ml) and extracted with 1 M HCl (3 × 20 ml). The combined aqueous extracts are washed with ethyl acetate, and carefully basified with solid sodium carbonate. The aqueous solution is extracted with chloroform (3 × 20 ml), and the combined extracts are dried ($MgSO_4$), filtered and concentrated, to give the crude product (3.23 g), which can be recrystallized from cyclohexane/toluene as fine yellow needles (2.33 g, 60%), m.p. 155–157°C. Similarly prepared are 4-acetyl-2-methyl- (46%) and 4-acetyl-2-hexylimidazoles (26%).

The β-formylacrylate (**11**) is conveniently prepared by acid hydrolysis of the corresponding ethyl α-phenyl-β-formylacrylate dimethylhydrazone methiodide, which in turn is available in quantitative yield when the corresponding *N,N*-dimethylhydrazone ester is quaternized with

4.3. FORMATION OF 1,5 AND 3,4 BONDS

methyl iodide. Reaction with an amidine converts (11) into an α-imidazolylphenylacetic acid (12) (Scheme 4.3.5). With benzamidine, the 2-phenylimidazole is formed in 66% yield, and there are other examples reported [31].

Amidines (and guanidines) react with α-cyanoepoxides (13) to give 4-amino-5-carbethoxy-, 5-cyano-4-hydroxy- and 4-carbethoxy-5-phenylimidazoles, depending on the reaction medium, and on the degree of steric hindrance in the epoxides (Scheme 4.3.6). If R^2 =H in the epoxide, mixtures of products are obtained. With added triethylamine an imidazoline is isolated, but it can be aromatized by heating in the presence of acetic acid. Although these reactions are reasonably chemoselective, the exotic natures of the starting materials reduce the appeal of the method somewhat [32].

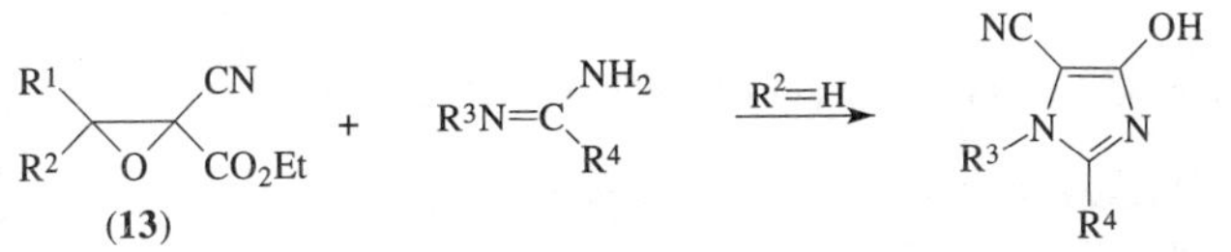

(13) + $R^3N=C(NH_2)R^4$ → [R^2=H] → imidazole with NC, OH, R^3, R^4

($R^3=R^4=Ph$, 70%)
($R^3=Ph, R^4=NHPh$, 75%)

Scheme 4.3.6

When a hydroxylamine reacts at room temperature with ethyl cyanoformate the products are either carbethoxyamino nitrones (14) or the corresponding aminooximes (15). Yields are good (62–76%) (Scheme 4.3.7). Either of these species will react with a propiolate ester to give an imidazole-2,4-dicarboxylic ester (16). The process involves formation of an initial Michael adduct, a vinyl oxyamidine, which can sometimes be isolated. When boiled in xylene this undergoes a [3,3]-sigmatropic rearrangement to the imidazole (16) [33]. The reaction has potential utility in view of the regiospecific way in which the amidine cyclizes giving 1,2,4 rather than 1,2,5 substitution; the latter alternative would require a [1,3]-rearrangement.

α-Amino-α-carbethoxy-N-methylnitrones (general method) [33]

The appropriate hydroxylamine (0.15–0.20 mol) in dry dichloromethane is treated with ethyl cyanoformate (1.2–1.5 eq.) at room temperature, monitoring the reaction by TLC. On completion, the solvent is evaporated *in vacuo* and the residue is washed repeatedly with petroleum ether until the solid crystallizes, e.g. (14) (R = Me, R^1 = Et) is obtained in 65% yield, m.p. 93–95°C, from dichloromethane–petroleum ether. Similarly prepared are (14) (R, R^1, given): iPr, Et, 73%; Me, Me, 68%; Ph, Et, 79%.

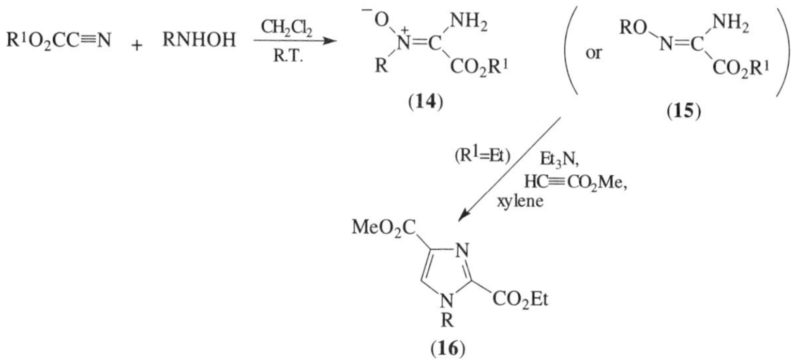

R¹O₂CC≡N + RNHOH →(CH₂Cl₂, R.T.)

⁻O–N⁺(R)=C(NH₂)(CO₂R¹) (14)

(or RO–N=C(NH₂)(CO₂R¹) (15))

(R¹=Et) Et₃N, HC≡CO₂Me, xylene ↓

MeO₂C–[imidazole ring with N–R]–CO₂Et
(16)

Scheme 4.3.7

2-Carbethoxy-4-carbomethoxyimidazole **(16)** *(R = H)* [33]

Ethyl α-aminooximinoacetate (15) (R = H, R¹ = Et) (0.08 g, 0.6 mmol) in xylene (6 ml) is mixed with triethylamine (1.2 eq.), and then treated with methyl propiolate (0.06 ml, 0.72 mmol). When reaction is complete (1 h), water is added, and the organic layer is separated and dried (Na₂SO₄). Removal of the solvent under reduced pressure gives the crude vinyl oxyamidine intermediate as a mixture of *cis* and *trans* isomers (m.p. 50–60°C). This is heated in xylene (6 ml) under reflux (14 h). Rotary evaporation of the solvent gives a solid which is purified by preparative TLC on plates precoated with 0.5 mm silica gel GF₂₅₄ developed with ethyl acetate–hexane (6:4). Crystallization of the purified product from ethyl acetate–hexane gives (16) (R = H) (0.08 g, 70%), m.p. 188–191°C. Similarly prepared without isolation of the intermediate is 2-carbethoxy-4-carbomethoxy-1-methylimidazole. The aminonitrone (14) (R = Me, R¹ = Et) (0.06 g) in benzene (4 ml) is mixed with triethylamine (1.2 eq.), then treated with methyl propiolate (0.04 ml, 0.49 mmol). When reaction is complete (1 h) the mixture is refluxed (15 min), the solvent is removed, and the residue recrystallized from ethyl acetate–hexane, to give the 1-methyl analogue of (16) (0.075 g), m.p. 155–156°C. Similarly prepared is 2,4-dicarbomethoxy-1-methylimidazole (68%).

Aminomalononitrile (readily made in 45–50% yield by reduction of the oximino compound with aluminium amalgam) condenses with formamidine acetate in ethanol to give 5-amino-4-cyanoimidazole in 35% yield [34].

Replacement of amidines by guanidines in a number of the foregoing procedures gives the analogous 2-aminoimidazoles. Thus, guanidines react with α-hydroxy- [13, 35] and α-halogenoketones [3, 6, 7, 15, 36, 54], and with α-diketones [37–39]. When α-diketones are used, the initial products formed

4.3. FORMATION OF 1,5 AND 3,4 BONDS

are 4-hydroxy-4H-imidazoles, which give 1H-imidazoles (**17**) on catalytic hydrogenation (Scheme 4.3.8).

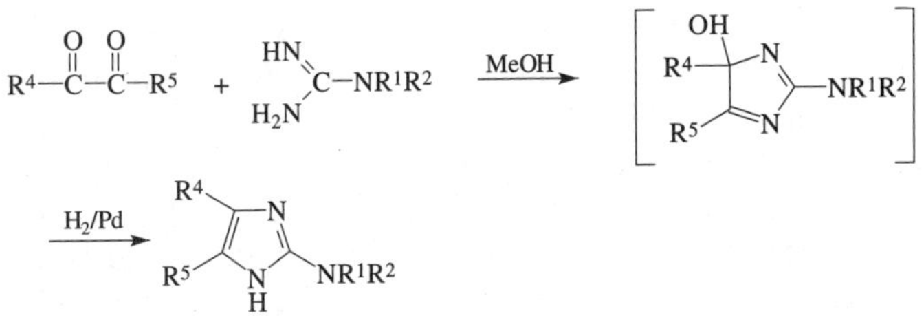

Scheme 4.3.8

5-Methyl-2-methylamino-4-phenylimidazole (**17**) *($R^1 = H$; R^2, $R^5 = Me$; $R^4 = Ph$)* [38]

Solutions of methylguanidine (0.14 g, 2 mmol) in methanol (10 ml) and 1-phenyl-1,2-propanedione (0.29 g, 2 mmol) in methanol (10 ml) are cooled to $-10°C$ and then mixed. The resulting solution is then hydrogenated in the presence of PtO_2 (0.1 g) while cooling with ice–water. After completion of the hydrogenation, concentrated HCl (0.1 ml) is added to the reaction mixture, the catalyst is filtered off, and the filtrate is rotary evaporated to dryness at about 30°C. The residue is dissolved in ethanol (2 ml), and about 50 ml of diethyl ether is added until cloudiness is observed. After 2 weeks of refrigeration, crystals (0.19 g, 37%) of the hydrochloride salt are obtained as colourless needles, m.p. 185–186°C.

There are many similar examples in which the 2-aminoimidazoles are obtained in good yields, e.g. (**17**) (R^1, R^2, R^4, R^5, yield given): H, H, Ph, Ph, 84%; H, H, p-ClC$_6$H$_4$, p-ClC$_6$H$_4$, 41%; H, H, Me, Me, 47%; H, H, Me, Et, 55%; Me, Me, Ph, Ph, 72%; Me, Me, p-ClC$_6$H$_4$, p-ClC$_6$H$_4$, 82%; Me, Me, p-MeOC$_6$H$_4$, p-MeOC$_6$H$_4$, 66% [39].

A recent simple synthesis of 2-aminoimidazoles has been accomplished by cyclization of α-halogenoketones with an N-acetylguanidine either at reflux in acetonitrile, or in DMF at ambient temperature. The initial products are the imidazol-2-ylacetamides, which are readily hydrolysed to give the 4,5-disubstituted 2-aminoimidazoles (**17**) (R^1, $R^2 = H$) in moderate to high yields [36].

General procedure [36]

To the acetylguanidine (3.0 eq.) in anhydrous DMF (or MeCN) is added an α-haloketone (typically a 0.3 mol l^{-1} solution). The reaction mixture is stirred at room temperature (96 h) (or with MeCN it is refluxed (16 h) under argon) before evaporation to dryness. The residue is washed with water, filtered, dried and either recrystallized or subjected to column chromatography on silica gel-60 or basic alumina, e.g. 2-acetamido-4-methylimidazole is purified by alumina column chromatography using 5% chloroform-methanol. 2-Acetamidoimidazoles prepared in this way include (4 and 5 substituent, yield given): Me, H, 32%; Et, H, 78%, tBu, H, 61%; Me, Me, 58%; Ph, H, 58%; Ph, Me, 65%; Ph, Ph, 61%. The acetamido derivatives are hydrolysed in the following manner. The 2-N-acetyl compound (0.1 g) is heated at reflux (24 h) in a 1:1 solution of methanol and water (5 ml) containing 5 drops of concentrated sulfuric acid. The reaction mixture is then evaporated, and the resulting sulfate salt recrystallized from water. Alternatively, the free base form can be generated by adjusting the pH to around 10 with 1% methanolic potassium hydroxide, and the product is then purified by column chromatography on Sephadex LH-20 with methanol. Under basic conditions (e.g. neat hydrazine at 70°C) the amide derivatives can also be hydrolysed in high yields (68–86%).

Symmetrical diarylguanidines condense with the mixed anhydride of acetic and chloroacetic acids (formally α-halocarbonyl species) to give 1-aryl-2-arylaminoimidazolin-4-ones (**18**) in a one-step reaction (Scheme 4.3.9). Yields are, however, only moderate [6].

Scheme 4.3.9

2-Anilino-1-phenylimidazolin-4-one (**18**) *(Ar = Ph)*

The mixed anhydride obtained by reaction of ketene with monochloroacetic acid (12 g) is added to a suspension of 1,3-diphenylguanidine (21 g) in acetone (60 ml) cooled to 0°C. The temperature of the reaction mixture rises to 10–15°C, and a precipitate is deposited. This is filtered and washed with acetone, giving the hydrochloride salt of (**18**) (Ar = Ph) (11.2 g, 40%), m.p. 235–236°C. Similarly prepared is (**18**) (Ar = *p*-tolyl) (38%).

As mentioned earlier, the other main NCN synthons which are commonly available are ureas, thioureas and isothioureas, which condense with

4.3. FORMATION OF 1,5 AND 3,4 BONDS

α-hydroxycarbonyl, α-halogenocarbonyl and α-dicarbonyl species to give imidazoles. Urea and methylurea will react with an acyloin or biacetyl in acidic solution to give 4-imidazolin-2-ones (45–85%) which are in tautomeric equilibrium with 2-hydroxyimidazoles. Both aliphatic and aromatic acyloins react, but 1,3-dimethylurea will not react under the same conditions with these CC substrates [40–42]. Thiourea reacts similarly to give the analogous 2-thiones [43–45], with the reactions preferably carried out in a high boiling solvent (e.g. n-hexanol) in the presence of catalytic quantities of HCl or dry HCl. Yields are usually moderate to good (50–60%).

The common reaction of an α-bromoketone with excess urea and ammonium acetate in aqueous acetic acid (refluxed for about 3 h) to give 4-imidazolin-2-ones in 50–75% yields can be simplified and adapted to a single-flask process from the original ketones. The ketone is initially brominated at 18–20°C in a mixture of acetic acid and urea, then, after the addition of 30% ammonia solution, the α-bromoketone which forms is heterocyclized in about 40% yield by further heating ($\sim$3 h) [42].

Cyanourea will react similarly with an α-haloketone to give 1-cyanoimidazolin-2-ones, which are readily hydrolysed to the 1-carbamoyl and ultimately to the N-unsubstituted analogues [46]. 4-Amino-2(2H)-imidazolones can be made from urea and an α-ketonitrile [47].

S-Methylisothiourea is formally analogous to an amidine or guanidine (see also reactions of amidine sulfinic acids earlier in this chapter). It reacts readily with acylvinylphosphonium salts in much the same way as amidines to give 2-methylthio-4-imidazolylphosphonium salts, which can be converted into multifunctional imidazoles with recovery of triphenylphosphine [28].

Carbodiimides react with dicarbonyl compounds (or their sulfur analogues) to give imidazoles [48, 49]. When diimmonium salts (**19**) are treated with guanidines or O-methylisoureas the initial products are 4,5-dihydroimidazoles, but these are readily aromatized by heating in the presence of triethylamine hydrochloride (Scheme 4.3.10). The mildly acidic conditions result in the loss of one of the amino functions from the intermediate [50]. Yields of 2,5-diaminoimidazoles are usually 60–80% overall.

Acetylamidrazones (**20**) can be made from imidate hydrochlorides and acetohydrazide in dry ethanol in the presence of triethylamine (Scheme 4.3.10). When treated with phenacyl bromide in boiling acetonitrile they are largely converted into 1-acetamidoimidazoles (**21**), although some acetylamidrazone hydrobromide also precipitates. Yields of (**21**) are around 46–64% with alkyl or aryl 2-substituents [51].

The reagent "betmip" (1-(triphenylphosphorylideneaminomethyl)benzotriazole) (**22**) [52] reacts with primary amines to form iminophosphoranes (**23**), which cyclize when treated with α-diketones to give (often) good yields of

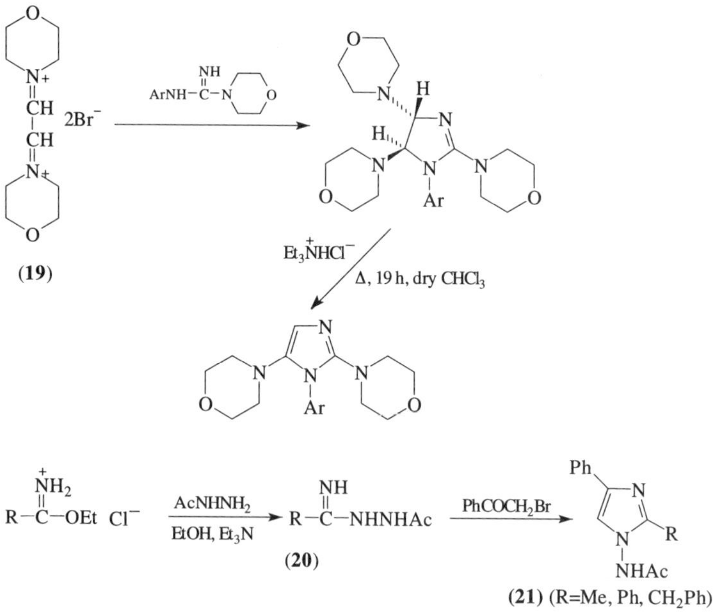

Scheme 4.3.10

1,4,5-trisubstituted imidazoles (Scheme 4.3.11). When the diketone is unsymmetrical the more nucleophilic iminophosphorane nitrogen atom attacks the more activated carbonyl function, giving only one regioisomer [53].

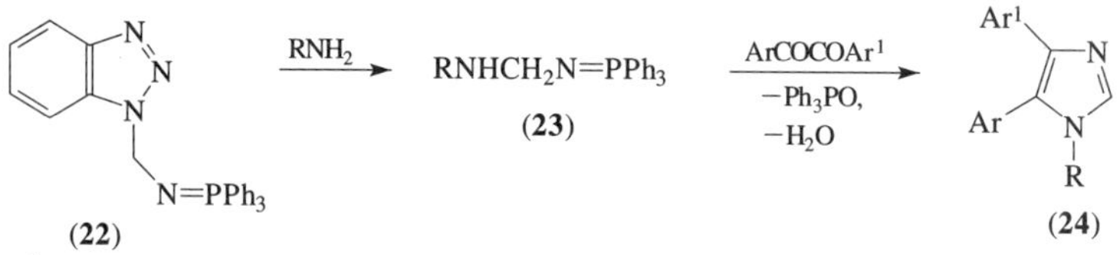

Scheme 4.3.11

1-Cycloheptyl-4,5-diphenylimidazole **(24)** *(R = cycloheptyl; Ar, Ar^1 = Ph)* [53]

To a solution of "betmip" [22] (2.5 g, 6.1 mmol) in THF (100 ml) is added cycloheptylamine (0.69 g, 6.1 mmol). The mixture is refluxed (12 h) before adding benzil (1.3 g, 6.1 mmol) and refluxing for a further 16 h. The mixture is cooled, diluted with diethyl ether (100 ml) and washed twice with 2 M KOH (25 ml) to remove benzotriazole. The crude imidazole product is purified by column chromatography on silica gel, eluting with ether to remove triphenylphosphine oxide. The product (1.57 g, 81%) is obtained as colourless needles m.p. 132–133.5°C. Similarly prepared are (**24**) (R, Ar, Ar[1], yield given): *p*-Me$_2$NC$_6$H$_4$, Ph, Ph, 84%; CH$_2$Ph, Ph, *p*-ClC$_6$H$_4$, 55%.

REFERENCES

1. M. R. Grimmett, in *Comprehensive Heterocyclic Chemistry* (ed. A. R. Katritzky and C. W. Rees). Pergamon Press, Oxford, 1984, Vol. 5, p. 457.
2. M. R. Grimmett, in *Comprehensive Heterocyclic Chemistry*. (ed. A. R. Katritzky and C. W. Rees). Elsevier, Oxford, 1996, Vol. 3.02 (ed. I. Shinkai) p. 77.
3. H. Beyer and S. Schmidt, *Liebigs Ann. Chem.* **748**, 109 (1971).
4. M. Hamaguchi and T. Ibata, *Chem. Lett.* 169 (1975).
5. M.-C. Dubroeucq, F. Rocquet and F. Weiss, *Tetrahedron Lett.* 4401 (1977).
6. Yu. V. Svetkin and A. N. Minlibaeva, *J. Org. Chem. USSR (Engl. Transl.)* **7**, 1339 (1971).
7. B. S. Drach, *Chem. Heterocycl. Compd. (Engl. Transl.)* **25**, 593 (1989).
8. H. Bredereck, R. Gompper, H. G. Schuh and G. Theilig, in *Newer Methods of Preparative Organic Chemistry* (ed. W. Foerst) Academic Press, New York, 1964, Vol. 3, p. 241.
9. H. Schubert, *J. Prakt. Chem.* **3**, 146 (1956).
10. F. C. Cooper and M. W. Partridge, *Org. Synth.* **36**, 64 (1956).
11. M. A. Ross, E. Iwanowicz, J. A. Reid, J. Liu and Z. Gu, *Tetrahedron Lett.* **33**, 5933 (1992).
12. G. C. Lancini and E. Lazzari, *J. Heterocycl. Chem* **3**, 152 (1966).
13. A. Kreutzberger, *J. Org. Chem.* **27**, 886 (1962).
14. A. F. Pozharskii, A. D. Garnovskii and A. M. Simonov, *Russ. Chem. Rev. (Engl. Transl.)* **35**, 122 (1966).
15. A. V. Ivashchenko, V. T. Lazareva, E. K. Prudnikova, S. P. Ivashchenko and V. G. Rumyantsev, *Chem Heterocycl. Compd. (Engl. Transl.)* **18**, 185 (1982).
16. R. K. Griffith and R. A. DiPietro, *Synthesis* 576 (1983).
17. J. L. Kelley, C. A. Miller and E. McLean, *J. Med. Chem.* **20**, 721 (1977).
18. P. Benjes and M. R. Grimmett, unpublished data, 1996.
19. B. Krieg, L. Brandt, B. Carl and G. Manecke, *Chem. Ber.* **100**, 4042 (1967).
20. C. Leschke, J. Altman and W. Schunack, *Synthesis* 197 (1993).
21. J. Altman, M. Wilchek, R. Lipp and W. Schunack, *Synth. Commun.* **19**, 2069 (1989).
22. D. Pocar, R. Stradi and B. Gioia, *Tetrahedron Lett.* 1839 (1976).
23. L. Citerio, D. Pocar, R. Stradi and B. Gioia, *J. Chem. Soc., Perkin Trans. 1* 309 (1978).
24. L. Citerio, D. Pocar, M. L. Saccarello and R. Stradi, *Tetrahedron* **35**, 2375 (1979).
25. H. O. House, L. J. Czuba, M. Gall and H. D. Olmstead, *J. Org. Chem.* **34**, 2324 (1969).
26. L. Citerio and R. Stradi, *Tetrahedron Lett.* 4227 (1977).
27. R. L. Webb, C. S. Labaw and G. R. Wellman, *ACS Congr. Abstr.* No. 162 (1979).
28. R. L. Webb and J. J. Lewis, *J. Heterocycl. Chem.* **18**, 1301 (1981).
29. C. A. Lipinski, T. E. Blizniak and R. H. Craig, *J. Org. Chem.* **49**, 566 (1984).
30. L. A. Reiter, *J. Org, Chem.* **49**, 3494 (1984).
31. C. Saturnino, M. Abarghaz, M. Schmitt, C.-G. Wermuth and J.-J. Bourguignon, *Heterocycles* **41**, 1491 (1995).
32. M. Guillemet, A. Robert and M. Baudy-Floc'h, *Tetrahedron Lett.* **36**, 547 (1995).
33. P. S. Branco, S. Prabhakar, A. M. Lobo and D. J. Williams, *Tetrahedron* **48**, 6335 (1992).
34. J. P. Ferris and L. E. Orgel, *J. Am. Chem. Soc.* **87**, 4976 (1965).
35. H. W. Schramm, M. Schubert-Zsilavecz, A. I. Saracoglu and Ch. Kratky, *Monatsh. Chem.* **122**, 1063 (1991).

36. T. L. Little and S. E. Webber, *J. Org. Chem.* **59**, 7299 (1994).
37. T. Nishimura, K. Nakano, S. Shibamoto and K. Kitajima, *J. Heterocycl. Chem.* **12**, 471 (1975).
38. T. Nishimura and K. Kitajima, *J. Org. Chem.* **41**, 1590 (1976).
39. T. Nishimura and K. Kitajima, *J. Org. Chem.* **44**, 818 (1979).
40. A. R. Butler and I. Hussain, *J. Chem. Soc., Perkin Trans. 2* 310 (1981).
41. H. M. Chawla and M. Pathak, *Tetrahedron* **46**, 1331 (1990).
42. S. I. Zav'yalov, I. V. Sitkareva, G. I. Ezhova, O. V. Dorofeeva, A. G. Zavozin and E. E. Rumyantseeva, *Chem. Heterocycl. Compd. (Engl. Transl.)* **26**, 708 (1990).
43. N. Kuhn and T. Kratz, *Synthesis* 561 (1993).
44. C. J. Broan and A. R. Butler, *J. Chem. Soc., Perkin Trans. 2* 1501 (1991).
45. P. M. Kochergin, V. E. Bogachev and M. G. Fomenko, Russian Patent 137,517 (1960); *Chem. Abstr.* **56**, 475 (1962).
46. H. Beyer and H. Schilling, *Chem. Ber.* **99**, 2110 (1966).
47. M. Srivastava and R. Lakhan, *Org. Prep. Proced. Int.* **25**, 708 (1993).
48. K. Hartke and A. Kumar, *Sulfur Lett.* **1**, 37 (1982); *Chem. Abstr.* **98**, 89 250 (1983).
49. R. Richter and E. A. Barsa, *J. Org. Chem.* **51**, 417 (1986).
50. L. Citerio, E. Rivera, M. L. Saccarello, R. Stradi and B. Gioia, *J. Heterocycl. Chem.* **17**, 97 (1980).
51. E. E. Glover, K. T. Rowbottom and D. C. Bishop, *J. Chem. Soc., Perkin Trans. 1* 2927 (1972).
52. A. R. Katritzky, J. Jiang and L. Urogdi, *Synthesis* 565 (1990).
53. A. R. Katritzky, J. Jiang and P. A. Harris, *Heterocycles* **31**, 2187 (1990).
54. G. Losse, A. Barth and R. Sachadae, *Chem. Ber.* **94**, 467 (1961).

—5—
Ring Syntheses which Involve Formation of Three or Four Bonds

5.1 FORMATION OF 1,2, 3,4 AND 1,5 BONDS OR 1,2, 2,3, 3,4 AND 1,5 BONDS

It is convenient to combine these two synthetic approaches because they are formally similar. Both condense an α-functionalized ketone or aldehyde (C-4–C-5 synthon) with an amine or ammonia (N-1, N-3) and an aldehyde (C-2). The alternative Bredereck modification uses formamide as the source of the C-2–N-3 bond and of N-1. The older (Radziszewski or Weidenhagen) methods give 4-mono-, 4,5-di- and 2,4,5-trialkyl or -triaryl imidazoles; the Bredereck "formamide" synthesis is largely restricted to the preparation of imidazoles with no 2 substituent.

Many of the classical methods grew out of the earliest synthesis of imidazole, which was achieved in 1858 by Debus [1] when he allowed glyoxal, formaldehyde and ammonia to react together. Although the earliest modifications of this method used α-diketones or α-ketoaldehydes as substrates [2], by the 1930s it was well established that α-hydroxycarbonyl compounds could serve equally well, provided that a mild oxidizer (e.g. ammoniacal copper(II) acetate, citrate or sulfate) was incorporated [3]. A further improvement was to use ammonium acetate in acetic acid as the nitrogen source. All of these early methods have deficiencies. There are problems associated with the synthesis of a wide range of α-hydroxyketones or α-dicarbonyls, yields are invariably rather poor, and more often than not mixtures of products are formed. There are, nevertheless, still applications to the preparation of simple 4-alkyl-, 4,5-dialkyl(diaryl)- and 2,4,5-trialkyl(triaryl)imidazoles. For example, pyruvaldehyde can be converted quite conveniently into 4-methylimidazole or 2,4-dimethylimidazole. However, reversed aldol reactions of pyruvaldehyde in ammoniacal solution lead to other imidazoles (e.g. 2-acetyl-4-methylimidazole) as minor products [4]. Such

TABLE 5.1.1
Imidazoles (2) made from α-dicarbonyl compounds, (an aldehyde) and ammonia (or ammonium acetate)

R^2	R^4	R^5	Yield (%)	Ref.
Ph	Ph	Ph	94	6
p-BrC$_6$H$_4$	p-MeOC$_6$H$_4$	p-MeOC$_6$H$_4$	71	9
p-MeOC$_6$H$_4$	Ph	Ph	53	9
2,4-Cl$_2$C$_6$H$_3$	Ph	Ph	66	9
p-BrC$_6$H$_4$	Ph	Ph	93	9
H	Ph	Ph	61,91	6,9
Me	Ph	Ph	55,96	6,9
p-ClC$_6$H$_4$	Ph	Ph	97	6,9
CF$_3$	Ph	Ph	38	6,9
Ph	Ph	1-Naphthyl	81	10
2-Thienyl	Ph	Ph	71	11
2-Thienyl	Ph	p-ClC$_6$H$_4$	67	11
2-Thienyl	Ph	p-MeC$_6$H$_4$	72	11
2-Thienyl	Ph	p-MeOC$_6$H$_4$	62	11
Ph	Ph	2-Thienyl	57	11
p-MeC$_6$H$_4$	Ph	2-Thienyl	57	11
p-MeOC$_6$H$_4$	Ph	2-Thienyl	64	11
m-MeC$_6$H$_4$	Ph	2-Thienyl	61	11
p-Me$_2$NC$_6$H$_4$	Ph	2-Thienyl	47	11
m-NO$_2$C$_6$H$_4$	Ph	2-Thienyl	82	11
2-Furyl	Ph	Ph	62	12
2-Furyl	Ph	p-ClC$_6$H$_4$	50	12
2-Furyl	Ph	p-MeC$_6$H$_4$	60	12
2-Furyl	Ph	p-MeOC$_6$H$_4$	52	12
Ph	Ph	CF$_3$	47	8
p-MeC$_6$H$_4$	Ph	CF$_3$	28	8
2-Pyridyl	Ph	CF$_3$	35	8
CF$_3$	Ph	CF$_3$	46	8
Ph	Ph	CO$_2$Et	88	13
Pr	Ph	CO$_2$Et	76	13
Ph	Pr	CO$_2$Et	81	13
Pr	Pr	CO$_2$Et	66	13
H	Ph	CO$_2$Et	90	13

alkaline fission of the two-carbon synthon serves to increase the likelihood of minor product proliferation and reduces overall yields, but one can use a large excess of the C-2 synthon to reduce these problems somewhat. In the synthesis of 2-acetyl-4-(1',2',3',4'-tetrahydroxybutyl)imidazole from D-glucosone, pyruvaldehyde and ammonium acetate, it was found convenient to use excess pyruvaldehyde ethylenedithioketal to prevent alkaline cleavage of the α-ketoaldehyde [5]. Examples of reactions which have usefully combined α-diketones and ammonia with or without added aldehyde

5.1. FORMATION OF 1,2, 3,4 AND 1,5 BONDS OR 1,2, 2,3, 3,4 AND 1,5 BONDS

include the high-yielding synthesis of lophine (2,4,5-triphenylimidazole) from benzil, benzaldehyde and ammonia [6], a rather elegant synthesis of 4,5-di-t-butylimidazole [7], and preparation of a series of 2-aryl-5-trifluoromethyl-4-phenylimidazoles from 3,3,3-trifluoro-1-phenylpropane-1,2-dione monohydrate, an aldehyde and ammonium acetate [8]. Table 5.1.1 lists some typical examples. The scope of the reaction has been extended recently by incorporation of an ester group into the α-dicarbonyl by use of an appropriate tricarbonyl monohydrate (1) ($R^5 = CO_2R$) and an alkyl or aryl aldehyde [13]. The appropriate tricarbonyl substrates (1) ($R^5 = CO_2R$) can be prepared in two steps from an acid chloride by the Wasserman method which proceeds via a ketophosphorane (3) [14], or by the Dess–Martin oxidation (with an iodine(V) reagent) of β-hydroxy- and β-ketoamides, -esters and -ketones (Scheme 5.1.1) [15]. The cyclocondensation of these tricarbonyls with an aldehyde and ammonium acetate serves to introduce an ester group at C-4(5) of the imidazole (2) (Scheme 5.1.1) [13].

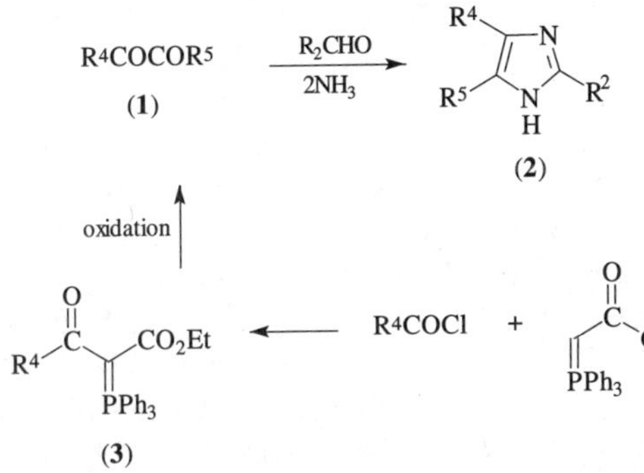

Scheme 5.1.1

Although 1,2-dicarbonyl substrates (especially unsymmetrical benzils) are often difficult to make, there are a number of approaches which may be appropriate. Propane-1,3-dithiol reacts with aldehydes to give cyclic thioacetals (in 52–91% yields) which form stable dithiane anions when treated with butyllithium. Subsequent quenching with an acid chloride followed by mercury(II) chloride treatment gives a 1,2-dicarbonyl species. Alternatively, substitution of an aldehyde for the acid chloride gives rise eventually to an α-hydroxycarbonyl derivative (Scheme 5.1.2) [16]. An alternative approach to α-ketoaldehydes (82–86% yields) reacts an α-ketonitrate ester with sodium acetate in DMSO [17]. Aryl α-diketones can be made from α-ketoanils, which are in turn made by cyanide ion-catalysed transformation of aromatic aldimines [18], and the range of unsymmetrical benzils has been increased by

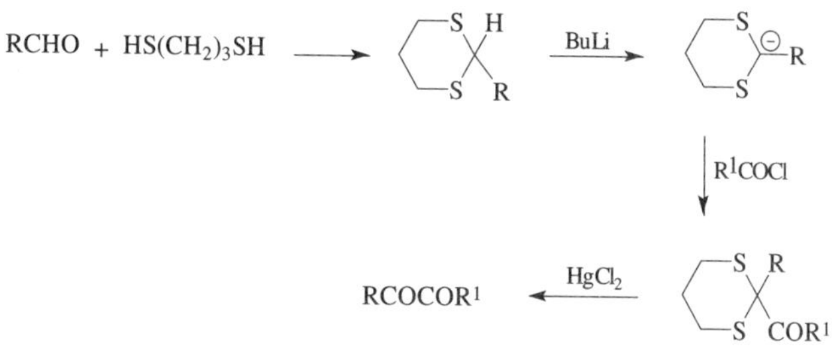

Scheme 5.1.2

RCHO + HS(CH$_2$)$_3$SH → [dithiane-CHR] —BuLi→ [dithiane-C$^{\ominus}$R]

↓ R^{1}COCl

RCOCOR^1 ←HgCl$_2$— [dithiane-CR(COR^1)]

Friedel–Crafts reaction of arenes with α-chloro-α-(methylthio)acetophenones [19]. When N-phenylbenzimidoyl chlorides are aroylated with an aryl halide in the presence of a catalytic amount of an azolium salt (such as 1,3-dimethylimidazolium iodide) and sodium hydride, followed by acid treatment in THF, unsymmetrical benzils can be isolated in 80–92% yields [20]. Added to this method is the sequence ketone → enaminoketone → α-diketone, which cleaves a carbon–carbon double bond in the final stage by photooxygenation [21].

Investigation of the mechanism of these reactions has suggested ways in which the yields can be improved. Acidic conditions (pH ~ 2) will prevent Cannizzaro rearrangement of any glyoxal-type species and also serve to hydrolyse any Schiff bases which result from side reactions of aldehyde and amine. Conditions should be adjusted so that the rate of hydrolysis of "linear" products is equal to the rate of cyclocondensation, allowing accumulation of the imidazole products. From glyoxal, formaldehyde and ammonium chloride the yield of imidazole can be increased to 85% by careful control of the conditions. With an appropriate alkylammonium chloride, 1-substituted imidazoles are also accessible (e.g. 1-methyl (56%), 1-isopropyl (46%), 1-cyclohexyl (49%), 1-n-butyl (55%), 1-t-butyl (25%)). The process may have some applications, but yields drop off with branched alkyl compounds [22]. Imidazolium salts are also available under similar conditions when two molar equivalents of a primary alkylamine are used [23].

2-(4′-Bromophenyl)-4,5-bis(4″-methoxyphenyl)imidazole (2)
(R^2 = p-Br−C$_6$H$_4$ R^4, R^5 = p-MeO−C$_6$H$_4$) [9]

A mixture of 4,4′-dimethoxybenzil (1.9 g, 7 mmol), 4-bromobenzaldehyde (1.5 g, 8.1 mmol) and anhydrous ammonium acetate (6.5 g, 84 mmol) in glacial acetic acid (50 ml) is heated. The solution turns yellow, and the reaction is shown to be complete (TLC) after 1.5 h. The solution is then slowly poured into a solution of 0.88 ammonia (100 ml) in ice–water (100 ml). The solid product is filtered and recrystallized from isopropanol to give the imidazole (2.2 g, 71%), m.p. 160–161°C.

Ethyl 2,4-diphenylimidazole-5-carboxylate (2) (R^2, R^4 = Ph, R^5 = CO_2Et) [13]

To a slurry of ammonium acetate (0.86 g, 11.2 mmol) in glacial acetic acid (3 ml) is added the tricarbonyl monohydrate (1) (R^4 = Ph, R^5 = CO_2Et) (0.23 g, 1.1 mmol) followed by benzaldehyde (0.236 g, 2.2 mmol). The mixture is heated to 65°C and stirred for 30 min, when TLC analysis indicates complete consumption of the tricarbonyl. After cooling to room temperature the acetic acid is evaporated, to give an oily residue which is dissolved in ethyl acetate and washed in turn with saturated sodium bicarbonate, water and brine. The organic phase is dried ($MgSO_4$), filtered and concentrated under reduced pressure. Chromatography on silica gel (hexanes–ethyl acetate 9:1) gives the above imidazole product (0.287 g, 88%), m.p. 165–167°C.

In the above reaction sequences, when the ammonia or amine is replaced by hydroxylamine, one can prepare 1-hydroxyimidazole 3-oxides, e.g. 1-hydroxyimidazole 3-oxide (68%) and its 2-methyl (74%) and 2-ethyl derivatives (70%). Subsequent hydrogenation with palladium on carbon in aqueous hydrochloric acid gives the 1-hydroxyimidazoles. It should be noted that the oxides decompose violently when heated [24]. Titanium(III) chloride in aqueous solution at room temperature is a suitable reducing agent [25].

1-Hydroxy-2,4,5-trimethylimidazole 3-oxide [26]

A solution of diacetyl (4.3 g, 50 mmol) and ethanol (2.2 g, 50 mmol) in methanol (400 ml) is treated with hydroxylamine hydrochloride (2 g, 100 mmol) and stirred at room temperature (20 h). The reaction mixture is neutralized with sodium carbonate (15 g), filtered, and concentrated under reduced pressure. The residue crystallizes from benzene as tan needles (5.5 g, 83%), m.p. 199–200°C. Similarly prepared is the 2-phenyl analogue (90%).

4,5-Dichloroimidazoles can be made by treatment of an arylaldehyde with cyanogen and hydrogen chloride in ethereal solution. The sole by-product appears to be oxamide, which is readily separated because of its sparing solubility in common solvents. Yields of 2-aryl-4,5-dichloroimidazoles lie in the range 36–72% [27]. The dichloroimidazoles are very base-sensitive, and even dilute alkalies convert them into *N*-methylbenzamides.

It has been reported that 1-hydroxyimidazoles (rather than imidazoles [28]) are accessible in high yields from reaction of nitrosonium fluoroborate with acetonitrile and the appropriate alkene; titanium(III) chloride reduces the hydroxyl function [24].

Replacement of the α-dicarbonyl compound with an α-hydroxyaldehyde or -ketone under oxidizing conditions is the cornerstone of the early imidazole syntheses based on reactions of reducing carbohydrates with ammonia. Yields in such reactions are seldom good, and the method was limited largely to the synthesis of 4-methyl- and 4-hydroxymethylimidazoles, for which more convenient methods are now available. If, however, it is necessary to make an imidazole with a polyhydroxyalkyl side chain of specific configuration at the 4- or 5-position, then the use of a carbohydrate starting material is indicated, e.g. in the synthesis of L-*erythro*-β-hydroxyhistidine from D-glucosamine [29]. There are, however, many examples in the literature in which simpler ketols or ketol acetates have served as imidazole sources [3, 30–35]. The method was originally devised by Weidenhagen [3]. Provided that an excess of the acyloin is used in an alcoholic solution of copper(II) acetate and ammonia, ring closure can lead to 2-imidazolylketones, e.g. 2-benzoyl-4-phenylimidazole (32%) [32], but the method is much more commonly applied to the synthesis of imidazoles unsubstituted at C-2, or simple 2-alkyl or 2-aryl analogues. Examples of products prepared by the general method include 2,4-bis-(*p*-methoxyphenyl)- (67%), 4-(*p*-phenoxyphenyl)- (59%), 2-(*p*-phenoxyphenyl)- (8%) [31], 2,4-di-α-furyl- (40%), 2-phenyl-4,5-di-α-furyl- (41%), 2-α-furyl-4,5-diphenyl- (30%) [36], 2-cyclohexyl-4-phenyl- (48%), 2-cyclohexyl-4,5-diphenyl- (34%) and 5-cyclohexyl-2,4-diphenylimidazoles (32%) [34]. Some of the yields quoted are for the copper salts. A complication of the interaction of α-acyloxyketones with ammonium acetate is the isolation of mixtures of oxazoles and imidazoles [37], and product complexity can be too great to offset the simplicity of the method. The use of excess ammonia serves to depress oxazole formation, and the ability to convert oxazoles into imidazoles (see Chapter 6) can also be a mitigating factor. A novel synthesis of acyloins in 55–95% yields increases the scope of these cyclizations [38], but attempts to make acyloins from epoxide precursors [39, 40] may only give modest yields, and it seems better to brominate a methyl ketone in methanol [41], and then convert the α-bromoketone into the acetoxyketone using potassium acetate before applying slightly modified Weidenhagen conditions. The copper(II) acetate is dissolved in 25% aqueous ammonia, and this solution is added dropwise to a solution of the acyloin (or acetoxyketone) and aldehyde in 25% aqueous ammonia at 0°C. The temperature is then gradually increased to 100°C. Improved yields of 2,4-disubstituted imidazoles have been obtained in this way [35].

4-Isopropyl-2-methylimidazole (**2**) *($R^2 = Me$, $R^4 = iPr$, $R^5 = H$)* [35]

1-Acetoxy-3-methyl-2-butanore (0.375 ml, 2.6 mmol) is dissolved in 3 M aqueous KOH (5 ml) at 0°C. To this a mixture of copper(II) acetate (5.1 g), 25%

5.1. FORMATION OF 1,2, 3,4 AND 1,5 BONDS OR 1,2, 2,3, 3,4 AND 1,5 BONDS

ammonia solution (70 ml) and ethanol (4.1 ml, 13.2 mmol) is added slowly at 0°C. The resulting mixture is shaken vigorously and then the temperature is raised in stages to 100°C until a blue-green precipitate appears. The suspension is allowed to cool to room temperature, then chilled to 0°C and kept at that temperature (6 h), then filtered. The solid product is washed once with ethanol and air dried to give the blue-white copper salt of the imidazole (0.454 g, 2.39 mmol). A hot aqueous suspension of this salt is treated with gaseous hydrogen sulfide (15–20 min), filtered, washed with hot water, and allowed to cool. Concentration of the filtrate *in vacuo* gives 4-isopropyl-2-methylimidazole (0.307 g, 88%), m.p. 91–94°C. Similarly prepared from hydroxy- or acetoxyketones are (**2**) (R^5 = H; R^2, R^4, yield given): Me, Me, 82%; Me, Pr, 36%; Me, iPr, 88%; Me, Bu, 41%; Me, iBu, 69%; H, iBu, 84%; Me, tBu, 98%; H, tBu, 94%; Me, Ph, 72%.

Rather less common are condensations between α-hydroxyketones, ammonia and an imidate. Equivalent amounts of the hydrochloride salt of the imidate and the α-substituted carbonyl compound (α-halogeno- and α-acetoxyketones also take part) are heated (40–70°C) in liquid ammonia (3–48 h) at elevated pressures. Yields of imidazoles vary between 10 and 90%, but the requirements of working with liquid ammonia in an autoclave may make this approach unappealing if alternatives are available [42, 43]. Examples include 4-hydroxymethyl- (35%), 4-hydroxymethyl-2-phenyl- (73%), 4-hydroxymethyl-2-(*p*-tolyl)- (68%) and 2-benzyl-4-hydroxymethylimidazoles (85%) [44].

In 1953, Bredereck and Theilig introduced their "formamide synthesis" in which α-diketones, α-hydroxyketones, α-halogenoketones, α-aminoketones, α-ketoesters or (under reducing conditions) α-oximinoketones react with formamide rather than ammonia (Scheme 5.1.3). Yields of imidazoles (**4**) can be as high as 90% and are reported seldom to drop below 40% [45–49]. In our hands, however, we have commonly found it difficult to achieve such consistently high yields, and they have been frequently in the range 20–30%. In spite of this, the process has sufficient simplicity to make it an appealing choice for a range of 4-substituted and 4,5-disubstituted

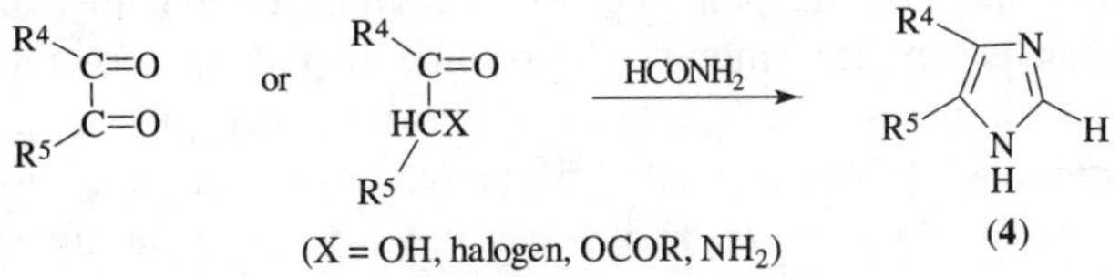

Scheme 5.1.3

TABLE 5.1.2
Imidazoles (4) made by "formamide syntheses"

X	R⁴	R⁵	Yield (%)	Ref.
OH	Me	Me	55	45
OH	Et	Et	67	45
OH	Pr	Pr	81	45
OH	iPr	iPr	85	45
OH	Bu	Bu	72	45
OH	iBu	iBu	61	45
OH	Ph	Ph	91	45
OH	Ph	Et	70	45
OH	p-FC$_6$H$_4$	p-FC$_6$H$_4$	63	50
Br	Ph	H	90	45
Br	Ph	iPr	69	45
Br	Et	Pr	49	45
Br	Me	Me	47[a]	45
Br	p-MeOC$_6$H$_4$	H	62	45
Br	Me	(CH$_2$)$_2$OH	94[b]	51
Br	p-ClC$_6$H$_4$	H	18	52
Br	p-NO$_2$C$_6$H$_4$	H	23	52
NH$_2$	Ph	Ph	90	48
OCOMe	Ph	Me	85[c]	53
OCOMe	Ph	Et	80[c]	53
OCOMe	Ph	H	75[c]	53

[a] Plus 22% 4,5-dimethyloxazole.
[b] The imidazole product also has a 2-methyl substituent.
[c] α-Bromoketal as substrate.

imidazoles (Table 5.1.2.). In the presence of added aldehyde it can be adapted to the preparation of 2,4,5-trisubstituted imidazoles [45, 49]. Replacement of ammonia by formamide reduces decomposition of α-diketones or α-hydroxyketones.

Highly branched α-hydroxyketones fail to react, and the formation of 4,5-disubstituted oxazoles as by-products (which predominate when concentrated sulfuric acid is the condensing medium) can be reduced if the temperature is kept in the range 180–200°C, and a large excess of formamide is used. The two-carbon synthon of choice in these reactions is often an α-bromoketone which can be prepared *in situ*. Thus, boiling the appropriate ketone with an equimolar amount of bromine in excess formamide gives good yields of simple 4- and 4,5-disubstituted imidazoles (4), e.g. 4-phenyl (61%), 4,5-diphenyl (67%), 4-methyl-5-phenyl (80%) and 4-ethyl-5-phenyl (81%) [54]. Where unsymmetrical ketones are involved, careful control of the solvent may allow selective orientation of the bromination [41]. One example added bromine in methanol to 4,5-dihydro-2-methylfuran to form the ketal which was refluxed in formamide to give a high yield to 4-methyl-5-hydroxyethylimidazole [51], and 4-hydroxypropylimidazole has been made similarly [55].

Some of the 4,5-dialkylimidazoles made by Bredereck [48] are thought to be complexes with formic acid, i.e. the reaction of acetoin with formamide gives a 1:1 hydrogen-bonded adduct of formic acid and 4,5-dimethylimidazole rather than the formate salt, and so an alternative synthesis of 4,5-dimethylimidazole has been proposed using the sequence 4-hydroxymethyl-5-methylimidazole 4-chloromethyl-5-methylimidazole   4,5-dimethylimidazole (80–90%) as the free base [56].

Extension of the method to the synthesis of 2-substituted imidazoles rests on the use of more complex amides then formamide, and this is a severe limitation. There are only a few examples of the use of acetamide which lead to 2-methylimidazoles [57].

4,5-Bis-(p-fluorophenyl)imidazole (4) (R^4, R^5 = p-FC_6H_4) [50]

A solution of 4,4'-difluorobenzoin (58.7 g, 0.23 mol) in formamide (300 ml) is refluxed (1 h), cooled and treated with water (50 ml). The resultant solid is filtered and washed in turn with water and cold acetonitrile to give the product (37 g, 63%), m.p. 247–250°C.

4-Phenylimidazole (4) (R^4 = Ph, R^5 = H) [52]

Freshly prepared phenacylbromide (20 g, 0.10 mol) is added to formamide (100 ml) with stirring. The temperature is raised to 150–180°C and maintained at that level (2 h). After cooling somewhat the solution is boiled with concentrated hydrochloric acid (200 ml) containing a small quantity of activated charcoal, cooled, filtered (Celite), and basified with aqueous ammonia, when the product crystallizes. Filtration, washing repeatedly with water, and recrystallization from water gives 4-phenylimidazole as colourless plates (3.6 g, 25%), m.p. 129–130°C.

Although α-aminoketones readily form imidazoles when they cyclize in the presence of formamide [48], preparative difficulties (see Sections 2.1.1, 2.2.1 and 4.1) point to alternative use of their precursors, α-oximinoketones, which can be reduced by dithionite or using catalytic methods in formamide at 70–100°C. Subsequent ring closure is achieved merely by raising the temperature to 180°C.

Cyclization of α-oximinoketones with an aldehyde in the presence of ammonia or an amine is a common way of making 1-hydroxyimidazoles or imidazole *N*-oxides (5) (Scheme 5.1.4) [58–63]. Yields are often very high

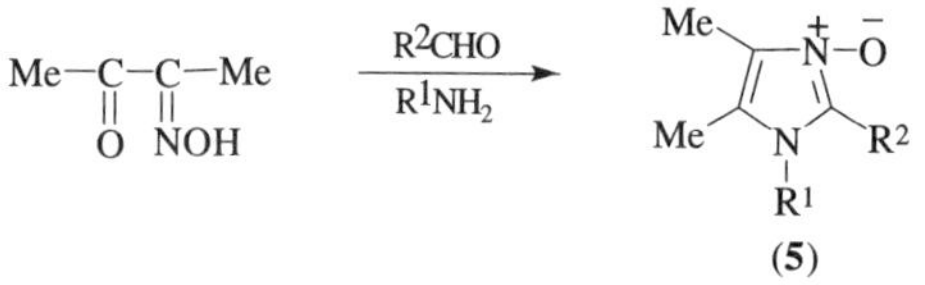

Scheme 5.1.4

and can be quantitative. There are formal similarities to the reactions between α-diketones, hydroxylamine and aldehydes (see above).

(R)-1(α-Carboxy-γ-methylthio)propyl-4,5-dimethylimidazole 3-oxide (5)
($R^1 = CH(CO_2H)(CH_2)_2SMe$, $R^2 = H$) [59]

A solution of diacetyl monoxime (1.01 g, 10 mmol), D(R)-methionine (0.89 g, 10 mmol) and 30% formalin solution (1.5 g) in ethanol (100 ml) is refluxed (3 h), then cooled and filtered. Rotary evaporation of the filtrate, treating the residue with dry acetone, again filtering, then addition of dry ether precipitates the product, which can be recrystallized from chloroform–hexane and added methanol as colourless crystals (2.35 g, 96%), m.p.178–181°C.

The old Maquenne synthesis of imidazoles [64] is seldom used nowadays. It was originally used to make imidazole-4,5-dicarboxylic acids from tartaric acid dinitrate and either an aliphatic aldehyde or formaldehyde precursor in the presence of ammonium ions at pH 3.5–6.5. The procedure is not an entirely pleasant experience, and has been happily superseded by other approaches, but it may be useful on occasion to make compounds such as vinyl-, hydroxy- and methoxy-substituted 2-arylimidazole-4,5-dicarboxylic acids [65]. Use of dialkyltartrate dinitrates gives the corresponding imidazole-4,5-dicarboxylates in 45–65% yields [31].

Another way of assembling the imidazole ring from three separate fragments cyclizes a mixture of an α-aminonitrile, amine and triethyl orthoformate (Scheme 5.1.5) [66, 67]. The process provides an alternative to some of the methods based on DAMN for making imidazoles with amino and nitrile

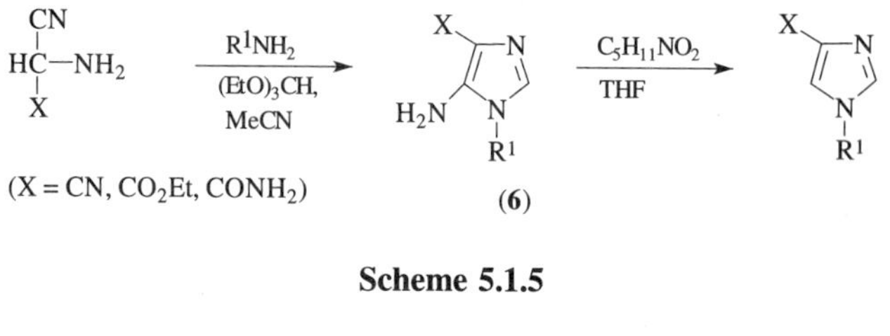

Scheme 5.1.5

5.1. FORMATION OF 1,2, 3,4 AND 1,5 BONDS OR 1,2, 2,3, 3,4 AND 1,5 BONDS

functions in the 4- and 5-positions (see Sections 2.1.1, 2.2.1 and 3.1.1), and the amino group can be later removed [67].

erythro-5-Amino-1-(2-hydroxy-3-nonyl)imidazole-4-carbonitrile (6) (R^1 = 2-hydroxy-3-nonyl, X = CN) [67]

Dry ammonia is bubbled (30 min) through a stirred solution of aminomalonitrile *p*-toluenesulfonate (commercially available from Aldrich) (4.5 g, 17.7 mmol) in dry acetonitrile (200 ml). The solid which separates is filtered off, and the filtrate is concentrated to 100 ml before addition of triethyl orthoformate (2.97 ml, 17.7 mmol) and refluxing (15 min). To the cooled mixture is added *erythro*-3-amino-2-nonanol (2.81 g, 17.7 mmol), and the solution is stirred overnight at room temperature. Evaporation of the solvent and flash chromatography of the residue on a silica gel column eluted with chloroform–methanol (96:4) gives (6) (2.0 g, 45%), m.p. 99–103°C. Similarly prepared are (6) (R^1, X, yield given): CH_2CH_2 NHEt, $CONH_2$, 51%; CH_2CH_2NHEt, CN, 72% [66].

erythro-1-(2-Hydroxy-3-nonyl)imidazole-4-carbonitrile [67]

To a boiling mixture of isoamyl nitrite in THF a solution of (6) (R^1 = 2-hydroxy-3-nonyl, X = CN) (1.0 g, 4.4 mmol) in THF (20 ml) is added dropwise over 1 h. After the addition in complete, the reaction mixture is refluxed (3 h), cooled and rotary evaporated. Flash chromatography of the residue on silica gel eluted with cyclohexane–chloroform–methanol (55:40:5) gives the nitrile (0.46 g, 50%) as an oil.

A number of miscellaneous methods which involve multibond formation are included here for completeness in spite of the fact that their synthetic utilities do not appear to be great. A reaction which closely resembles the α-halogenoketone-amide method is the cyclization of an α-chloro-α-phenylthioketone (prepared from the diazoketone) with ammonia and a carboxylic acid. It has been used to prepare 2-ethyl-4-methyl-5-phenylthioimidazole in 32% yield [68].

2,4,5-Triarylimidazoles have been isolated from reactions of alkenes, carbon monoxide and ammonia in the presence of a rhodium catalyst, while benzylamines react with catalytic quantities of metal carbonyls to form the same compounds [69, 70]. 4-Aminoimidazolium salts have been made by assembling iminochloro sulfides, benzaldimines and isocyanides in a process believed to involve a transient *N*-imidobenzylideniminium halide intermediate. Yields of 25–76% are reported [71].

162 5. RING SYNTHESES WHICH INVOLVE FORMATION OF THREE/FOUR BONDS

Recently, a method has been devised to overcome the problems of making 2,4,5-triarylimidazoles (**8**) with different aryl groups. Essentially, reactions are between α,α-dilithioarylnitromethanes and aryl nitriles; an unusual cyclization–elimination mechanism is believed to be operating (Scheme 5.1.6). The chemistry is reported to work best when the aromatic ring of the arylnitromethane dianion (**7**) contains electron-donating substituents. If it is desired to prepare an imidazole with three different aryl groups one can add in turn two different aryl cyanides to a THF–HMPA solution of the dianion. Although yields are frequently only modest, the method appears to offer considerable potential [72]. The phenylnitromethane starting materials are made either from phenylacetic acid or phenylacetonitrile precursors in high yields [73]. Despite the fact that an imidazole with three different aryl substituents is not the sole product isolated (see experimental data below), the general method should have applications.

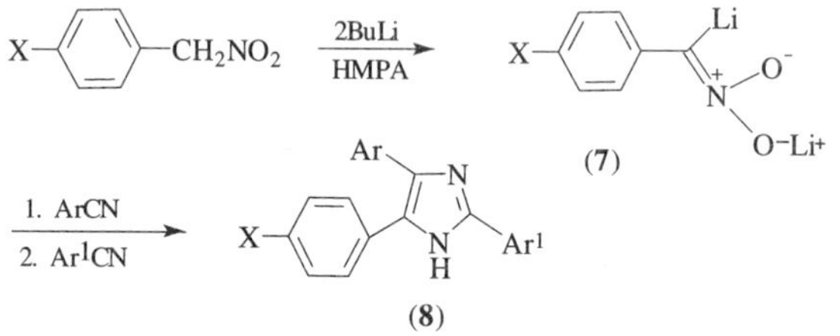

Scheme 5.1.6

General procedure for 2,4,5-triarylimidazoles (**8**) [72]

To a solution of arylnitromethane (10 mmol) in THF (50 ml per gram of substrate) is added HMPA (5.2 ml, 30 mmol), and the solution is cooled to −80°C before addition of n-butyllithium (8.4 ml of a 2.5 M solution in hexane; 21 mmol). The resulting solution is stirred (100 min) and gradually allowed to warm up to −20°C during this period. The mixture is then recooled to −70°C, and aryl cyanide (21 mmol) in THF (2 ml per gram of substrate) is added. The mixture is allowed to come to room temperature and is stirred (18 h) before work-up, which commonly involves concentration *in vacuo* followed by column chromatography on silica gel. Prepared in this way are (**8**) (X, Ar = Ar^1, yield, chromatography solvent mixture listed): F, Ph, 67%, ether–hexane (2:8–3:7); F, p-CF$_3$C$_6$H$_4$, 37%, ether–hexane (3:7); F, p-MeOC$_6$H$_4$, 10%, ether–hexane (1:1); H, Ph, 64%, ether–hexane–Et$_3$N (3:7:0.1); H, p-CF$_3$C$_6$H$_4$, 39%, ether–hexane (3:7); MeO, Ph, 60%, ether–hexane (1:1); MeO, p-CF$_3$C$_6$H$_4$, 62%, ether–hexane–Et$_3$N (40:59:1); MeO, p-MeOC$_6$H$_4$, 15%, ether–hexane (7:3).

2-(4'-Pyridyl)-4-(4'-fluorophenyl)-5-phenylimidazole [72]

To a solution of p-fluorophenylnitromethane (0.5 g, 3.22 mmol) in THF (25 ml) is added HMPA (1.68 ml, 9.67 mmol). The solution is cooled to $-70°C$ and n-butyllithium (2.71 ml of 2.5 M solution in hexane; 6.78 mmol) is added. The solution is stirred at $-70°C$ to $10°C$ (60 min), then cooled to $-80°C$ and benzonitrile (0.329 ml, 3.23 mmol) is added. The reaction mixture is stirred at $-80°C$ to $-20°C$ (40 min) and then at $-20°C$ to room temperature (20 min). The temperature is then lowered again to $-80°C$, and 4-cyanopyridine (0.335 g, 3.22 mmol) in THF (3.0 ml) is added. After allowing it to come to room temperature the solution is stirred (16 h) and concentrated *in vacuo* to dryness. The residue is chromatographed on silica gel using ethyl acetate as the eluent to give (**8**) (X = F, Ar, Ar^1 = Ph) (0.13 g, 13%), m.p. 256–257°C, and (**8**) (X = F, Ar = Ph, Ar^1 = 4-pyridyl) (0.31 g, 31%) as a white crystalline powder, m.p. 291°C (dec.)

REFERENCES

1. H. Debus, *Liebigs Ann. Chem.* **107**, 204 (1858).
2. B. Radziszewski, *Chem. Ber.* **15**, 2706 (1882).
3. R. Weidenhagen and R. Herrmann, *Chem. Ber.* **68**, 1953 (1935).
4. M. R. Grimmett and E. L. Richards, *J. Chem. Soc.* 3751 (1965).
5. J. G. Sweeney, E. Ricks, M. C. Estrada-Valdes, G. A. Iacobucci and R. C. Long, *J. Org. Chem.* **50**, 1133 (1985).
6. D. Davidson, M. Weiss and M. Jelling, *J. Org. Chem.* **2**, 319 (1938).
7. H. Wynberg and A. D. Groot, *J. Chem. Soc., Chem. Commun.* 171 (1965).
8. J. G. Lombardino, *J. Heterocycl. Chem.* **10**, 697 (1973).
9. J. G. Lombardino and E. H. Wiseman, *J. Med. Chem.* **17**, 1182 (1974).
10. A. A. Bardina, B. S. Tanaseichuk and A. A. Khomenko, *J. Org, Chem. USSR (Engl. Transl.)* **7**, 1307 (1971).
11. B. S. Tanaseichuk and S. V. Yartseva, *J. Org. Chem. USSR (Engl. Transl)*, **7**, 1299 (1971).
12. S. V. Yartseva and B. S. Tanaseichuk, *J. Org. Chem. USSR (Engl. Transl.)* **7**, 1558 (1971).
13. M. F. A Brackeen, J. A. Stafford, P. L. Feldman and D. S. Karanewsky, *Tetrahedron Lett.* **35**, 1635 (1994).
14. H. H. Wasserman, D. S. Ennis, C. A. Blum and V. M. Rotello, *Tetrahedron Lett.* **33**, 6003 (1992).
15. M. J. Batchelor, R. J. Gillespie, J. M. C. Golec and C. J. R. Hedgecock, *Tetrahedron Lett.* **34**, 167 (1993).
16. E. J. Corey and D. Seebach, *Angew. Chem., Int. Ed. Engl.* **4**, 1075 (1965).
17. N. Kornblum and H. W. Frazier, *J. Am. Chem. Soc.* **88**, 865 (1966).
18. J. S. Walia, J. Singh, M. S. Chattha and M. Satyanarayana, *Tetrahedron Lett.* 195 (1969).
19. H. Ishibashi, K. Matsuoka and M. Ikeda, *Chem. Pharm. Bull.* **39**, 1854 (1991).
20. A. Miyashita, H. Matsuda and T. Higashino, *Chem. Pharm. Bull.* **40**, 2627 (1992).
21. H. H. Wasserman and J. L. Ives, *J. Am. Chem. Soc.* **98**, 7868 (1976).
22. A. A. Gridnev and I. M. Mihaltseva, *Synth. Commun.* **24**, 1547 (1994).
23. A. Arduengo (Du Pont), US Patent 5,077,414 (1992); *Chem. Abstr.* **116**, 106 289 (1992).
24. G. Laus, J. Stadlwieser and W. Klötzer, *Synthesis* 773 (1989).

25. B. H. Lipschutz and M. C. Morey, *Tetrahedron Lett.* **25**, 1319 (1984).
26. K. Akagane and G. G. Allan, *Chem. Ind. (London)* 38 (1974).
27. D. Günther and D. Bosse. *Angew. Chem., Ind. Ed. Engl.* **19**, 130 (1980).
28. M. L. Scheinbaum and M. B. Dines, *Tetrahedron Lett.* 2205 (1971).
29. S. M. Hecht, K. M. Rupprecht and P. M. Jacobs, *J. Am. Chem. Soc.* **101**, 3982 (1979).
30. M. R. Grimmett, *Adv. Heterocycl. Chem.* **27**, 241 (1980).
31. H. Schubert and H. Ladish, *J. Prakt. Chem.* **18**, 199 (1962).
32. H. Schubert, *J. Prakt. Chem.* **8**, 333 (1959).
33. C. F. Huebner, US Patent 2,744,899 (1956); *Chem. Abstr.* **51**, 486 (1957).
34. H. Schubert, *J. Prakt. Chem.* **3**, 146 (1956).
35. B. H. Lipschutz and M. C. Morey, *J. Org. Chem.* 48, 3745 (1983).
36. H. Schubert, E. Hagen and G. Lehmann, *J. Prakt. Chem.* **17**, 173 (1962).
37. P. P. E. Strzybny, T. van Es and O. G. Backeberg, *J. Org. Chem.* **28**, 3381 (1963).
38. E. L. M. van Rozendaal, S. J. T. Kuster, E. T. Rump and H. W. Scheeren, *Synth. Commun.* **24**, 367 (1994).
39. T. Tsuji, *Tetrahedron Lett.* 2413 (1966).
40. T. M. Santosusso and D. Swern, *Tetrahedron Lett.* 4261 (1968).
41. M. Gaudry and A. Marquet, *Bull. Soc. Chim. Fr.* 1849 (1967).
42. V. I. Kelarev and V. N. Koshelev, *Russ. Chem. Rev (Engl. Transl.)* **64**, 317 (1995).
43. G. Estenne, G. Leclerc, P. Dodey and P. Renaut, *J. Heterocycl. Chem.* **31**, 1121 (1994).
44. P. Dziuron and W. Schunack, *Arch. Pharm. (Weinheim, Ger.)* **306**, 347 (1973).
45. H. Bredereck, R. Gompper, H. G. V. Schuh and W. Theilig, *Angew. Chem.* **71**, 753 (1959).
46. M. R. Grimmett, *Adv. Heterocycl. Chem.* **12**, 103 (1970).
47. H. Bredereck, R. Gompper, H. G. V. Schuh and W. Theilig, in *Newer Methods of Preparative Organic Chemistry* (ed. W. Foerst). Academic Press, New York, 1964, Vol. 3, p. 241.
48. H. Bredereck and G. Theilig, *Chem. Ber.* **86**, 88 (1953).
49. H. Bredereck, R. Gompper and D. Hayer, *Chem. Ber.* **92**, 338 (1959).
50. T. R. Sharpe, S. C. Cherkofsky, W. E. Hewes, D. H. Smith, W. A. Gregory, S. B. Haber, M. Leadbetter and J. G. Witney, *J. Med. Chem.* **28**, 1188 (1985).
51. J. R. Pfister, W. Kurz and I. T. Harrison, *J. Heterocycl. Chem.* **18**, 831 (1981).
52. P. Benjes and M. R. Grimmett, unpublished; P. Benjes, PhD thesis, University of Otago, 1994.
53. A. Novelli and A. de Santis, *Tetrahedron Lett.* 265 (1967).
54. H. Bredereck, F. Effenberger, F. Marquez and K. Ockewitz, *Chem. Ber.* **93**, 2083 (1960).
55. G. A. A. Kivits and J. Hora, *J. Heterocycl. Chem.* **12**, 577 (1975).
56. A. D'Sa and L. A. Cohen, *J. Heterocycl. Chem.* **28**, 1819 (1991).
57. F. Marquez, *Anales Real Soc. Espan. Fis. Quim. (Madrid), Ser. B* **57**, 723 (1961); *Chem Abstr.* **57**, 12 467 (1962).
58. R. Breslow and S. Chung, *Tetrahedron Lett.* **30**, 4353 (1989).
59. H. Lettau, P. Nuhn, R. Schneider and P. Stenger, *Pharmazie* **45**, 830 (1990); *Chem. Abstr.* **114**, 185 371 (1991).
60. R. Hossbach, H. Lettau, P. Nuhn, R. Schneider, P. Stenger and B. Stiebitz, *Pharmazie* **46**, 412 (1991); *Chem. Abstr.* **116**, 128 783 (1992).
61. H. Lettau, *Z. Chem.* **10**, 462 (1970).
62. H. Lettau, *Z. Chem.* **10**, 431 (1970).
63. H. Lettau, *Z. Chem.* **10**, 211 (1970).
64. M. Maquenne, *Ann. Chim. Phys.* **24**, 525 (1891).
65. B. Krieg, R. Schlegel and G. Manecke, *Chem. Ber.* **107**, 168 (1975).
66. P. R. Birkett, C. B. Chapleo and G. Mackenzie, *Synthesis* 822 (1991).
67. G. Cristulli, A. Eleuteri, P. Franchetti, M. Grifantini, S. Vittori and G. Lupidi, *J. Med. Chem.* **34**, 1187 (1991).

REFERENCES

68. H. J. Bestmann and E. Singer, in *Newer Methods of Preparative Organic Chemistry* (ed. W. Foerst). Academic Press, New York, 1964, Vol. 3, p. 487.
69. Y. Iwashita and M. Sakuraba, *J. Org. Chem.* **36**, 3927 (1971).
70. Y. Mori and J. Tsuji, *Tetrahedron* **27**, 4039 (1971).
71. Y. Malvaut, E. Marchand and G. Morel. *J. Org. Chem.* **57**, 2121 (1992).
72. J. F. Hayes, M. B. Mitchell and C. Wicks, *Heterocycles* **38**, 575 (1994).
73. A. P. Black and F. H. Babers, *Org. Synth. Coll. Vol. 2* 512 (1943).

–6–
Syntheses From Other Heterocycles

Such synthetic approaches can only be valid if the "other heterocycles" are readily available, or if their transformations lead to imidazoles difficult to make by other means. It is certainly important to be able to aromatize imidazolines since a number of ring-synthetic procedures lead to reduced imidazoles. 4-Aminoisoxazoles are sources of α-acylaminoenaminones which cyclize with bases to give 4-acylimidazoles. Oxazole–imidazole conversion has largely historical importance, but it is also implicated in some ring-synthetic procedures (e.g. the Bredereck method, see Chapter 5). Transformations of benzofuroxans into 2-substituted benzimidazole N-oxides have some synthetic importance. Few, if any, ring contractions appear to have major application.

Although some conversions of other heterocyclic compounds into benzimidazoles are of little more than academic interest, there may be synthetic value if the reactions occur under neutral conditions (the usual cyclizations of o-phenylenediamines are in acidic media).

6.1 IMIDAZOLES

6.1.1 From three-membered rings

The few examples in which azirines or aziridines ring expand in the presence of a suitable CN synthon do not constitute an important synthetic route to imidazoles, but the high yields frequently reported merit a brief summary here.

2H-Azirines or their salts are converted into imidazoles (**2**) when treated with nitriles, a process which is promoted by perchloric acid [1] or boron trifluoride etherate [2, 3] (Scheme 6.1.1). These reactions are essentially [3 + 2] addition reactions of an intermediate aza-allyl cation (formed by ring opening of the azirine) and the nitrile; the 1,2 and 3,4 bonds are formed during the cyclization. Similar reactions with 1,3,4-thiadiazol-2-ones are carried out in isopropanol under nitrogen at room temperature when R = mesyl, or at 80°C for R = methoxy. Yields of imidazole-2-carboxamides (**1**) (R^4 = NMePh) are in the range 65–96% for a variety of substituents (R^5 = Me, Et, Ph) (Scheme 6.1.1). Parabanic acid reacts similarly with such azirines [4]. The cyclizations of 2,3-diphenyl-2H-azirine with nitriles are carried out at elevated temperatures to give a wide range of 2-substituted 4,5-diphenylimidazoles, e.g. 2-methyl (150°C, 71%), 2-t-butyl (240°C, 42%), 2-phenyl (130°C, 82%), 2-vinyl (240°C, 51%), 2-benzyl (135°C, 79%) [2, 3]. There are, however, more common methods for making such compounds. Photolysis of some azirines leads to nitrile ylides which can be recyclized in the presence of reagents such as ethyl cyanoformate to give imidazoles, but such reactions have little obvious synthetic utility [5, 6].

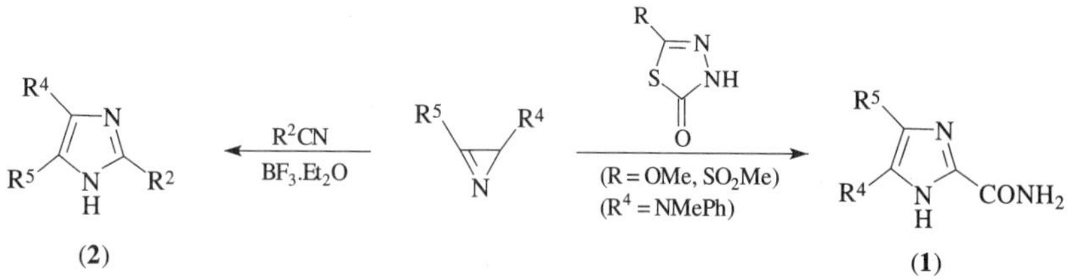

Scheme 6.1.1

6.1.2 From five-membered rings

6.1.2.1 Pyrazoles

Pyrazoles rearrange under photochemical conditions to imidazoles with yields as high as 30%. Frequently, however, mixtures of isomers are formed, decreasing the synthetic utility of the approach. Thus, photolysis of 1,5-dimethylpyrazole gives a mixture of 1,2-, 1,4- and 1,5-dimethylimidazoles [7]; 3-cyanopyrazole gives 2- (25%) and 4-cyanoimidazoles (11%) [8]; 4-cyano-1-methylpyrazole gives 1-methyl-4-cyano- (25%) and 1-methyl-5-cyanoimidazoles (10%) [8, 9]; 3-methylpyrazole gives a mixture of 2- (16%) and 4-methylimidazoles (16%); and 1-methylpyrazole gives 1-methylimidazole (29%) [10]. Control of temperature during the photolyses may be used to restrict isomer formation to some extent [7]. The usual consequence of these photolyses is the interchange of N-2 and N-3 in the pyrazole, but complications of tautomerism arise in N-unsubstituted pyrazoles.

6.1.2.2 Dehydrogenation of imidazolines

There are a number of synthetic procedures which give reduced imidazoles as the primary products, and it is usually not too difficult to aromatize these. For example, 2-alkyl-2-imidazolines can be made by heating acylethylenediamines in the presence of magnesium powder [11]. Barium manganate in dry

6.1. IMIDAZOLES

dioxane serves to dehydrogenate these imidazolines, and it is a useful reagent in a variety of cases [12], but it is also a very toxic compound. Imidazolines have been dehydrogenated with sulfur [13], and by catalytic methods using Raney nickel, platinum or palladium, often at high temperature, in addition to the use of other hydrogen acceptors (selenium, copper(II) oxide, cyclohexanone) [14–16]. The best and most reliable methods use active manganese dioxide [17, 18], permanganate [12, 19] or palladium on carbon [16]. It is important to be able to oxidize 2-alkylimidazolines to 2-alkylimidazoles in an efficient and economical manner because the latter compounds are important as synthetic intermediates in the synthesis of nitroimidazole drugs such as metronidazole, tinidazole and dimetridazole [12]. The use of palladium on carbon is a mild method suitable for converting 2- and 2,4-disubstituted imidazolines into the corresponding imidazoles. Results with this reagent are rather more variable with 4-substituted imidazolines, e.g. 4-phenyl (51%) and 4-benzyl (unsuccessful) [16].

Methyl 1-methylimidazole-4-carboxylate [17]

To a solution of methyl 1-methyl-2-imidazoline-4-carboxylate (7.5 g, 50 mmol) in chloroform (150 ml) is added active manganese dioxide [20, 21] (30 g). The suspension is stirred at room temperature (16 h), filtered and washed with hot chloroform. The filtrate and washings are rotary evaporated to dryness to give essentially pure product (5.32 g, 72%) which can be sublimed ($b_{0.05}$ 80°C) to give pure crystals, m.p. 97–98°C.

2-Methylimidazole [12]

To 2-methyl-2-imidazoline (0.5 g, 5.95 mmol) dissolved in dry dioxane (15 ml) under nitrogen is added, in small portions with stirring, potassium permanganate (1.5 g, 9.5 mmol). The mixture is then refluxed (12 h), allowed to cool to room temperature, and filtered through Celite. The filtrate is evaporated under reduced pressure to a yellowish solid, which is recrystallized from toluene (0.352 g, 72%), m.p. 140–142°C. Similar dehydrogenations of the 2-ethyl- and 2-propyl-2-imidazolines were accomplished with 48 and 50% efficiencies, respectively.

2,4-Diphenylimidazole [16]

A suspension of 10% Pd/C (0.3 g) and 2,4-diphenylimidazoline (0.3 g) in toluene (7.5 ml) is refluxed under argon (40 h). After recovery of the catalyst by filtration, the filtrate is rotary evaporated to give a solid, which on recrystallization from dichloromethane–hexane forms needles (0.2 g, 85%), m.p. 161–162°C.

6.1.2.3 Triazoles and tetrazoles

Sodium hydride in DMF at room temperature induces rearrangement of some 1,4-dialkyl-4H-1,2,4-triazolium salts to 5-aminoimidazoles; the reaction can be generalized to permit synthesis of 4-acyl-5-aminoimidazoles from readily accessible starting materials, but yields are only moderate (21–60%) [22].

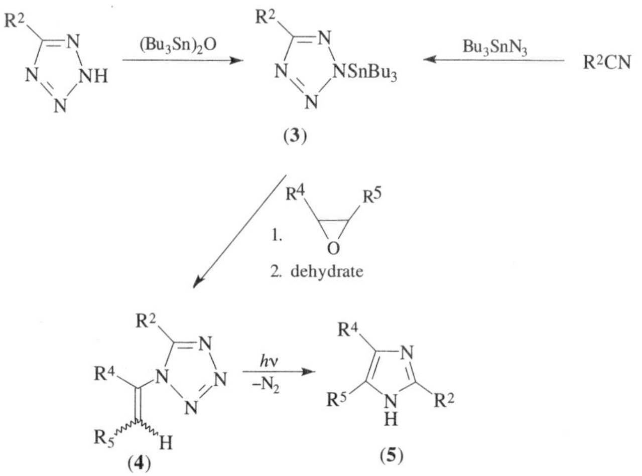

Scheme 6.1.2

Photolysis of 1-vinyltetrazoles (**4**) gives moderate yields of imidazoles, presumably via cyclization of imidoyl nitrenes which are generated when nitrogen is lost (Scheme 6.1.2). The process relies on efficient preparation of 1-vinyltetrazoles from 2-tributylstannyltetrazoles (**3**) (quite easily made from tetrazoles or from nitriles) which are N-1-alkylated at room temperature by ring-opened epoxides. The tributylstannyl group is cleaved by gaseous hydrogen chloride, and the alcohols are readily dehydrated with methyl triphenoxyphosphonium iodide in HMPA or DMF (often preferable) at room temperature, followed by stirring for a few hours with 10% aqueous caustic soda. Irradiation of (**4**) is carried out in a variety of solvents (usually light petroleum (b. 60–80°C) or ethanol), and it has been found that yields are increased if two equivalents of trifluoracetic acid are added to the ethanol. Yields of imidazoles (**5**) are 32–73%, e.g. (**5**) (R^2, R^4, R^5, yield given): H, H, H, 32%; Ph, H, H, 61%; H, Me, H, 62%; Me, Me, H, 73%; Ph, Me, H, 66%; Ph, Ph, H, 66% [23, 24]. Problems with the method include the formation of mixtures of 1- and 2-isomers at the alkylation stage, although the 1-isomer:2-isomer alkylation ratio is usually greater than 4:1, and chromatographic methods readily separate the isomers. The dehydration of the 2-hydroxyalkyl

6.1. IMIDAZOLES

substituents also gives a mixture of vinyl and allyl isomers, but the former usually predominate. Yields of the 1-alkenyltetrazoles (**4**) are usually 69–93% for a wide variety of substituents [24]. The reaction can be adapted to the synthesis of 4-acylimidazoles by reacting a conjugated alkyne, such as ethyl propiolate, with (**3**), giving rise to 1-alkenyltetrazoles bearing a conjugated acyl substituent. When these are irradiated the products are 4-acylimidazoles. This is potentially a useful method since 4-acylimidazoles are often difficult to make by other methods, e.g. (**5**) (R^2, R^4, R^5, yield given): Ph, CO_2Me, H, 61%; Ph, CO_2Me, CO_2Me, 24%; Ph, COMe, H, 40%; Ph, COPr, H, 57%; H, CO_2Me, H, 10%; Me, CO_2Me, H, 63%; CO_2Me, CO_2Me, H, 63% [25].

2-Tri-n-butylstannyltetrazole (**3**) ($R^2 = H$) [24]

A mixture of $1H$-tetrazole (0.212 g, 3.03 mmol) and tri-n-butylstannyl oxide (0.902 g, 1.51 mmol) in ethanol (2.5 ml) is refluxed under nitrogen (2 h), and the solvent is removed to give the product as an oil.

General procedure: ring-opening of epoxides [24]

A solution of (**3**) and the epoxide (1.0–2.5 mol eq.) in diethyl ether (1–2 ml per millimole of tetrazole) is stirred at room temperature until no more (**3**) remains (TLC, ^{1}H NMR). Excess gaseous HCl or glacial acetic acid is added, the mixture is stirred (1–2 h), and the solvent is removed. The products are isolated by column chromatography on silica gel H (type 60) or alumina 60H (basic; type E). It may be necessary to prewash with cold light petroleum (b. 60–80°C) to remove stannyl salts.

1-(2-Hydroxyethyl)tetrazole. The 2-tri-n-butylstannyltetrazole (4.60 g, 12.8 mmol) and ethylene oxide (1.6 ml, 32 mmol) are stirred in ether (10 ml) for 43 h. Quenching with acetic acid (0.73 ml, 12.8 mmol) followed by washing with petroleum, and chromatography on alumina, gives the product (0.415 g, 28%) as an oil.

1-Vinyltetrazole (**4**) (R^2, R^4, $R^5 = H$) [24]

1-(2-Hydroxyethyl)tetrazole (0.690 g, 6.05 mmol) is allowed to react with methyltriphenoxyphosphonium iodide (3.42 g, 7.56 mmol) in DMF (10 ml) at room temperature (4 h). The solution is then poured into 10% aqueous sodium hydroxide and again stirred at room temperature (27 h) before continuous ether extraction and chromatography on silica gel, to give pure 1-vinyltetrazole (0.400 g, 69%).

Photolysis of 1-vinyltetrazole [24]

A solution of 1-vinyltetrazole (0.191 g, 1.99 mmol) in ethanol (150 ml) is irradiated with light of 254 nm in a quartz vessel with nitrogen passing through the solution (40 min). The solvent is removed *in vacuo* and the residue distilled at 75°C/0.33 mmHg to give imidazole (0.043 g, 32%), m.p. 86-89°C. Similarly prepared are 4-methyl- (62%), 2,4-dimethyl- (73%), 2-phenyl- (61%), 4-methyl-2-phenyl- (66%) and 2,4-diphenylimidazoles (66%).

Suitably substituted tetrazolium salts are transformed by aqueous base into 4-imidazolones, e.g. 1,5-dimethyl-4-phenacyltetrazolium bromide, heated in sodium bicarbonate for 15-20 min at 80-100°C gives 1,2-dimethyl-5-phenyl-4-imidazolone in 48% yield [26].

6.1.2.4 Benzimidazoles

A seldom used route to imidazoles which may have the occasional application is that which oxidizes a benzimidazole to give the imidazole-4,5-dicarboxylic acid. Either chromic acid or 30% hydrogen peroxide is effective, but N-substituted benzimidazoles cannot be converted in the same way [27, 28].

Imidazole-4,5-dicarboxylic acid [28]

To concentrated sulfuric acid (70 ml) and water (55 ml) at 90°C is added benzimidazole (5 g), followed by powdered potassium dichromate (37 g). After 10-15 min the reaction is quenched carefully with water. The product is filtered, and washed with water in about 70% yield, m.p. ~ 290°C (dec.). Similar oxidation of 2-methylbenzimidazole gives 2-methylimidazole-4,5-dicarboxylic acid (52%).

6.1.2.5 Isoxazoles

Under conditions of thermolysis, photolysis and hydrogenation certain isoxazoles can be converted into imidazoles. For example, merely heating 5-amino-3,4-dialkylisoxazoles at 180-190°C gives 40-65% yields of 4,5-dialkylimidazolin-2-ones in what initially appears to be a Dimroth-type rearrangement. Compounds such as 4-methyl-5-propyl-, 4-butyl-5-propyl-, 5-benzyl-4-methyl-, 4,5-dimethyl- and 4-ethyl-5-methylimidazoles can be formed in the same way, but only if urea is present and if the reaction is carried out in the condensed phase. Without the added urea (or an arylamine) the yields are only 40-65%; with urea they reach 70-90% [29-31].

Of rather more synthetic importance are reactions which hydrogenate amides of 4-aminoisoxazoles (**6**) to form α-(acylamino)enaminones (**7**) which cyclize in the presence of bases to give 4-acylimidazoles (**8**) (Scheme 6.1.3). In terms of the final cyclization step, this reaction could well be classified under 1,2 bond formation (see Section 2.1.1). Its importance lies in its general applicability to the synthesis to 4-acylimidazoles (Tables 6.1.1 and 6.1.2); indeed, it

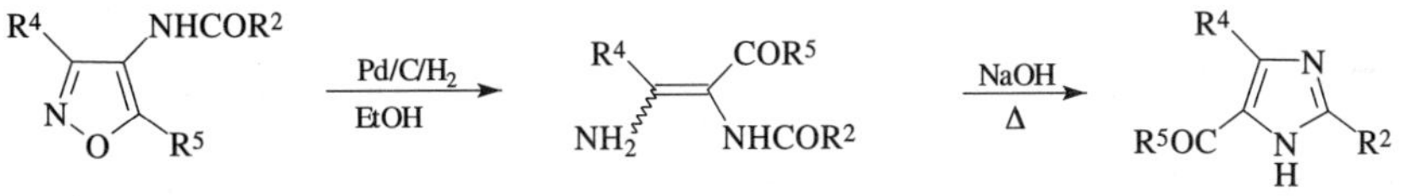

6.1. IMIDAZOLES

Scheme 6.1.3

TABLE 6.1.1
4(5)-Acylimidazoles (**8**) made from isoxazole amides (**6**) [32, 33]

R^2	R^4	R^5	Yield (%)
H	H	H	79
Me	H	H	45
H	H	Me	98
Me	H	Me	78
CH_2OMe	H	Me	70
CH_2NMe_2	H	Me	52
Et	H	Me	93
iPr	H	Me	95
tBu	H	Me	98
Ph	H	Me	80
CF_3	H	Me	83
H	Me	H	86
Me	Me	H	74
Ph	Me	H	67
Me	Ph	Ph	91

TABLE 6.1.2
1-Substituted 4(and 5)-acylimidazoles [32]

R^1	R^2	R^4	R^5	Yield (%)
CH_2Ph	H	H	H	53
Me	Me	H	Me	55
CH_2Ph	Me	H	Me	62
CH_2Ph	H	Me	H	85
CH_2Ph	Me	Me	H	92
CH_2Ph	H	CHO	H	66
CH_2Ph	Me	CHO	Me	18

appears to be the method of choice. The required 4-aminoisoxazole starting materials can be made quite simply by reduction of 4-nitroisoxazoles (made by direct nitration of isoxazoles [32]). They are then acylated, hydrogenated, and the open chain products (**7**) heated with base, usually without prior isolation. If the amino group of the original 4-aminoisoxazole carries a substituent, the ultimate products will be 1-substituted-5-acylimidazoles, a substitution pattern not readily achieved by alternative methods (except via 5-lithioimidazoles, and from some 4-acylimidazoles by sequential N-acylation, N-alkylation and deacylation of the 3-acyl-1-alkylimidazolium salt [32]). The 2-substituent is derived from the acylating agent, while the substituents at C-4(5) and on the adjacent acyl group come from the original 3- and 5-substituents of the isoxazole. The 1-substituent is limited only by what groups can be placed on the amino group of the 4-aminoisoxazole. Simple alkyl and aralkyl groups can be readily introduced, but hindered alkyl and aryl groups will cause some difficulty. The method can be adapted to give 1,4 regiochemistry of the acyl group if C-5 of the imidazole is unsubstituted. This is similar to the usual product of alkylation of a 4(5)-acylimidazole (see Section 7.2). The nitrogen substituent in this instance must be derived from a primary amine sufficiently nucleophilic to add to the β-position of a β-amino-α,β-unsaturated ketone. Only 1,5-disubstituted and 1,2,5-trisubstituted 4-acylimidazoles cannot be made using this chemistry [32, 33].

N-(5-Methyl-4-isoxazolyl)acetamide (**6**) *($R^4 = H$; R^2, $R^5 = Me$)* [32]

4-Amino-5-methylisoxazole hydrochloride (1.35 g, 10 mmol) is treated with sodium acetate (0.820 g, 10 mmol) and excess acetic anhydride (5 ml) in glacial acetic acid (45 ml) at room temperature (2 h). Concentration at reduced pressure gives a residue which is taken up in diethyl ether, dried (MgSO$_4$) and filtered. The filtrate is rotary evaporated before removal of traces of acetic acid under high vacuum to give a white solid (1.38 g, 98%), m.p. 94–97°C.

General procedure for isoxazole–imidazole interconversion [32]

The 4-(acylamino)isoxazole (**6**) is hydrogenated at 40 psi (580 bar) over 10% palladium on carbon (25–50% w/w) in ethanol (~10 ml per millimole of reactant). After about 1 h the reaction is usually complete (TLC, MeOH–CHCl$_3$ (1:9)), when the catalyst is filtered off and washed with ethanol. If the reaction is found to be incomplete after 1 h a second portion of 10% palladium on carbon is added and hydrogenation is continued for another hour. The filtrate containing (**7**) is treated with sodium hydroxide pellets (1.1 eq.) at reflux (1 h), solid ammonium chloride (1.2 eq.) is added, the reaction is allowed to cool to ambient temperature, and the ethanol is removed *in vacuo*. The residue is slurried in acetone, the mixture is filtered, and the filtrate is concentrated to give the crude product, usually as a solid. Flash chromatography [34] or recrystallization gives the imidazoles (**8**). (See Table 6.1.1.)

6.1. IMIDAZOLES

N-*(4-Isoxazolyl)formamide* [32]

4-Aminoisoxazole hydrochloride (1.33 g, 11 mmol) is refluxed (24 h) in a mixture of formic acid (11 ml) and ethyl formate (110 ml). After concentration of the mixture *in vacuo* the resulting oil is diluted with toluene and re-evaporated to give a dark tan solid. Flash chromatography [34] (MeOH–CHCl$_3$, 5:95) gives the product (1.09 g, 89%), m.p. 125–129°C. Further purification can be achieved by recrystallization from cyclohexane–ethyl acetate.

Imidazole-4-carbaldehyde (**8**) *(R^2, R^4, $R^5 = H$)* [32]

This is made from *N*-(4-isoxazolyl)formamide by the above general procedure with the modification that the intermediate α-amino-α,β-unsaturated aldehyde (**7**) (R^2, R^4, $R^5 = H$) precipitates. In consequence, the catalyst is not filtered off until after the sodium hydroxide treatment and ammonium chloride quench. The yield after flash chromatography [34] (MeOH–CHCl$_3$, 1:9) is 79%, m.p. 170–171°C.

4-(N-Benzylamino)isoxazole [32]

4-Aminoisoxazole hydrochloride (0.603 g, 5 mmol), sodium acetate (0.410 g, 5 mmol) and benzaldehyde (0.584 g, 5.5 mmol) are dissolved together in methanol (50 ml). After 15 min at room temperature, sodium cyanoborohydride (0.346 g, 5.5 mmol) is added in one portion and the mixture is allowed to stand (4 h) before concentration *in vacuo*. The residue is taken up in saturated sodium bicarbonate solution (25 ml), and extracted with chloroform (3 × 20 ml). The extracts are dried (MgSO$_4$), filtered and rotary evaporated. The residue, dissolved in ethanol, is treated with 12 M hydrochloric acid (1 ml) and evaporated to give a solid which recrystallizes from isopropanol (10 ml) as off-white crystals (0.242 g, 23%), m.p. 159–161°C. A second crop can be obtained on concentration of the filtrate, drying the residue under high vacuum, and slurrying it in ether to yield further product (0.646 g, 61%).

N-*Benzyl*-N-*(4-isoxazolyl)formamide* [32]

This is prepared from 4-(*N*-benzylamino)isoxazole by the same procedure as for *N*-(4-isoxazolyl)formamide above. Thus, reaction of 4-(*N*-benzylamino)isoxazole (0.843 g, 4 mmol) gives, after flash chromatography [34] (ethyl acetate–isopropyl ether, 2:8), a yellow solid (0.714 g, 88%), m.p. 71–73°C.

1-Benzylimidazole-5-carbaldehyde [32]

This is prepared from *N*-benzyl-*N*-(4-isoxazolyl)formamide by the general procedure to give a yellow oil (after flash chromatography) which solidifies on standing (53%). The product is recrystallized from pentane, m.p. 53–55°C.

1-Benzylimidazole-4-carbaldehyde [32]

N-(4-Isoxazolyl)formamide (0.672 g, 6 mmol) is hydrogenated at 45 psi (650 bar) over 10% palladium on carbon (0.350 g) in methanol (60 ml) at room temperature. If after 1 h the reaction is incomplete, a second portion of catalyst is added and hydrogenation is repeated (1 h). The catalyst is filtered off, washed well with methanol, and the filtrate is treated with benzylamine (6.43 g, 60 mmol) and stirred at room temperature (20 h). Sodium hydroxide (0.264 g, 6.6 mmol) is added, and the mixture is warmed to reflux and maintained at that temperature (2 h), before addition of ammonium chloride (0.385 g, 7.2 mmol). After cooling to room temperature the methanol is rotary evaporated, and the residue is dissolved in ethyl acetate (50 ml) and washed with saturated ammonium chloride solution (5 × 25 ml) to remove unreacted benzylamine. The ethyl acetate solution is dried ($MgSO_4$), filtered and concentrated *in vacuo* to an orange oil which still contains a considerable quantity of the benzylimine of the desired carbaldehyde. Thus, the oil is dissolved in 3 M hydrochloric acid (10 ml) and stirred at room temperature (16 h), then neutralized with solid sodium carbonate, and extracted with ethyl acetate (3 × 25 ml). The extract is dried, filtered and again concentrated to an orange oil, which on flash chomatography [34] (ethyl acetate–isopropyl ether, 1:1 and then ethyl acetate) gives the product (0.738 g, 66%) as a pale yellow oil.

4-Acetylimidazole (**8**) *(R^2, R^4 = H; R^5 = Me)* [32]

4-Amino-5-methylisoxazole hydrochloride (0.777 g, 5.77 mmol) is heated at 90°C (4 h) in triethyl orthoformate (6 ml). After standing overnight at room temperature, the solvent is removed and the residue is flash chromatographed (methanol–chloroform, 2.5:97.5) to give methyl *N*-(5-methyl-4-isoxazolyl)methanimidoate (0.421 g, 52%) as a yellow oil. This imidate is then hydrogenated at 15 psi (220 bar) over 10% palladium on carbon (0.100 g) in methanol. After 0.5 h and 2.5 h additional portions (0.100 g) of catalyst are added. At the end of 3.5 h of hydrogenation the catalyst is removed by filtration and the filtrate is refluxed (18 h). Concentration of the mixture *in vacuo* and flash chromatography (methanol–chloroform,1:9) of the residue gives 4-acetylimidazole (0.078 g, 23%) as an off-white solid, m.p. 153–156°C.

6.1. IMIDAZOLES

N-*(5-Methyl-4-isoxazolyl)acetamide* **(6)** *($R^4 = H$; R^2, $R^5 = Me$)* [32]

4-Amino-5-methylisoxazole hydrochloride (1.35 g, 10 mmol) is treated with sodium acetate (0.820 g, 10 mmol) and excess acetic anhydride (5 ml) at room temperature (2 h). The mixture is then concentrated under reduced pressure, and the residue is taken up in diethyl ether, dried (MgSO$_4$), filtered, and the filtrate concentrated *in vacuo* before removal of traces of acetic anhydride under high vacuum. The product is isolated as a white solid (1.38 g, 98%), m.p. 94–97°C.

4-Acetyl-2-methylimidazole **(8)** *($R^4 = H$; R^2, $R^5 = Me$)* [32]

This is made by the general procedure above in 37% yield, m.p. 128–130°C, recrystallized from ethyl acetate–isopropyl ether.

1,4-Diacetyl-2-methylimidazole [32]

4-Acetyl-2-methylimidazole (2.50 g, 20.1 mmol) and triethylamine (2.24 g, 22.1 mmol) are combined in chloroform (100 ml), and acetyl chloride (1.74 g, 22.1 mmol) is added dropwise. After 4.5 h the reaction mixture is washed with water (3 × 50 ml) and dried (MgSO$_4$). Filtration and concentration of the filtrate gives a light yellow solid (2.39 g, 71%). ^{1}H NMR: 2.50 (s, 3H), 2.60 (s, 3H), 2.65 (s, 3H), 7.77 (s, 1H).

5-Acetyl-1,2-dimethylimidazole [32]

The above crude product (2.39 g, 14.4 mmol) is dissolved in chloroform (30 ml) and stirred at room temperature (24 h) with trimethyloxonium tetrafluoroborate (2.34 g, 15.8 mmol). A further portion (1.17 g, 7.9 mmol) of methylating agent is then added. After 20 h the mixture is concentrated *in vacuo* and the residue is dissolved in water. The solution is made basic with solid sodium carbonate, and extracted with chloroform (5 × 25 ml). The combined extracts are dried (MgSO$_4$), concentrated to a brown oil, and flash chromatographed (methanol–chloroform, 5:95) to give a pale yellow solid (1.30 g, 65%), m.p. 80.5–81.5°C. The same product can also be made in 55% yield rather more directly from 5-methyl-4-(methylamino)isoxazole.

When 3,4-disubstituted 4-aminoisoxazolin-5(4*H*)-ones **(9)** are hydrogenated, and then subsequently recyclized, high yields of imidazoles **(10)** are obtained (Scheme 6.1.4). The starting materials can be readily made from isoxazolin-5(4*H*)-ones by sequential bromination, reaction with sodium azide and conversion of azide into amino by reaction with hydrobromic acid in acetic acid. The amines are then converted into formamidine or acetamidine derivatives with DMF or dimethylacetamide under reflux in phosphoryl chloride–chloroform. Yields at this stage are 50–93%. Catalytic hydrogenolysis of the N—O bond is followed by decarboxylation, cyclization and elimination of dimethylamine under the influence of hydrobromic and acetic acids. The recyclization ultimately forms the 1,2 bond. A wide variety of imidazoles (**10**) with R^2, R^4 = H, alkyl and R^5 = aryl, alkyl can be made by this method, though the 2-substituent seems limited to hydrogen or methyl at this stage [35].

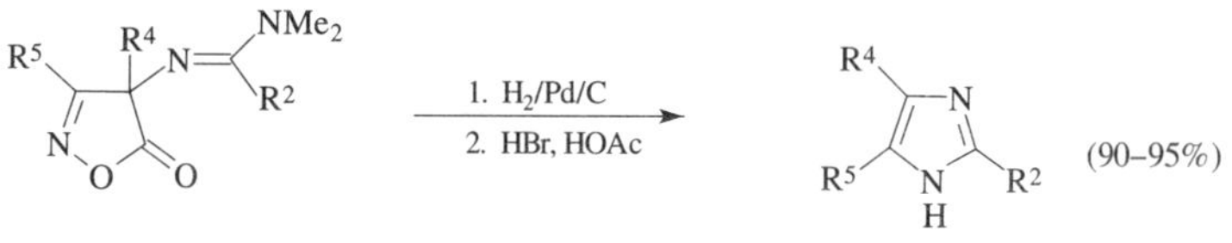

Scheme 6.1.4

6.1.2.6 Oxazoles and oxazolines

There is a long history of oxazole–imidazole interconversion [36]. The oxazoles can be heated with ammonia, formamide, hydrazine (to give 1-aminoimidazoles) and primary amines, often in the presence of a Brønsted acid [37–39]. Yields are improved if formamide is used rather than ammonia, or with mixtures of these reagents at elevated temperatures (140–210°C) in an autoclave [36, 40]. The reaction fails with 2,4,5-triethyloxazole and with benzoxazole, rendering it normally unsuitable for benzimidazole synthesis (but see below). Aliphatic groups in the 2- and 5-positions of oxazole apparently cause some steric hindrance to the reaction since such substrates react only with difficulty with formamide at 140°C, but elevation of the temperature to 200–210°C and the use of a mixture of formamide and ammonia seems to deal satisfactorily with such recalcitrant oxazoles [40, 41]. Aryl and halogenoalkyl groups in the 4 position assist the transformations presumably by lowering electron density at the annular oxygen. The mechanism probably involves amine attack at C-2 of the oxazole with ring opening of the C(2)—O bond [15, 42, 43].

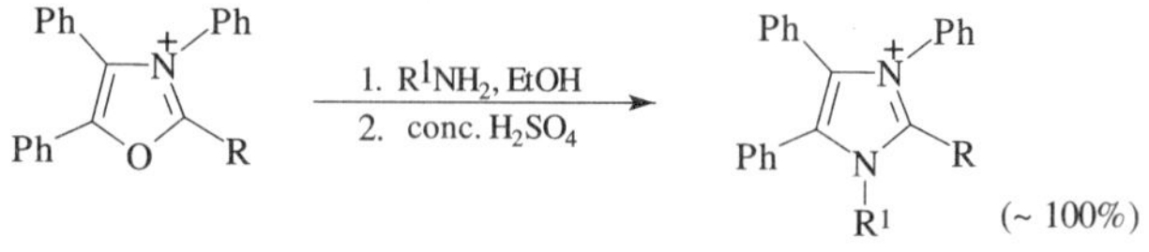

Scheme 6.1.5

Oxazolium salts, too, are similarly converted into imidazoles under relatively mild conditions (Scheme 6.1.5) [44–47]. Simple primary aliphatic

6.1. IMIDAZOLES

amines need only to be stirred in ethanol for half an hour to form the 4,5-dihydroimidazolium salt, which is aromatized by treatment with concentrated sulfuric acid. More hindered amines react more slowly to form open-chain benzamidines which are readily cyclized by the acid [44]. 4-Carbomethoxyoxazoles are also a source of α-acyl-α-amino acid esters, which react with thiocyanates to give imidazoles [48] (see Section 4.1).

Imidazoles can also be made by heating 4-tosyloxazolines in saturated methanolic ammonia or monoalkylamines. These reactions proceed through intermolecular condensation of α-aminoketones and amidines and intramolecular cyclization of α-amidinoketones, respectively [49] (see Section 4.2). When N-unsubstituted 4-oxazolin-2-ones are added to isocyanates, the 2-oxo-4-oxazoline-3-carboxamide products cleave under the influence of strong acids and heat. Subsequent ring closure gives 4-imidazolin-2-ones in good yields (44-98%) [50]. Base-catalysed rearrangement of 5-methylaminobenzoxazole gives 4-hydroxy-1-methylbenzimidazole (85%). This reaction depends on ring-opening of the C—O bond and recyclization with the 4-methylamino group [51].

2,4,5-Triphenylimidazole [41]

2,4,5-Triphenyloxazole (5 g), formamide (40 g) and liquid ammonia (100 ml) are heated in an autoclave at 200-210°C (5 h). The brownish reaction product is poured into water, and the flocculent precipitate is filtered, washed with water and recrystallized from ethanol (4.3 g, 85%), m.p. 273°C. Similarly prepared are 2,5-diethyl-4-phenyl- (25%), 2-methyl-4,5-dipropyl- (70%) [41], 4-ethyl-5-phenyl- (50%), 4-phenyl-5-propyl- (40%) and 4-benzyl-5-ethyl imidazoles (5%) [40]. From 2-methyloxazole-4-carboxylic acid boiled at 150°C in a sealed tube with aqueous ammonia is obtained 2-methylimidazole (22%); boiling with aniline gives 2-methyl-1-phenylimidazole (67%) [52].

3-Benzyl-1,4,5-triphenylimidazolium perchlorate [44]

Benzylamine (0.28 g, 2.5 mmol) is added dropwise to a suspension of 3,4,5-triphenyloxazolium perchlorate (0.56 g, 1.4 mmol) in ethanol at 20°C. The solution is stirred (2 h), then evaporated *in vacuo*. The resulting gum is warmed with concentrated sulfuric acid (15 min) before addition of cold water, to precipitate the product, which is recrystallized from isopropanol (40%), m.p. 207-212°C.

6.1.2.7 Thiazoles

5-Amino-1,3-thiazoline-2-thiones undergo Dimroth rearrangement to give imidazole-2,5-dithiols in a reaction which does not appear to have much synthetic potential [53]. A similar rearrangement of 5-aminothiazoles in a basic medium gives imidazole-5-thiols by electrocyclic ring opening and ring closure of the dianion which forms [54]. With mild alkali, 5-amino-2-methylaminothiazoles give 5-amino-1-methylimidazolin-2-thiones, e.g. in 10% aqueous sodium carbonate at reflux for a few minutes 5-amino-2-methylaminothiazole is converted into 5-amino-1-methylimidazolin-2-thione (65%). In contrast, 2-acylamino-5-aminothiazoles rearrange to 5-acylaminoimidazolin-2-thiones [55]. Aryl isocyanates convert 2,4-disubstituted thiazole N-oxides into imidazole-5-thiols (the 2,5 isomers are unreactive). This latter reaction has some application for making imidazole-5-thiols; the initial reaction products are bis-(5-imidazolyl)disulfides, which are converted by lithium aluminium hydride into the thiols. Air oxidation reverses the process [56]. Reaction of a 5-amino-4-tosylthiazole with more than one equivalent of butyllithium gives the dianion, which rearranges and reacts with acid to give a 4-tosylimidazole-5-thiol in 85–95% yield [56].

6.1.2.8 Oxadiazoles or thiadiazoles

Ring transformation of 2-amino-3-phenacyl-1,3,4-oxadiazolium halides with amines, liquid ammonia or heterocyclic bases gives 1-acylamino-2-amino-4-arylimidazoles [57]. When treated with an α-halogenoketone in ethanol 2,5-diamino-1,3,4-thiadiazoles form salts, which react with hydrazine hydrate to give 1-aminoimidazolin-2-thiones [58].

6.1.3 Ring contractions

There are considerable data available on imidazole formation by ring contractions of pyrimidines, pyrazines and triazines [15, 43, 59–61]. Few of the reactions, however, have synthetic potential except perhaps for the thermolytic conversions of azidopyrimidines and azidopyrazines into 1-cyano-substituted imidazoles, and the reactions of chloropyrimidines and chloropyrazines with potassium amide in liquid ammonia to give 4- and 2-cyanoimidazoles, respectively. Ring contractions of quinoxaline 1-oxides may also have some applications.

6.1.3.1 Pyrimidines

When treated with potassium amide in liquid ammonia, suitably substituted and activated pyrimidines are converted into imidazoles, a ring contraction which has been shown by ^{14}C labelling to involve cleavage of the 5,6 bond (Scheme 6.1.6). 5-Amino-4-chloro-2-phenylpyrimidine is converted under the reaction conditions into 4-cyano-2-phenylimidazole in 30–35% yield, rather better than is achieved by Cornforth's method (using aminoacetonitrile with

6.1. IMIDAZOLES

an imidate [62]). 4-Chloro-5-methyl-2-phenylpyrimidine is similarly converted into 4-ethynyl-2-phenylimidazole (30–35%) (Scheme 6.1.6) [63, 64].

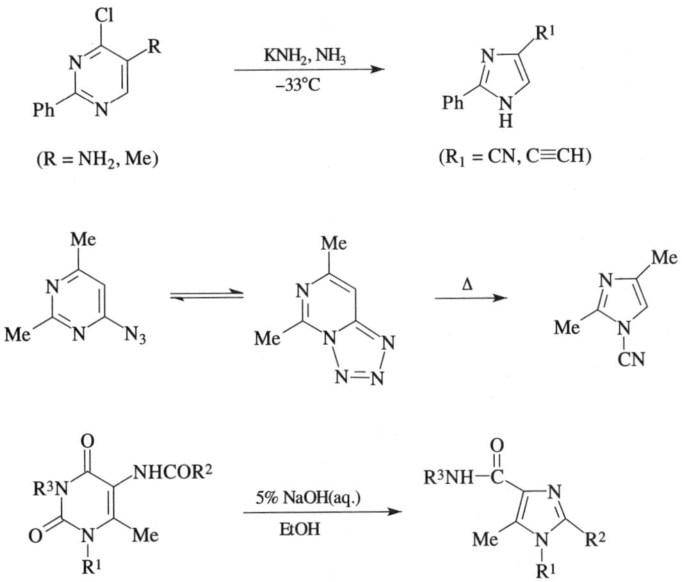

Scheme 6.1.6

When 4- or 6-azidopyrimidines are pyrolysed in the gas phase they are transformed into 1-cyanoimidazoles, usually in 80–90% yields. Since the azidopyrimidines are easily prepared from the chloro precursors, this method has some synthetic interest. Examples of 1-cyanoimidazoles made in this way include the 2,4-dimethoxy (12%) and 4-methyl-2-methylthio (88%) derivatives (Scheme 6.1.6) [65–67].

The few examples of photochemical ring contractions of 1,3-diazines into imidazoles have no obvious synthetic importance [68].

Ethanolic sodium hydroxide converts 5-acylaminopyrimidin-4(3H)-ones (including acylated uracils) into 4-carbamoylimidazoles (**11**) in quite good yields (Scheme 6.1.6), e.g. (**11**), (R^1, R^2, R^3, yield given): Ph, Me, Me, 81%; Me, Ph, Me, 70%; Me, Ph, Ph, 44%; Me, Me, Me, 44%; Ph, Me, Ph, 34%. If there is no R^1 substituent on the uracil, N-unsubstituted (**11**) are formed (51–84%) [69, 70].

A variety of oxidative metal salts in alcoholic solution are capable of converting 5-amino-6-methyl-3-phenylpyrimidin-4(3H)-one into 2-alkoxy-imidazoles (51–71% yields) [71, 73].

2,5-Dimethyl-4-methylcarbamoyl-1-phenylimidazole (**11**) *($R^1 = Ph$; R^2, $R^3 = Me$)* [70]

A mixture of 5-acetamido-3,6-dimethyl-1-phenyluracil (0.500 g, 1.83 mmol), 5% aqueous sodium hydroxide (5 ml), and ethanol (30 ml) is refluxed (3 h), then cooled and neutralized with 5% aqueous hydrochloric acid solution. Extraction with chloroform, drying the extracts (MgSO$_4$), and concentration *in vacuo* gives the product, which is recrystallized from benzene (0.339 g, 81%), m.p. 154–156°C.

2-Methoxy-5-methyl-4-phenylcarbamoylimidazole (**11**) *($R^1 = H$, $R^2 = OMe$, $R^3 = Ph$)* [73]

Anhydrous zinc chloride (0.2 mol eq.) and phenyliodine(III) diacetate [72] (1.2 mol eq.) are added to a solution of 5-aminopyrimidin-4(3*H*)-one (0.100 g, 0.5 mmol) in dry methanol (10 ml), and the mixture is stirred (1 h) at room temperature. The reaction mixture is then poured into water and extracted with dichloromethane. The extracts are dried (MgSO$_4$), filtered and evaporated. The residue is purified by column chromatography on silica gel using hexane–ethyl acetate (2:1) to give the product (0.118 g, 57%).

6.1.3.2 Pyrazines

The photochemical rearrangement of pyrazine *N*-oxides and 2,3-dihydropyrazines to imidazoles seems to have no useful synthetic application [74–76]. Of more interest are the conversions of 2-azidopyrazines and their *N*-oxides. Thus, 2-azidopyrazine 1-oxide heated in benzene gives 2-cyano-1-hydroxyimidazole (83%); 2-azidopyrazine heated at 220°C in acetic acid gives 1-cyanoimidazole (70%) [77–81]. The reactions probably proceed through nitrene intermediates. Azidopyrazines are readily made by heating the chloro analogues with sodium azide in DMF; yields are usually good (62–81%). Photochemical transformations give lower yields and mixtures.

Chloropyrazines undergo ring contraction when treated with potassium amide in liquid ammonia. The reactions resemble those of pyridines and pyrimidines, and may be useful for making 2-cyanoimidazoles despite the formation of product mixtures and low yields. For example, 2-chloropyrazine is converted into a mixture of 2-aminopyrazine (15%), imidazole (14–15%) and 2-cyanoimidazole (30–36%) [82]. In a similar experiment, 2-cyano-4-phenylimidazole (19%) can be obtained [83].

2,5-Dialkyl-1-cyanoimidazoles [79, 80]

A 2-azido-3,6-dialkylpyrazine (3–5 g) is heated at 240°C on a metal bath (30–60 s). After cessation of nitrogen evolution, the crude 1-cyanoimidazole is distilled under reduced pressure as a pale yellow solid or colourless oil. Prepared in this general way are the following 1-cyanoimidazoles (2-, 4- and 5-substituents, yield given): Me, H, Me, 89%; Et, H, Et, 93%; Pr, H, Pr, 91%; iPr, H, iPr, 96%; Bu, H, Bu, 90%; sBu, H, sBu, 88%; H, Ph, Me, 92%; Me, H, Ph, 89%; Ph, H, Me, 74%; Me, Ph, H, 90%; Ph, Me, H, 76%; H, Ph, Ph, 92%; Ph, H, Ph, 91%; Ph, Ph, H, 91%.

2,5-Dialkyl-4-chloro-1-cyanoimidazoles [81]

A monoazidomonochloropyrazine (0.100 g) is heated in a test tube at 220°C in a metal bath. The reaction is complete in a moment with explosive generation of nitrogen. After cooling, the product is purified by silica gel column chromatography (5 g), eluting with hexane containing an increasing amount of ethyl acetate. Made in this way are the following 4-chloro-1-cyanoimidazoles: 2,5-dimethyl (80%), 2,5-diethyl (39%), 2,5-dipropyl (77%), 2,5-dibutyl (68%), 2,5-diisobutyl (89%) and 2,5-diphenyl (61%).

6.1.3.3 Other six- and seven-membered rings

Under reducing regimes some 1,3,5- and 1,2,4-triazines ring contract to imidazoles [84–86], but there is no obvious synthetic importance. When α-halogenooximes react with amidines, they appear to form 4H-1,2,5-oxadiazines, which are deoxygenated by iron carbonyls to give 2,4-disubstituted imidazoles in good yields [87]. Thermal rearrangement of 1H-1,4-diazepine-7(6H)-thiones occurs on brief heating at 120°C in DMSO to give 1-vinylimidazole-2(3H)-thiones in 92–95% yields [88].

REFERENCES

1. P. Beak and J. L. Miesel, *J. Am. Chem. Soc.* **89**, 2375 (1967).
2. H. Bader and H. J. Hansen, *Chimia* **29**, 264 (1975).
3. H. Bader and H. J. Hansen, *Helv. Chim. Acta* **61**, 286 (1978).
4. J. M. Villalgordo, A. Linden and H. Heimgartner, *Helv. Chim. Acta* **75**, 2270 (1992).
5. B. Jackson, H. Märky, H. J. Hansen and H. Schmidt, *Helv. Chim. Acta* **55**, 919 (1972).
6. V. P. Semenov, A. N. Studenikov and A. A. Potekhin, *Chem. Heterocycl. Compd (Engl. Transl.)* **15**, 467 (1979).
7. R. E. Connors, D. S. Burns, E. M. Kurzweil and J. W. Pavlik, *J. Org. Chem.* **57**, 1937 (1992).
8. J. A. Barltrop, A. C. Day, A. G. Mack, A. Shahrisa and S. Wakamatsu, *J. Chem. Soc., Chem. Commun.* 604 (1981).
9. S. Wakamatsu, J. A. Barltrop and A. C. Day, *Chem. Lett.* 667 (1982).
10. H. Tiefenthaler, W. Dörscheln, H. Goth and H. Schmid, *Helv. Chim. Acta* **50**, 2244 (1967).
11. H. C. Chitwood and E. E. Reid, *J. Am. Chem. Soc.* **57**, 2424 (1935).
12. J. L. Hughey, S. Knapp and H. Schugar, *Synthesis* 489 (1980).
13. E. Jassmann and H. Schulz, *Pharmazie* **18**, 461 (1963).

14. R. E. Klem, H. F. Skinner, H. Walba and R. W. Isensee, *J. Heterocycl. Chem.* **7**, 403 (1970).
15. M. R. Grimmett, *Adv. Heterocycl. Chem.* **12**, 103 (1970).
16. Y. Amemiya, D. D. Miller and F.-L. Hsu, *Synth. Commun.* **20**, 2483 (1990).
17. P. K. Martin, H. R. Matthews, H. Rapoport and G. Thyagarajan, *J. Org. Chem.* **33**, 3758 (1968).
18. E. Duranti and C. Balsamini, *Synthesis* 815 (1974).
19. M. E. Campos, R. Jiménez, F. Martinez and H. Salgado, *Heterocycles* **40**, 841 (1995).
20. J. Attenburrow, A. F. B. Cameron, J. H. Chapman, R. M. Evans, B. A. Hems, A. B. A. Jansen and T. Walker, *J. Chem. Soc.* 1094 (1952).
21. L. F. Fieser and M. Fieser, in *Reagents for Organic Synthesis*. Wiley, New York, 1967, p. 637.
22. C. N. Rentzea, *Angew. Chem., Int. Ed. Engl.* **25**, 652 (1986).
23. M. Casey, C. J. Moody and C. W. Rees, *J. Chem. Soc., Chem. Commun.* 714 (1982).
24. M. Casey, C. J. Moody and C. W. Rees, *J. Chem. Soc., Perkin Trans. 1* 1933 (1984).
25. M. Casey, C. J. Moody, C. W. Rees and R. G. Young, *J. Chem. Soc., Perkin Trans. 1* 741 (1985).
26. D. Moderhack and A. Lembke, *Chem.-Ztg* 109,432 (1985); *Chem. Abstr.* **106**, 18 433 (1987).
27. H. V. Euler, H. Hasselquist and O. Heidenberger, *Arkiv Kemi* **14**, 419 (1958); *Chem. Abstr.* **54**, 12 156 (1960).
28. L. S. Efros, N. V. Khromov-Borisov, L. R. Davidenkov and M. M. Nedel, *Zh. Obshch. Khim.* **26**, 455 (1956); *Chem. Abstr* **50**, 13 881 (1956).
29. H. Kano, *Yakugaku Zasshi*, **72**, 150 (1952); *Chem. Abstr.* **46**, 11 180 (1952).
30. H. Kano, *Yakugaku Zasshi* **72**, 1118 (1952); *Chem. Abstr.* **47**, 6936 (1953).
31. T. Nishiwaki, *Synthesis* 20 (1975).
32. L. A. Reiter, *J. Org. Chem.* **52**, 2714 (1987).
33. L. A. Reiter, *Tetrahedron Lett.* **26**, 3423 (1985).
34. W. C. Still, M. Kahn and A. Mitra, *J. Org. Chem.* **43**, 2923 (1978).
35. E. M. Beccalli, A. Marchesini and T. Pilati, *Synthesis* 127 (1991).
36. T. Nishiwaki, *Synthesis* 20 (1975).
37. W. Hafner and H. Prigge, German Patent 1,923,643 (1970); *Chem. Abstr.* **74**, 22 838 (1971).
38. K. Fitzi, Swiss Patent 561,718 (1975); *Chem. Abstr.* **83**, 131 598 (1975).
39. H. Sasaki and T. Kitagawa, *Chem. Pharm. Bull.* **36**, 1593 (1988).
40. H. Bredereck, R. Gompper and F. Reich, *Chem. Ber.* **93**, 723 (1960).
41. H. Bredereck, R. Gompper and H. Wild, *Chem. Ber.* **88**, 1351 (1955).
42. A. F. Pozharskii, A. D. Garnovskii and A. M. Simonov, *Russ. Chem. Rev. (Engl. Transl.)* **35**, 122 (1966).
43. M. R. Grimmett, in *Comprehensive Heterocyclic Chemistry* (ed. A. R. Katritzky and C. W. Rees). Pergamon Press, Oxford, 1984, Vol. 5 (ed. K. T. Potts), p. 457.
44. A. R. Katritzky and A. Zia, *J. Chem. Soc., Perkin Trans. 1* 131 (1982).
45. Y. Kikugawa and L. A. Cohen, *Chem. Pharm. Bull.* **24**, 3205 (1976).
46. R. Gompper, *Chem. Ber.* **90**, 374 (1957).
47. A. Hetzheim, O. Peters and H. Beyer, *Chem. Ber.* **100**, 3418 (1967).
48. S. Maeda, M. Suzuki, T. Iwasaki, K. Matsumoto and Y. Iwasawa, *Chem. Pharm. Bull.* **32**, 2536 (1984).
49. D. A. Horne, K. Yakushijin and G. Büchi, *Heterocycles* **39**, 139 (1994).
50. B. Krieg and H. Lautenschläger, *Liebigs Ann. Chem.* 208 (1976).
51. A. R. Katritzky, R. P. Musgrave, B. Rachwal and C. Zaklika, *Heterocycles* **41**, 345 (1995).
52. J. W. Cornforth and R. H. Cornforth, *J. Chem. Soc.* 96 (1947).
53. G. L'abbé, W. Meutermans and M. Bruynseels, *Bull. Soc. Chim. Belg.* **95**, 1129 (1986).

REFERENCES

54. S. P. J. M. van Nispen, J. H. Bregman, D. G. van Engen, A. M. van Leusen, H. Saikachi, T. Kitagawa and H. Sasaki, *J. R. Neth. Chem. Soc.* **101**, 28 (1982).
55. A. H. Cook, J. D. Downer and I. Heilbron, *J. Chem. Soc.* 2028 (1948).
56. N. Honjo, T. Niiya and Y. Goto, *Chem. Pharm. Bull.* **30**, 1722 (1982).
57. A. Hetzheim, O. Peters and H. Beyer, *Chem. Ber.* **100**, 3418 (1967).
58. A. Sitte, H. Paul and G. Hilgetg, *Z. Chem.* **7**, 341 (1967).
59. M. R. Grimmett, *Adv. Heterocycl. Chem.* **27**, 241 (1980).
60. M. R. Grimmett, in *Comprehensive Heterocyclic Chemistry*. (ed. A. R. Katritzky and C. W. Rees), Elsevier, Oxford, 1996, Vol. 3.02 (ed. I. Shinkai) p. 77.
61. H. C. van der Plas, in *Ring Transformations of Heterocycles*. Academic Press, New York, 1973, Vol. 2.
62. J. W. Cornforth, E. Fawaz, L. J. Goldsworthy and R. Robinson, *J. Chem. Soc.* 1549 (1949).
63. H. W. van Meeteren and H. C. van der Plas, *Recl. Trav. Chim. Pays-Bas, Belg.* **87**, 1089 (1968).
64. H. W. van Meeteren, H. C. van der Plas and D. A. de Bie, *Recl. Trav. Chim. Pays-Bas, Belg.* **88**, 728 (1969).
65. W. D. Crow and C. Wentrup, *J. Chem. Soc., Chem. Commun.* 1082 (1968).
66. C. Wentrup, C. Thétaz and R. Gleiter, *Helv. Chim. Acta* **55**, 2633 (1972).
67. C. Wentrup and C. Thétaz, *Helv. Chim. Acta* **59**, 256 (1976).
68. R. E. van der Stoel, H. C. van der Plas and H. Jongejan, *J. R. Neth. Chem. Soc.* **102**, 364 (1983).
69. T. Ueda, I. Matsuura, N. Murakami, S. Nagai, J. Sakakibara and M. Goto, *Tetrahedron Lett.* **29**, 4607 (1988).
70. I. Matsuura, T. Ueda, N. Murakami, S. Nagai and J. Sakakibara, *J. Chem. Soc., Perkin Trans. 1* 2821 (1991).
71. I. Matsuura, T. Ueda, N. Murakami, S. Nagai and J. Sakakibara, *J. Chem. Soc., Chem. Commun.* 1688 (1991).
72. L. F. Fieser and M. Fieser, *Reagents for Organic Synthesis*. Wiley, New York, 1967, Vol. 1, p. 508.
73. I. Matsuura, T. Ueda, N. Murakami, S. Nagai and J. Sakakibara, *Chem. Pharm. Bull.* **41**, 608 (1993).
74. N. Ikekawa, Y. Honma and R. Kenkyusho, *Tetrahedron Lett.* 1197 (1967).
75. N. J. Leonard and B. Zwanenberg, *J. Am. Chem. Soc.* **89**, 4456 (1967).
76. T. Matsuura and Y. Ito, *Bull. Chem. Soc. Jpn* **47**, 1724 (1974).
77. H. J. J. Loozen and E. F. Godefroi, *J. Org. Chem.* **38**, 3495 (1973).
78. R. A. Abramovitch and B. W. Cue, *J. Am. Chem. Soc.* **98**, 1478 (1976).
79. A. Ohta, T. Watanabe, J. Nishiyama, K. Uehara and R. Hirate, *Heterocycles* **14**, 1963 (1980).
80. T. Watanabe, J. Nishiyama, R. Hirate, K. Uehara, M. Inoue, K. Matsumoto and A. Ohta, *J. Heterocycl. Chem.* **20**, 1277 (1983).
81. T. Watanabe, I. Ueda, N. Hayakawa, Y. Kondo, H. Adachi, A. Iwasaki, S. Kawamata, F. Mentori, M. Ichikawa, K. Yuasa, A. Ohta, T. Kurihara and H. Miyamae, *J. Heterocycl. Chem.* **27**, 711 (1990).
82. P. J. Lont, H. C. van der Plas and A. Koudijs, *Recl. Trav. Chim. Pays-Bas, Belg.* **90**, 207 (1971).
83. P. J. Lont and H. C. van der Plas, *Recl. Trav. Chim. Pays-Bas, Belg.* **92**, 449 (1973).
84. A. H. Cook and D. G. Jones, *J. Chem. Soc.* 278 (1941).
85. C. Grundmann and A. Kreutzberger, *J. Am. Chem. Soc.* **77**, 6559 (1955).
86. R. Metze and G. Scherowsky, *Chem. Ber.* **92**, 2481 (1959).
87. S. Nakanishi, J. Nantaku and Y. Otsujix, *Chem. Lett.* 341 (1983).

88. J. Barluenga, R. P. Carlon, J. Gonzalez, J. Joglar, F. L. Ortiz and S. Fustero, *Bull. Soc. Chim. Fr.* **129**, 566 (1992).

6.2 BENZIMIDAZOLES

6.2.1 From imidazoles

Reactions which ring close on to a preformed imidazole ring are quite uncommon since they normally require at least one aldehyde group on the imidazole substrate at C-4 and/or C-5. Reactions of this type have been used to make 7-alkylbenzimidazoles [1], and 5,6-diacylbenzimidazoles have been made in good to poor yields by condensing an imidazole-4,5-dicarbaldehyde with a suitable sulfide or sulfinyl reagent. Base-catalysed sulfur extrusion from an intermediate imidazothiepin is probably involved in this latter example [2].

6.2.2 From 1-aryltetrazoles

There are reports that benzimidazoles can be made by cyclization of *N*-arylimidoyl nitrenes generated by photolysis of sulfimides of 1-aryltetrazoles [3, 4]. These should be compared to the related imidazole synthesis in Section 6.1.2.3. Examples include the formation of 2-phenylbenzimidazole (42%) from 1,5-diphenyltetrazole [5], and 5-phenoxybenzimidazole (36%) from 5-phenoxy-1-phenyltetrazole [6]. These reactions have little apparent synthetic importance.

6.2.3 From 1-(2-nitroaryl)-1,2,3-triazolines

A seemingly practical entry to 1-alkyl-2-aminobenzimidazoles (**2**) cyclizes 5-amino-1-(2-nitroaryl)-1,2,3-triazolines (**1**) in the presence of triethylphosphite (Scheme 6.2.1). The starting triazolines are readily made from an aldehyde, secondary amine and arylazide in an inert solvent. In the one-pot reaction the triethylphosphite serves to reduce the nitro group to an amino group as well as assisting the cyclization [7].

General procedure for synthesis of 1-alkyl-2-aminobenzimidazoles (**2**) [7]

A mixture of the triazoline (10 mmol) and triethyl phosphite (20 ml) is refluxed (10–30 h) under an inert atmosphere. When the amidine intermediate is no longer detectable (TLC, 40% ethyl acetate–cyclohexane) the solution is rotary evaporated, and the crude residue is chromatographed on a silica gel column using the same solvent as for TLC. The first fraction (minor) is unchanged amidine; the second fraction contains the benzimidazole (**2**) in 60–95% yield.

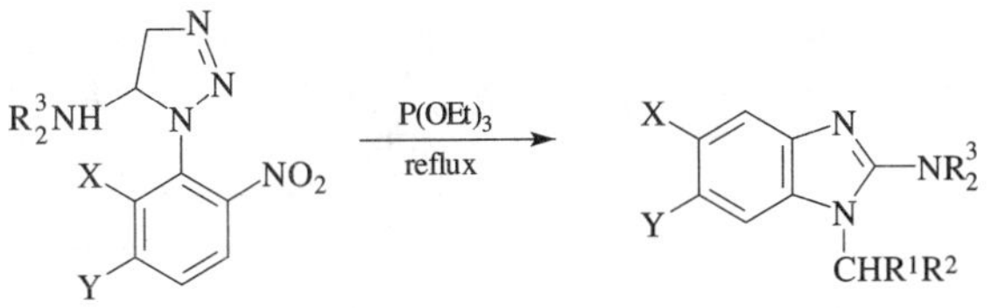

6.2. BENZIMIDAZOLES

Scheme 6.2.1

Prepared in this way are the following (**2**) (X, Y, R^1, R^2, $NR^3{}_2$, yield given): MeO, H, Me, Me, $N(CH_2Ph)_2$, 69%; MeO, H, Me, Me, morpholino, 65%; H, H, Et, H, morpholino, 73%; H, H, Et, H, NMe_2, 60%; H, H, Me, H, morpholino, 62%; H, H, Me, Me, pyrrolidino, 83%; H, H, Ph, H, morpholino, 95%.

6.2.4 From aryl-1,2,4-oxadiazol-5-ones

Pyrolysis of 3,4-diaryl-1,2,4-oxadiazol-5-ones at 190–260°C gives 2-arylbenzimidazoles, usually in moderate to good yields. The starting oxadiazolones can be made quite easily from N-hydroxy-N'-arylamidines and ethyl chloroformate, thereby giving some value to this approach to benzimidazole synthesis [8, 9]. Yields of 2-arylbenzimidazoles are usually in the range 70–90%, e.g. 2-phenyl-4-methyl- (91%), 2-phenyl-4-nitro- (78%), 2-phenyl-5-nitro- (75%), 2-(p-nitrophenyl)- (74%) and 2-(p-methoxyphenyl)-benzimidazoles (14%) [8]. Modifications of the method have included peroxide-induced thermolysis and photolysis of oxadiazolones and the analogous thiones in dioxane solution [10, 11].

6.2.5 From benzo five-membered heterocycles

6.2.5.1 Indazoles (cf. Section 6.1.2.1)

Photolysis of indazole converts it into benzimidazole in 27% yield. 2-Alkylindazoles form 1-alkylbenzimidazoles under the same conditions, but 1-alkylindazoles give 2-alkylamino benzonitriles [12, 13]. The reactions appear to have little synthetic significance.

6.2.5.2 Benzoxazoles (cf. Section 6.1.2.6)

4-Hydroxy-1-methylbenzimidazoles (**4**) can be made isomerically pure in 33% overall yield from 4-aminobenzoxazoles (**3**), which are readily available by condensation of 2-amino-3-nitrophenol with triethyl orthoformate and reduction of the nitro group. The sequence alkylates the amino group of (**3**) and then rearranges the methylaminobenzoxazole in base (Scheme 6.2.2). The 7-hydroxy isomer is formed in a sequence which reduces 4-nitrooxazole to 3-nitro-2-methylaminophenol, hydrogenation to the 1,2-diamine, and cyclization with triethyl orthoformate (1,2 and 2,3 bond formation) [14].

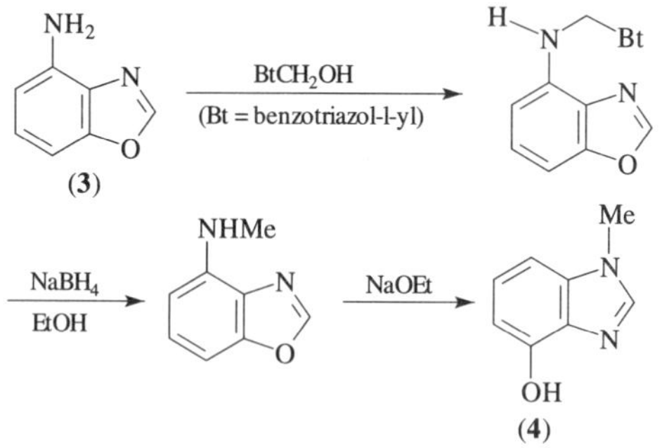

Scheme 6.2.2

6.2.5.3 Benzofuroxans (benzofurazan 1-oxides, 2,1,3-benzoxadiazole 1-oxides)

When benzofuroxans (**5**) are treated with suitable nucleophiles (nitroalkane carbanions, barbituric acid, 2,4,6-trialkylhexahydro-1,3,5-triazines, cyanacetamide anions) nucleophilic attack at a ring nitrogen transforms them into 1-hydroxybenzimidazoles and their 3-oxides (**6**) [23] (Scheme 6.2.3). Benzofuroxan (**5**) (X = H) is the cyclic tautomer of *o*-dinitrosobenzene, and, as such, it should be able to react with an active methylene compound to give an *o*-nitrosoanil, and ultimately a benzimidazole *N*-oxide. Matters are somewhat more complicated than this because the nitroso groups can act either as electron acceptors or donors, but what is found is that if the active methylene group has a good leaving group (nitrile, nitro, sulfone) the product will be a 1-hydroxybenzimidazole 3-oxide. If the methylene is activated by a keto function, then quinoxaline 1,4-dioxides are likely to be the major products [15, 16]. In other instances the *o*-nitrosoanil may merely give the 1-hydroxybenzimidazole. With a malonyl derivative such as barbituric acid as the active methylene reagent the sole product is the 1-hydroxybenzimidazole-2-carboxylic acid [15, 17, 18]. Nitroalkanes or other strong alkylating agents such as alkyltrifluoromethyl sulfonates give 2-alkyl-1-hydroxybenzimidazole 3-oxides [19–21]. Phenylsulfonylacetic acid and phenylsulfonylacetophenone give the same products [20–22], while cyanacetamides lead to 2-carbamoyl analogues [24], and formaldehyde gives 1,3-dihydroxybenzimidazol-2-ones

6.2. BENZIMIDAZOLES

[25]. Extension of the reactions to a variety of substituted benzofuroxans is less appealing since synthesis of a range of (**5**) will be necessary, and these will exist in solution as mixtures of tautomers, which will give rise to mixtures of benzimidazole N-oxides.

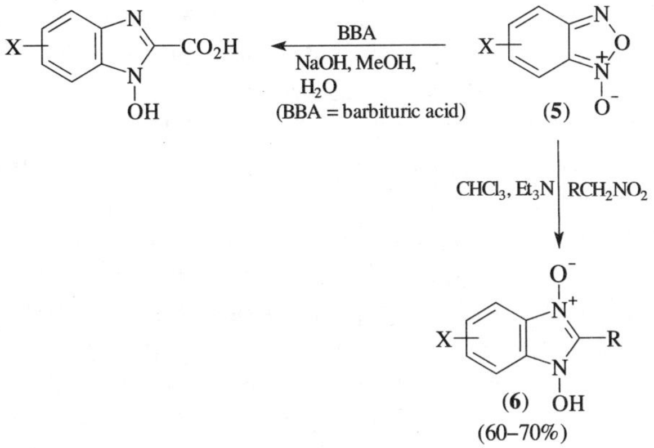

Scheme 6.2.3

1-Hydroxybenzimidazole 3-oxides (**6**) *(X = H)* [26]

Benzofurazan 1-oxide (**5**) (X = H) (2.0 g, 15 mmol) in ethanol (50 ml) is cooled to 0°C and saturated with gaseous ammonia. The primary nitroalkane (20 mmol) is then added, and the mixture is kept at room temperature (72 h) before evaporation of the solvent below 40°C and recrystallization. Prepared in this way are (**6**) (X = H, R, yield given): H, 62%; Me, 64%; Ph, 52% [26]; Et, 66%; $CH_2CH_2CONH_2$, 70%; CO_2Et, 35% [20].

2-Aminocarbonyl-1-hydroxybenzimidazole 3-oxide (**6**) *(X = H, R = $CONH_2$)* [24]

Benzofuroxan (**5**) (X = H) (13.6 g, 0.1 mol) and sodium hydroxide (20 g, 0.5 mol) are suspended and dissolved respectively in water (100 ml). Then, with ice cooling to maintain the reaction temperature between 10 and 15°C, a stream of cyanacetamide (25.2 g, 0.3 mol) is added over 10 min. After brief concentration the sodium salt of (**6**) separates as a dense yellow precipitate, which is collected by suction filtration, redissolved in water, and acidified with dilute hydrochloric acid to give (**6**) (X = H, R = $CONH_2$) as pale yellow crystals (15 g, 78%) which decompose violently at 219°C. Similarly prepared are (**6**) (X, R, yield given): H, CONHMe, 59%; H, CONHEt, 72%; H, CONH-cyclohexyl, 92%; H, CONHPh, 41%; 6-Cl, $CONH_2$, 53%; 6-Me, $CONH_2$, 32%; 6-OMe, $CONH_2$, 37%.

6.2.6 From benzo six-membered heterocycles

6.2.6.1 Quinoxalines

Potassium amide in liquid ammonia converts 2-halogenoquinoxalines into benzimidazole. The reaction, which resembles that of the analogous pyrazines (see Section 6.1.3.2), has no synthetic value [27]. When boiled with aqueous methanol 2,3-diphenylquinoxaline 1-oxide ring contracts to form 1-benzoyl-2-phenylbenzimidazole [28], while methyl quinoxaline-2-carbamate 1-oxide gives a quantitative yield of methyl benzimidazole-2-carbamate when photolysed in methanol or dichloromethane in the presence of triethylamine; similar photolysis in acetonitrile gives the 1-formyl-2-carbamate [29]. A variety of 3-substituted quinoxaline 1-oxides form 2-substituted benzimidazole 1-oxides in rather poor yields when treated with 30% hydrogen peroxide and potassium hydroxide [30, 31]. Thermal decomposition of 2-azidoquinoxaline 1-oxides (or the 1,4-dioxides) can be achieved merely by refluxing in benzene solution for 30 min. If the azidoquinoxaline oxide is unsubstituted in the 3-position a good yield of 2-cyano-1-hydroxybenzimidazole (92%) is obtained [32, 33].

6.2.6.2 Miscellaneous ring contractions

4-Azidoquinazolines yield 1-cyanobenzoyimidazoles almost quantitatively when pyrolysed in the gas phase [34].

"Mebendazole" (methyl 5-benzoylbenzimidazol-2-yl carbamate) has been made in good yield by treating the appropriate benzo-2,1,4-thiadiazine with either mineral acid or triphenylphosphine. The starting materials are easily made by dithionite reduction of o-nitro-arylthiocarbamoylcarbamates, and the general method can be extended to other mebendazole analogues [35, 36]. Reductive ring closures of 3-(4-thiazolyl)benzotriazine 1-oxides, using zinc in acetic acid or platinum oxide in ethanol, give thiabendazole derivatives [37]. There is a well-documented acid-catalysed contraction of 1,5-benzodiazepines to benzimidazoles [38, 39].

REFERENCES

1. H. J. Loozen and E. F. Godefroi, *J. Org. Chem.* **38**, 3495 (1973).
2. E. F. Godefroi, A. Corvers and A. De Groot, *Tetrahedron Lett.* 2173 (1972).
3. T. L. Gilchrist, C. J. Moody and C. W. Rees, *J. Chem. Soc., Perkin Trans. 1* 1964 (1975).
4. R. N. Butler, *Adv. Heterocycl. Chem.* **21**, 323 (1977).
5. R. M. Moriarty and J. M. Kliegman, *J. Am. Chem. Soc.* **89**, 5959 (1967).
6. F. L. Bach, J. Karliner and G. E. Van Lear, *J. Chem. Soc., Chem. Commun.* 1110 (1969).
7. E. Erba, G. Mai and D. Pocar, *J. Chem. Soc., Perkin Trans. 1* 2709 (1992).
8. T. Bacchetti and A. Alemagna, *Atti. Accad. Nazl. Lincei. Cl. Sci. Fis., Mat. Natur. Rend.* **28**, 824 (1960); *Chem. Abstr.* **56**, 7304 (1962).

REFERENCES

9. T. Bacchetti and A. Alemagna, *Atti. Accad. Nazl. Lincei. Cl. Sci. Fis., Mat. Natur. Rend.* **22**, 637 (1957); *Chem. Abstr.* **52**, 15 511 (1958).
10. R. L. Ellsworth, D. F. Hinkley and E. F. Schoenewaldt, French Patent 2,014,402 (1970); *Chem. Abstr.* **74**, 76 422 (1971).
11. J. H. Boyer and P. J. A. Frints, *J. Heterocycl. Chem.* **7**, 59 (1970).
12. J. P. Ferris, K. V. Prabhu and R. L. Strong, *J. Am. Chem. Soc.* **97**, 2835 (1975).
13. J. P. Ferris and F. R. Antonucci, *J. Chem. Soc., Chem. Commun.* 126 (1972).
14. A. R. Katritzky, R. P. Musgrave, B. Rachwal and C. Zaklika, *Heterocycles* **41**, 345 (1995).
15. W. Durckheimer, *Liebigs Ann. Chem.* **756**, 145 (1972).
16. S. S. Sabri, M. M. El-Abadelah and H. A. Yasin, *J. Heterocycl. Chem.* **24**, 165 (1987).
17. F. Sen, K. Ley and K. Wagner, *Synthesis* 703 (1975).
18. G. V. Nikitina and M. S. Pevzner, *Chem. Heterocycl. Compds USSR (Engl. Transl.)* **29**, 127 (1993).
19. A. J. Boulton, A. C. G. Gray and A. R. Katritzky, *J. Chem. Soc. (B)* 911 (1967).
20. M. J. Abu El-Haj, *J. Org. Chem.* **37**, 2519 (1972).
21. D. W. S. Latham, O. Meth-Cohn and H. Suschitzky, *J. Chem. Soc., Chem. Commun.* 1040 (1972).
22. D. P. Claypool, A. R. Sidani and K. J. Flanagan, *J. Org. Chem.* **37**, 2372 (1972).
23. M. J. Haddadin and C. H. Issidorides, British Patent 1,305,138 (1973); *Chem. Abstr.* **78**, 136 339 (1973).
24. F. Seng and K. Ley, *Synthesis* 606 (1972).
25. F. Seng and K. Ley, *Angew. Chem. Int. Ed. Engl.* **11**, 1009 (1972).
26. D. W. S. Latham, O. Meth-Cohn, H. Suschitzky and J. A. L. Herbert, *J. Chem. Soc., Perkin Trans. 1* 470 (1977).
27. P. J. Lont and H. C. van der Plas, *Recl. Trav. Chim. Pays-Bas, Belg.* **91**, 850 (1972).
28. C. Kaneko, I. Yokoe, S. Yamada and M. Ishikawa, *Chem. Pharm. Bull.* **14**, 1316 (1966).
29. R. A. Burrell, J. M. Cox and E. G. Savins, *J. Chem. Soc., Perkin Trans. 1* 2707 (1973).
30. E. Hayashi and C. Iijima, *Yakugaku Zasshi* **82**, 1093 (1962); *Chem. Abstr.* **58**, 4551 (1963).
31. E. Hayashi and Y. Miura, *Yakugaku Zasshi* **87**, 648 (1967); *Chem. Abstr.* **67**, 90 775 (1967).
32. R. A. Abramovitch and B. W. Cue, *Heterocycles* **1**, 227 (1973).
33. J. P. Dirlam, B. W. Cue and K. J. Gombatz, *J. Org. Chem.* **43**, 76 (1978).
34. C. Wentrup and C. Thetaz, *Helv. Chim. Acta* **59**, 256 (1976).
35. A. C. Barker and R. G. Foster, German Patent 2,246,605 (1973); *Chem. Abstr.* **79**, 5341 (1973).
36. H. Loewe, J. Urbanietz, R. Kirsch and D. Duewel, German Patent 2,332,486 (1975); *Chem. Abstr.* **83**, 43 324 (1975).
37. R. L. Ellsworth, D. F. Hinkley and E. F. Schoenewaldt, French Patent 2,014,422 (1970); *Chem. Abstr.* **74**, 76 423 (1971).
38. D. Lloyd, R. H. McDougall and D. R. Marshall, *J. Chem. Soc.* 3785 (1965).
39. Y. Okamoto and K. Tagaki, *J. Heterocycl. Chem.* **24**, 885 (1987).

-7-
Aromatic Substitution Approaches to Synthesis

Although substituents can be introduced prior to ring formation, many precursors are difficult or impossible to prepare, and other substituents may not survive the cyclization procedures. Only carbon, oxygen, nitrogen and sulfur substituents are commonly introduced before cyclization, and even these can only be introduced to some ring positions. For these reasons it is useful to be able to introduce a wide range of groups with good selectivity to the preformed rings. There is a review article which discusses in general terms the introduction of substituents into azoles [1], focusing on the various strategies which use activating, directing and protecting groups to ensure that mono and regioselective substitution is achieved. Imidazoles and benzimidazoles are quite prone to electrophilic substitution reactions, and benzimidazoles undergo some nucleophilic substitutions quite readily. Nucleophilic and radical substitutions are, however, less well known in these compounds, but there are exceptions which may be synthetically useful.

Electrophilic attack in imidazole is usually most facile at an annular nitrogen, and there are many examples of N-alkylation, -protonation, -acylation, cyanation, -arylation and -silylation. N-Nitration is much less common; N-oxidation is virtually non-existent. When an annular nitrogen becomes substituted, tautomerism in the molecules is blocked, and mixtures of isomers are usually formed with substituted benzimidazoles and 4(5)-substituted imidazoles.

Electrophilic substitution on carbon in imidazole is largely restricted to nitration, sulfonation and halogenation. Acidic electrophiles tend to protonate the azole, producing the much less reactive imidazolium species. For this reason, Friedel–Crafts reactions are not known, and C-alkyl and C-acyl imidazoles must be made by alternative methods. Most electrophilic reactions in imidazole (**1**) (except for diazo coupling, some acylations, and deuterium exchange) occur initially at the 4(or 5)-position. 2-Substitution is much more difficult to achieve, and in 1-substituted imidazoles C-5 attack is preferred. Strong electron donors, such as the methoxy group, at C-2 or C-4 of imidazole activate C-5 to electrophilic attack, while a methoxy group at C-5 activates both of the other ring carbon positions. Nitro groups analogously activate groups α and γ to themselves to nucleophilic attack.

In benzimidazole (2), electrophilic substitution occurs most readily in the fused benzene ring, commonly at C-5; a second substituent usually enters at C-6, although in all cases groups already present on the molecules can substantially modify the usual orientation of substitution. Thus, in benzimidazoles, a strongly electron-releasing group at C-5 will direct subsequent attack to C-4; electron-withdrawing substituents lead to subsequent attack at C-4 or C-6. Increased reactivity is achieved if the molecules can be induced to react in their anionic forms.

Nucleophilic substitution reactions in imidazoles are confined largely to displacements of halogen or suitable sulfo groups, but even these need to be activated by an electron-withdrawing group elsewhere in the molecule, or by quaternization. In imidazole and benzimidazole the 2-position is the most prone to nucleophilic attack; 2-halogenoimidazoles are almost as reactive as 2-halogenopyridines. A number of nucleophiles cause ring opening of the molecules, especially under vigorous conditions. Radical substitutions are, however, little known, and do not appear to have much synthetic potential.

In summary, substitution processes on the preformed rings usually lead to 1-, 2-, 4- and 5-substitution in imidazole, and 1-, 2-, 5- and 6-substitution in benzimidazole [2–8].

REFERENCES

1. M. Begtrup, *Bull. Soc. Chim. Belg.* **97**, 573 (1988).
2. P. N. Preston, in *Benzimidazoles and Congeneric Tricyclic Compounds* (ed. P. N. Preston). Wiley-Interscience, New York, 1981.
3. A. F. Pozharskii, A. D. Garnovskii and A. M. Simonov, *Russ. Chem. Rev. (Engl. Transl.)* **35**, 122 (1966).
4. K. Schofield, M. R. Grimmett and B. R. T. Keene, *Heteroaromatic Nitrogen Compounds. The Azoles.* Cambridge University Press, Cambridge, 1976.
5. M. R. Grimmett, in *Comprehensive Heterocyclic Chemistry* (ed. A. R. Katritzky and C. W. Rees). Pergamon Press, Oxford, 1984, Vol. 5 (ed. K. T. Potts), p. 457.
6. M. R. Grimmett, *Comprehensive Heterocyclic Chemistry*. (ed. A. R. Katritzky and C. W. Rees), Elsevier, Oxford, 1996, Vol. 3.02 (ed. I. Shinkai) p. 77.
7. M. R. Grimmett, *Adv. Heterocycl. Chem.* **12**, 103 (1970).
8. M. R. Grimmett, *Adv. Heterocycl. Chem.* **27**, 241 (1980).

7.1 SYNTHESIS OF N-SUBSTITUTED IMIDAZOLES BY ELECTROPHILIC SUBSTITUTION

In addition to syntheses which are aimed at N-substituted imidazoles and benzimidazoles as the ultimate products, there are many examples of processes which introduce a group on to an annular nitrogen either as a "directing agent" or as a "blocking agent". Such groups, then, need to be easy to attach and also easy to remove without affecting the newly introduced substituent. The use of acyl blocking groups for regiospecific synthesis of 1,5-disubstituted imidazoles (see Section 7.1.4), and the variety of blocking groups used in lithiation processes (see Sections 7.1.1 and 7.2.2) are the most important examples of these [1].

7.1.1 N-Alkylimidazoles [2]

Imidazole alkylations can be carried out under a variety of reaction conditions. For conventional N-alkylations which are unlikely to be complicated in terms of regiochemistry, it is preferable to alkylate the imidazole anion (an S_E2cB process). Such reactions are faster, higher yielding and less prone to azole salt formation than those in "neutral" conditions. The anion is generated best by the use of sodium in ethanol or liquid ammonia, with sodium or potassium hydroxide or carbonate, or by use of sodium hydride in dry DMF [3]. Addition of the alkylating agent to the deprotonated substrate completes the reaction.

Recently, Begtrup and Larsen [4] have developed a new approach which reportedly allows better control of alkylation (and also acylation and silylation). The deprotonation and nucleophilic displacement steps are separated so that reaction conditions can be better controlled, and reaction capability values (or reaction potentials) have been allocated to alkylating agents and solvents. These reaction capabilities (which also include temperature and time) are additive and are normalized to pK_a units so that they can be related to the substrate acidity. The alkylation capability is considered satisfactory if the yield of alkylazole reaches 75% or more under standard conditions. For successful imidazole alkylation the relationship below must hold (methyl iodide is given the arbitrary P_a of 7). The advantages of separating the process into two steps are that one can choose separate solvents for each step, the destruction of the alkylating agent by the base is all but eliminated, and the nucleophilicity of the imidazole anion can be solvent enhanced.

$$[20 - (P_a(\text{alkylating agent}) + P_a(\text{solvent}) + P_a(\text{temperature and time})$$
$$\leq pK_a \text{ (acidic) imidazole}]$$

The deprotonation can be achieved using solid or aqueous alkali metal hydroxides, or sodium hydride in DMF or acetonitrile. The approach has, however, only limited synthetic application to unsymmetrically substituted imidazoles which give mixtures on alkylation. Quaternization (see later) can be avoided by keeping the sum of the normalized reaction potentials below the pK_a (basic) value of the 1-alkylimidazole, but if this is not possible, salt formation can be minimized by using only one equivalent of the alkylating agent and a solvent which readily dissolves the anion.

Deprotonation [4]

(a) Aqueous sodium hydroxide (33%, 10 mmol as determined by titration) and the imidazole (10 mmol) are heated to 100°C for 10 min. The solution is cooled to 20°C, and evaporated at 0.5 mmHg until a solid remains. This is dried at 200°C (0.5 mmHg) to give the hygroscopic azole sodium salt as a crystalline powder (97–98%).

(b) Solid potassium hydroxide (10 mmol as determined by titration) and the imidazole (10 mmol) are heated at 200°C (0.5 mmHg) for 0.5 h to produce the hygroscopic potassium salt as crystalline lumps (97–98%).

(c) Dry acetonitrile (20 ml) is added with stirring and cooling in an ice bath to a mixture of a 55% suspension of sodium hydride in mineral oil (1.57 g) and the imidazole (30.0 mmol) under nitrogen. After stirring (24 h) the acetonitrile is removed at 20°C (1 mmHg), and the residue is washed with hexane (3 × 4 ml) to give a quantitative yield of the imidazole sodium salt.

1-Methylimidazole [4]

Imidazole sodium salt (0.90 g, 10 mmol), acetonitrile (4.3 ml) and methyl iodide (0.62 ml) are mixed with stirring at −25°C in a closed reaction vessel [5]. The vessel is kept in the bath while its temperature is raised to 20°C in about 1 h. Stirring is then continued at 20°C (72 h). The solvent is removed *in vacuo* at 20°C, the residue is extracted into dichloromethane (3 × 8 ml), the solvent is again removed. The resulting product (0.80 g, 98%) is distilled in a ball-tube, b_{10} 200°C. Similarly prepared is 1-benzylimidazole (94%).

1-n-Propylimidazole [6]

Imidazole (13.60 g, 0.20 mol) is carefully added to a solution of sodium (5 g, 0.22 mol) in liquid ammonia (300 ml) containing a few crystals of ferric nitrate. 1-Iodopropane (35.7 g, 0.21 mol) is then added over a period of 30 min with continuous stirring. The ammonia is allowed to evaporate overnight, water (100 ml) is added, and the solution is exhaustively extracted with chloroform. The extracts are dried (Na_2SO_4), concentrated and distilled to give 1-n-propylimidazole (15.0 g, 68%), $b_{1.0}$ 70°C.

1-Benzyl-2,4,5-tribromoimidazole [7]

A stirred mixture of 2,4,5-tribromoimidazole (50.0 g, 0.16 mol), benzyl chloride (20.74 g, 0.16 mol), sodium carbonate (17.37 g, 0.16 mol) and DMF

7.1. SYNTHESIS OF N-SUBSTITUTED IMIDAZOLES

(100 ml) is heated under reflux overnight. The cooled mixture is filtered and the solvent rotary evaporated. To the residue is added water (50 ml), and the resulting oil is triturated until solidification. The water is decanted and the residue recrystallized from ethanol to give the product (57.0 g, 90%), m.p. 67-68°C.

Both liquid-liquid and solid-liquid phase transfer catalytic methods have been successfully adapted to imidazole and benzimidazole alkylation [8, 9]. The advantages which accrue are largely a consequence of being able to separate the alkylating agent from the basic media used to generate the imidazole anion. This helps circumvent the problem of competing base-induced eliminations from the alkylating agents (e.g. t-butylchloride is largely converted into 2-methylpropene) and leads to increased yields of 1-alkylimidazoles. This approach has been comprehensively reviewed [10].

1-Benzylimidazole (using phase transfer catalysis) [8]

In a round-bottomed flask provided with a refrigerant, with a calcium chloride tube, and magnetic stirring, is introduced, in order, xylene (200 ml), imidazole (2.04 g, 30 mmol), anhydrous potassium carbonate (4.14 g, 30 mmol), powdered potassium hydroxide (1.68 g, 30 mmol), tetrabutylammonium bromide (0.48 g, 1.5 mmol) and benzyl chloride (3.79 g, 30 mmol). After 14-20 h of heating under reflux, the hot reaction mixture is filtered and the residue is washed with warm xylene (2 × 25 ml). After drying (Na_2SO_4), the solution is evaporated *in vacuo* and the residue subjected to column chromatography on silica gel, eluting with chloroform to give 1-benzylimidazole (3.60 g, 76%), m.p. 70-72°C.

Other alkylating agents which have been used include diazomethane [2], trialkyl phosphates [11], alkoxyphosphonium salts [12], dimethylformamide dialkyl acetals [13], trialkyl orthoformates [14], alkyl cyanoformates [15], dialkyl carbonates (less toxic) [16] and, where more powerful reagents are needed, alkyl fluorosulfonates, trialkyloxonium tetrafluoroborates [17], alkyl triflates and mesylates [18, 19].

When unsymmetrical imidazoles are alkylated, mixtures of isomeric products are usually formed, e.g. methylation of 4(5)-methylimidazole gives about equal quantities of the 1,4- and 1,5-dimethylimidazoles under a variety of reaction conditions. The orientation of N-alkylation of such unsymmetrical imidazoles is controlled by a number of factors (reaction conditions, polar and steric natures of the substituent(s), the nature of the alkylating agent, solvent, tautomerism in the substrate) [2]. An understanding of these factors can help with the preferential synthesis of one of the possible isomers (**6**) and

(**7**) (Scheme 7.1.1). One can also use sterically demanding blocking groups to direct alkylation to a less-preferred position. 3-Substituted L-histidines have been made in this way by blocking the 1-position with trityl, phenacyl, benzyloxymethyl or t-butoxymethyl groups, quaternizing the protected imidazole, and then removing the blocking group [18, 20].

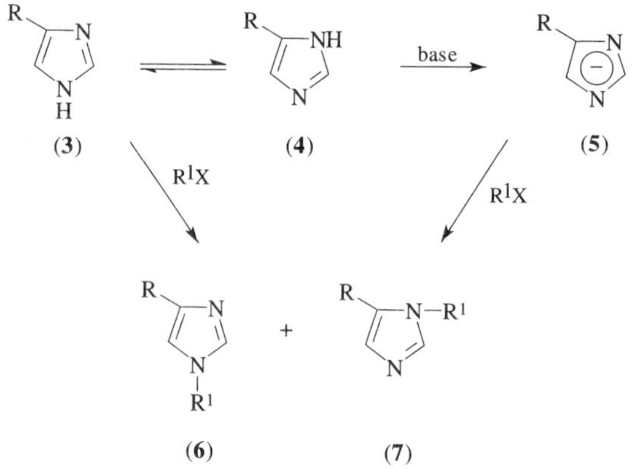

Scheme 7.1.1

As a rule of thumb one can assume that reactions which are carried out in basic media involve the anion (**5**), which N-alkylates to give mainly the 1,4 isomer (**6**) if R is bulky or electron-withdrawing. In free base alkylations the tautomeric mixture (**3**) ⇌ (**4**) reacts. An electron-withdrawing substituent (R) will ensure that much of the reaction will be with the major tautomer (**3**) even if that tautomer is the less reactive. The product mixture formed will have mainly 1,5 regiochemistry (**7**), e.g. 4(5)-nitroimidazole methylated in basic medium gives mainly 1-methyl-4-nitroimidazole; in neutral media 1-methyl-5-nitroimidazole forms almost exclusively. Sterically demanding groups, however, favour the 1,4 isomer. Table 7.1.1 lists some other examples.

When isomeric mixtures are formed they can usually be separated quite readily by column or radial chromatography, for example: mixtures of 4- and 5-halogeno-1-methylimidazoles separate on an alumina column eluted with 2% ethyl acetate in dichloromethane; 1,4- and 1,5-dimethylimidazoles use methanol–ethyl acetate–dichloromethane–hexanes (1:4:5:40); and 1-methyl-4- and -5-nitroimidazoles are separated by 4% methanol in dichloromethane. Further purification is achieved by vacuum distillation in a Kugelrohr apparatus [21].

The range of alkyl and substituted-alkyl blocking groups used in lithiation processes include benzyl, p-methoxybenzyl, t-butyl, trityl, alkoxymethyl, hydroxymethyl, SEM ([2-(trimethylsilyl)ethoxy]methyl) and

7.1. SYNTHESIS OF N-SUBSTITUTED IMIDAZOLES

TABLE 7.1.1
Ratios (percentage yields) of (6):(7) (R^1 = Me) on methylation [21]

R	Neutral conditions[a]	Basic conditions
Me	1.4:1 (49)	1.2:1 (58)[b]
Cl	1:17 (53)	3.0:1 (58)[b]
Br	1:11.5 (44)	2.6:1 (45)[b]
CHO	1:3.6 (34)	1.0:1 (24)[b]
NO_2	1:50 (41)	8.6:1 (37)[b]
CO_2 Me	1:6.5 (34)	1.5:1 (18)[b]
Ph	4.0:1 (56)	7.3:1 (47)[c]

[a] Me_2SO_4-EtOH.
[b] Me_2SO_4-NaOH-H_2O.
[c] Me_2SO_4-NaOEt-EtOH.

(8) R^1 = CH_2Ar, tBu; CPh_3; CH_2OR (R = H, alkyl), CH_2NR_2, $CH_2O(CH_2)_2SiMe_3$ (SEM)

dialkylaminomethyl (see structure (8); see also carboxyl, acyl, trimethylsilyl, dimethylaminosulfonyl, arylsulfonyl, dialkoxymethyl groups, discussed in Sections 7.1.4 and 7.1.5). Benzyl protecting groups are unsatisfactory because competitive lithiation occurs at the methylene group, and quite severe oxidative and reductive methods are needed later to remove the group. p-Methoxybenzyl and 3,4-dimethoxybenzyl groups, however, are rather more easily removed using oxidative methods or treatment with acid [22, 23]. t-Butyl groups, too, are difficult to remove (besides being difficult to introduce). 1-Triphenylmethylimidazole is only slightly soluble in diethyl ether, retarding the deprotonation step, and although it is more soluble in THF, quantitative lithiation is still difficult to achieve. Silyl groups may migrate from one nitrogen to another or to a ring carbon during reaction. N-Alkoxyalkylimidazoles require severe deprotection regimes, but the SEM group is readily introduced, is stable to lithiation, and is removed quite easily either by warming with dilute acid, or dry tetrabutylammonium fluoride. There are, however, problems which arise when the lithio derivatives react with some electrophiles (see Section 7.2.2). The most useful substituted alkyl blocking groups appear to be hydroxymethyl (see Section 7.1.2) and dialkylaminomethyl. The latter are made from the parent heterocycle under Mannich conditions, and subsequent lithiation is easily achieved in ether or THF solution. The protecting group is easily removed by acid-catalysed hydrolysis [24-26].

1-[(Dimethylamino)methyl]imidazole [24]

Imidazole (20.4 g, 0.3 mol) and 97% dimethylamine hydrochloride (26.0 g, 0.3 mol) are dissolved in water (50 ml), and concentrated hydrochloric acid is added until the pH is below 5. Aqueous formaldehyde solution (37%, 27 g, 0.33 mol) is added, and the mixture is allowed to stand at room temperature (48 h). The solution is made strongly alkaline with 20% potassium hydroxide solution, and the organic material is salted out with potassium carbonate, and extracted with chloroform. The combined organic layers are dried (K_2CO_3) and concentrated to give an oil, which is distilled under vacuum to give the pure product (29 g, 78%), b_2 100–102°C, $b_{1.5}$ 95°C.

1-(N,N-Dimethylsulfamoyl)imidazole [25, 27]

Dimethylchlorosulfonamide (13.0 ml, 0.12 mol) is stirred with imidazole (9.54 g, 0.14 mol) and triethylamine (18.0 ml, 0.13 mol) in benzene (160 ml) at room temperature (16 h). The mixture is then filtered, the precipitate is washed with benzene (100 ml), the filtrate and washings are combined, and the solvent is evaporated. Distillation of the residue gives the above product (19.94 g, 95%), $b_{0.4}$ 110°C. It solidifies on standing to give a white solid, m.p. 42–44°C.

1-Ethoxymethyl-2-phenylimidazole [28]

To 2-phenylimidazole (10.0 g, 70 mmol) in anhydrous THF (550 ml) at −20°C under nitrogen is slowly added n-butyllithium (1.6 M in hexanes; 48 ml, 77 mmol) with stirring. After 45 min chloromethylethyl ether (7.5 g, 94 mmol) is slowly added to the cloudy solution at −20°C, and the mixture is allowed to warm to ambient temperature overnight. The clear solution which is formed is poured into a saturated aqueous solution of ammonium chloride (700 ml), and the aqueous layer is separated and washed twice with diethyl ether (200 ml). The combined organic layers are washed with saturated aqueous sodium chloride, dried (Na_2SO_4), filtered and rotary evaporated to give a clear yellow oil (9.0 g). Chromatography (silica gel, ethyl acetate–hexanes) gives the pure product (8.4 g, 51%). Similarly prepared are 1-(1-ethoxyethyl)- (78%), 1-(1-ethoxyethyl)-2-methyl- (76%) and 1-(1-ethoxyethyl)-2-phenylimidazoles (86%) [29].

4-Iodo-1-tritylimidazole [30]

To a solution of 4-iodoimidazole (3.38 g, 17 mmol) in DMF (15 ml) is added triphenylmethyl chloride (5.56 g, 20 mmol) and triethylamine (1.5 ml). The solution is stirred at room temperature overnight, then poured into ice water

7.1. SYNTHESIS OF N-SUBSTITUTED IMIDAZOLES

(200 ml). The obtained solid is filtered and dried to give the product, which is recrystallized from ethyl acetate–cyclohexane (7.04 g, 95%), m.p. 224–225°C.

General procedure: 1-SEM-protected imidazoles [31]

Under a blanket of nitrogen, 50% sodium hydride (8.2 g, 0.17 mol) is washed with hexane. The flask is then charged with dry DMF (250 ml), and the imidazole derivative (0.175 mol) is added in small portions. After stirring at room temperature (1.5 h), [2-(trimethylsilyl)ethoxy]methyl chloride (SEM-Cl, available from Aldrich) (30.8 g, 0.185 mol) is added dropwise. The reaction mixture becomes slightly warm, and is stirred (1 h) before quenching with water and extracting with ethyl acetate (3 × 200 ml). The combined organic layers are washed with water (3 × 200 ml), dried (Na_2SO_4) and concentrated to a light brown oil, which is purified by distillation, e.g. 1-SEM-imidazole (65%, $b_{0.2}$ 94°C) and 1-SEM-benzimidazole (50%, $b_{0.2}$ 220°C).

Quaternary salts can be made by alkylation of 1-substituted imidazoles. They may also be formed directly from imidazoles treated with an excess of alkylating agent. Begtrup and Larsen predict quaternization in terms of quaternization potentials (P_q) using the relationship

$$20 - [P_q(\text{alkylating agent}) + P_q(\text{solvent}) + P_q(\text{temperature}) + P_q(\text{time})]$$
$$\leq pK_a \text{ (1-alkylimidazole)}$$

Standard reaction conditions are 20°C (72 h) in methanol, dichloromethane, acetonitrile or DMF. Under these conditions 1-methylimidazole gives a 30% yield of 1,3-dimethylimidazolium iodide (using methyl iodide) and a 99% yield of 1-benzyl-3-methylimidazolium chloride (using benzyl chloride) [3]. Heating promotes alkylation, but does not seem to have much effect or ease of quaternization, although it does increase yields.

Imidazole quaternary salts (**9**) are frequently quite hygroscopic and need to be kept dry if storage is intended. In their synthesis all apparatus must be scrupulously dry, solvents must be dried before use, and reactions and recrystallizations need to be carried out under dry nitrogen. Liquid reagents are best added using a syringe through a serum cap, and Schlenk apparatus is recommended for filtrations.

(**9**)

Quaternization (Begtrup method) [3]

The 1-alkylimidazole (10 mmol) in DMF, methanol, acetonitrile or dichloromethane (4 ml) and the alkylating agent (10 mmol) in a screw-cap sealed reaction vessel are kept at 20°C (72 h) or heated at 100°C (3 h). Removal of the solvent and washing with dry ether (3 × 5 ml) gives the imidazolium salt.

1,3-Dimethylimidazolium iodide (9) (R^1, R^3 = Me, X = I) [32]

Methyl iodide (7 g, 52 mmol) is added dropwise to a cold, stirred solution of 1-methylimidazole (4.0 g, 49 mmol) in dry benzene (55 ml) under dry nitrogen. After 20 min the mixture is refluxed (5 h), and the oily layer which separates is washed with dry benzene (20 ml) to induce crystallization. Removal of the solvent *in vacuo* gives the salt (5.8 g, 57%), m.p. 88°C. Similarly prepared are a number of other imidazolium iodides, bromides and chlorides (**9**) (R^1, R^3, X, yield given): Me, Et, I, 97%; Me, CH=CH$_2$, I, 96%; Me, Bu, I, 89%; Me, Ph, I, 62%; Me, CH$_2$Ph, I, 93%; Me, Et, Br, 82%; Me, Et, Cl, 78%; Et, Et, I, 79%; Pr, Bu, I, 68%; iPr, Bu, I, 65%.

1-Ethyl-3-methylimidazolium tetraphenylborate [32]

To 1-methyl-3-ethylimidazolium bromide (0.1 g) in dry methanol (5 ml) is added silver tetraphenylborate (0.2 g) in methanol (15 ml). The mixture is refluxed (2 h) and filtered from silver bromide, and then the solvent is evaporated to give the tetraphenylborate salt (36%), m.p. 140°C. Similarly prepared using silver perchlorate in dry acetone is 1-ethyl-3-methylimidazolium perchlorate (83%) as an oily liquid.

7.1.2 *N*-Alkylbenzimidazoles

Similar methods apply, and mixtures of products result when the original substrate is substituted in the fused benzene ring. Quaternization is more difficult because benzimidazole is less basic than imidazole. When 5(6)-substituted benzimidazoles are alkylated the product ratios depend on the resonance electronic effects of the substituent, e.g. methylation of 5(6)-nitrobenzimidazole gives a 1:1 ratio of 1,5 and 1,6 isomers. Substituents in the 4(7)-positions have increased electronic directing effects, and steric effects also come into play, e.g. methylation of 4(7)-nitrobenzimidazole in basic medium gives a 6:1 ratio of 1,4 to 1,7 isomers. And so, in designing a synthetic approach to a 1-alkyl-*C*-nitrobenzimidazole, all of these factors need to be taken into account. It may be more valid to nitrate a 1-alkylbenzimidazole than to alkylate a *C*-nitrobenzimidazole [2].

7.1. SYNTHESIS OF N-SUBSTITUTED IMIDAZOLES

In reactions which require the formation of 2-lithiobenzimidazoles, protection of nitrogen as the hemiaminal, sequential lithiation, quenching with electrophile, and deprotection can be achieved in a one-pot reaction. The 1-hydroxymethylbenzimidazoles are smoothly deprotected by mild acid hydrolysis using dilute hydrochloric acid or a silica gel column. The hydroxymethyl group activates adjacent ring positions in general [33].

1-Hydroxymethylbenzimidazole [33]

To benzimidazole (1.57 g, 13.3 mmol) in a two-necked flask is added THF (40 ml) at 20°C, to give a suspension. Formaldehyde (1 ml, 1.0 eq., 37% aqueous solution) is then added at 20°C to give a homogeneous solution. After 5 min TLC (silica gel–ethyl acetate) shows that all of the benzimidazole has been consumed and 1-hydroxymethylbenzimidazole has been formed. The solvent is rotary evaporated and the residue is dried *in vacuo* (24 h). Paraformaldehyde (1 eq.) in THF under argon can replace the formalin solution. The hydroxymethylbenzimidazole can be used without further purification in lithiation processes.

7.1.3 N-Arylimidazoles

Although there are a number of ring-synthetic methods which can be used to make N-arylimidazoles, there are occasions when it is useful to be able to carry out a direct N-arylation. Highly activated aryl halides such as 1-fluoro-2,4-dinitrobenzene react in benzene solution containing a little triethylamine, to give 1-dinitrophenylimidazoles in 77–92% yields [34]. Less reactive aryl halides need to be heated over an extended period with the imidazole in a high-boiling solvent such as nitrobenzene and in the presence of potassium carbonate and copper(I) bromide [5, 35]. This is a modifed Ullmann method characterized by generally poor yields. Alternative reaction conditions include the use of sodium carbonate in DMF, copper oxide and pyridine [36, 37], and quite recently the use of aryllead(IV) triacetates as "aryl cation equivalents". Under relatively mild conditions (1.1–1.5 eq. of *p*-tolyllead triacetate in the presence of a catalytic amount of copper diacetate at 90–100°C for 4–6 h in dichloromethane–DMF) excellent yields have been obtained [38, 39]. A wide range of aryllead triacetates can be made relatively simply [40]. Typically, a solution of the azole with a slight excess of the aryllead triacetate and a catalytic amount of copper(II) acetate is refluxed, and the copper species are removed by treatment with hydrogen sulfide [39].

1-*p*-Nitrophenylimidazole and its benzimidazole analogue have been made by arylation of the azole anions under phase transfer conditions with *p*-fluoronitrobenzene in an ultrasound bath. These conditions appear to give better yields than magnetic stirring [41].

1-Phenylimidazole [42]

A mixture of imidazole (10.2 g, 0.15 mol), bromobenzene (23.6 g, 0.15 mol), anhydrous potassium carbonate (20 g) and cuprous bromide (0.9 g) in nitrobenzene (230 ml) is heated under reflux (2 days). CARE — bumping occurs. After cooling and filtering, the filtrate is acidified, steam distilled to remove nitrobenzene, and the remaining solution is neutralized and extracted with chloroform to give 1-phenylimidazole (4 g, 18%). The crude product sublimes at 90°C/0.05 mmHg as hygroscopic crystals. Picrate m.p. 155–156°C.

Alternative modified Ullmann procedure [37]

A mixture of the azole (50 mmol), aryl halide (50 mmol), anhydrous potassium carbonate (7 g) and copper(II) oxide (0.25 g) in pyridine (10 ml) is refluxed for some hours. The cooled mixture is filtered, extracted with benzene or chloroform, and the combined extracts and filtrate are rotary evaporated to give a residue which is chromatographed on alumina, eluting with benzene or benzene–chloroform. Prepared in this way are the following 1-substituted imidazoles (1-substituent, heating time, yield given): *o*-nitrophenyl, 11 h, 64%; *m*-nitrophenyl, 19.5 h, 30%; *p*-nitrophenyl, 10 h, 54%; *o*-cyanophenyl, 50 h, 43%; *m*-cyanophenyl, 50.5 h, 13%; *p*-cyanophenyl, 50 h, 73%; *o*-acetylphenyl, 7 h, 33%; *m*-acetylphenyl, 48 h, 68%; *p*-acetylphenyl, 48 h, 82%; α-pyridyl, 19 h, 37%; β-pyridyl, 24 h, 51%; γ-pyridyl, 12 h, 30%.

General procedure for 1-arylimidazoles (and 1-arylbenzimidazoles) [39]

A solution or suspension of the starting azole (0.37–0.84 mmol), *p*-tolyllead acetate (1.1 eq.) and copper(II) acetate (0.01 g) in dichloromethane (2–5 ml) is refluxed. Solvent slowly boils off during the process, but in no case is the reaction allowed to evaporate to dryness. The green reaction mixture is diluted with chloroform (20 ml), and the solution is poured on to dilute aqueous H_2S (50 ml). The two-phase system is stirred vigorously (1 h), then filtered through Celite to remove inorganic sulfides. The pale yellow filtrate is separated, dried (Na_2SO_4) and evaporated, and the residue is chromatographed on silica gel to give the product. Prepared in this way are the following (product, yield, time, temperature, chromatography solvent given): 1-(*p*-tolyl)imidazole, 82%, 6 h, 90°C, chloroform; 1-(*p*-tolyl)benzimidazole, 98%, 4.5 h, 90°C, dichloromethane; 2-phenyl-1-(*p*-tolyl)benzimidazole, 75%, 4.5 h, 90°C, petroleum ether–ethyl acetate (99:1).

Phase transfer method [41]

A mixture of the azole (10 mmol), tetrabutylammonium bromide (1 mmol) and potassium hydroxide (20 mmol) are placed in an ultrasound bath for 2–3 h. A slight excess of *p*-fluoronitrobenzene (11 mmol) is then added, and the mixture is again subjected to ultrasonication (24 h). The reaction progress can be monitored by TLC (hexane–ethyl acetate, 1:1). The crude product is extracted with dichloromethane, the solution is dried (MgSO$_4$), and the solvent is evaporated, before recrystallization of the 1-arylazole. Prepared in this way are 1-(p-nitrophenyl)imidazole (recrystallized from hexane–ethyl acetate, 76%) and 1-(p-nitrophenyl)benzimidazole (recrystallized from ethanol, 77%).

7.1.4 N-Acylimidazoles and N-acylbenzimidazoles

N-Acylimidazoles are of importance in a number of biological processes such as transacylation. They also have considerable synthetic utility in a variety of acylation reactions [43]. Reagents such as 1-acetylimidazole (used as an acetylating agent) and 1,1′-carbonyldiimidazole (widely employed as a coupling agent in peptide synthesis) are commercially available; others can be made quite readily by reacting acyl halides or anhydrides with the free imidazoles. It is necessary to use a non-nucleophilic solvent and usually anhydrous conditions since many 1-acylimidazoles (azolides) react very readily with any nucleophiles, including water and alcohols. It is usual to use a 1:2 ratio of acid chloride and imidazole in an inert solvent at room temperature [44, 45], or to treat 1,1′-carbonyldiimidazole with a carboxylic acid at room temperature [46–48]. Other methods which have been used include the use of ketene in benzene solution [49, 50], isopropenyl acetate [51], acetic anhydride neat or in pyridine [52, 53], and treatment of an N-trimethylsilylimidazole with an acid chloride [54].

Weak bases can assist the reaction. Naturally, the only useful "N-acyl" substituents of use in lithiation sequences are the acetals (e.g. dialkoxymethyl), which can be made quite readily by treating the imidazole with a trialkyl orthoformate with p-toluenesulfonic acid as a catalyst. The groups are easily removed again, but have the drawback of also being unstable under dilithiation conditions [25, 55].

The fact that N-acylation and N-aroylation reactions are reversible means that the thermodynamic product is the only one obtained, i.e. a 4(5)-substituted imidazole gives only the 1-acyl(aroyl)-4-substituted isomer for steric reasons. This observation is utilized in regiospecific synthesis of 1,5-disubstituted imidazoles. The 1-acyl(aroyl)-4-substituted imidazole is first quaternized using a powerful alkylating agent such as a trialkyloxonium salt or methyl fluorosulfonate, and the acyl(aroyl) blocking group is then removed under mildly basic conditions (Scheme 7.1.2) [56, 57].

1-(p-Toluoyl)imidazole [46]

A solution of p-toluic acid (6.8 g, 50 mmol) in THF (80 ml) and 1,1′-carbonyldiimidazole (7.2 g, 50 mmol) in THF (80 ml) is allowed to stand

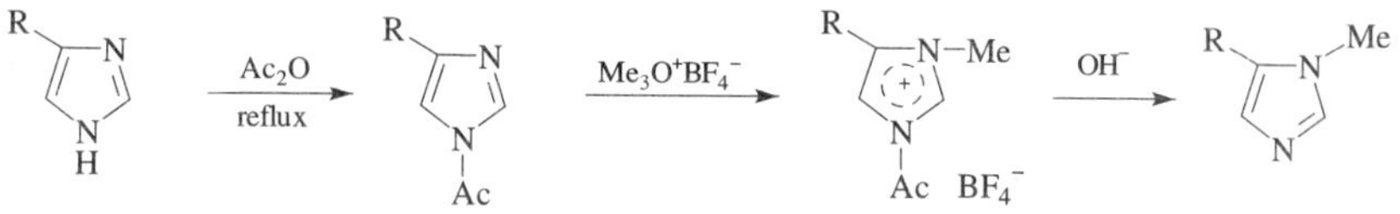

Scheme 7.1.2

(2 h) in a stoppered flask at 30–40°C. The solvent is removed *in vacuo*, and the residue is extracted with hot cyclohexane, leaving the imidazole almost completely behind. Repeated recrystallization from cyclohexane gives the product (7.3 g, 80%), m.p. 72.5–73.5°C.

1-Benzoyl-4-phenylimidazole [56]

Benzoyl chloride (2.4 g, 17 mmol) is slowly added to a stirred solution of 4-phenylimidazole (2.0 g, 14 mmol) and sodium hydroxide (1.2 g, 28 mmol) in acetone (10 ml) and water (40 ml). During the addition the product precipitates. After 30 min cold water (50 ml) is added to ensure complete precipitation. The solid is filtered, washed with water, dried and recrystallized as white platelets from chloroform–ether (3.3 g, 95%), m.p. 124.0–124.5°C.

1-Ethyl-5-phenylimidazole [56]

1-Benzoyl-4-phenylimidazole (4.0 g, 16 mmol) and triethyloxonium fluoroborate (commercially available from Aldrich) (3.06 g, 16 mmol) are dissolved in dichloromethane (20 ml) and stirred at room temperature (48 h). After rotary evaporation of the solvent the product is shown by NMR to be the expected salt, 1-benzoyl-3-ethyl-4-phenylimidazolium fluoroborate. This salt is then dissolved in water (25 ml). The acidic solution is made slightly basic with sodium carbonate, and the product is extracted with chloroform (4 × 50 ml). The extracts are dried (K_2CO_3) and filtered, the solvent removed, and the residue vacuum distilled to give 1-ethyl-5-phenylimidazole (2.50 g, 91%), $b_{0.4}$ 109–110°C.

1-(Diethoxymethyl)imidazole [55]

Imidazole (12.8 g, 0.2 mol), triethyl orthoformate (118.4 g, 0.8 mol) and *p*-toluenesulfonic acid (1 g) are heated at 130°C until no more ethanol distils from the mixture. Excess orthoformate is then removed *in vacuo*, solid sodium carbonate (1 g) is added, and the residue is fractionally distilled to give the acetal (28.2 g, 82%) as a colourless oil, $b_{0.02}$ 52°C.

1-Benzoylimidazole (Begtrup method) [3]

The imidazole sodium salt (10 mmol), dichloromethane (20 ml) and benzoyl chloride (10 mmol) are mixed with stirring and cooling in an ice bath under

7.1. SYNTHESIS OF N-SUBSTITUTED IMIDAZOLES

a nitrogen atmosphere. After being stirred at 35°C (24 h) in a screw-cap sealed reaction vessel the mixture is pressure filtered and extracted with dichloromethane (3 × 5 ml) under nitrogen. After removal of the solvent the 1-benzoylimidazole (90%) is obtained as an oil or low-melting point solid, m.p. 19-20°C.

7.1.5 Other N-substituted imidazoles

Frequently it is necessary to place a group on a ring nitrogen, either to direct subsequent C-substitution, or to protect that nitrogen from attack by a subsequent reagent which would normally react more readily at nitrogen than at carbon. Such protecting groups should ideally have some directional or activating effects, but it is more important for them to be easy to attach and easy to remove in the presence of other substituents. Indeed, it is advantageous to have blocking groups which can be removed specifically by acid, base, oxidation or reduction. Groups such as methyl and phenyl are of little value as they are too difficult to remove. Benzyl groups can be removed by oxidative or reductive methods, but alkyl and aryl groups are generally unsatisfactory substrates for metallation procedures because they can themselves be converted into anions.

1-Trimethylsilylimidazoles have been known for many years. The silyl group imparts increased volatility to the azoles, making them suitable for gas chromatographic separation, purification and analytical procedures. The compounds can be made quite simply by reaction of the imidazole with trimethylsilyl chloride, or by application of Begtup's two-stage method [3, 54, 58]. The trimethylsilyl group is easily hydrolysed, but gives poor results in lithiation procedures as it is subject to N → C rearrangements [25, 59].

Sulfonyl groups are easily introduced, but they are very readily hydrolysed and also seem to reduce the ease of lithiation when utilized as blocking groups in such procedures. The dimethylsulfamoyl group can be introduced using the sulfamoyl chloride with or without added triethylamine, and has considerable advantages as a blocking agent in lithiation reactions.

Recently, interest has been rekindled in the use of the vinyl group as a blocking agent which can be cleaved by ozonolysis [60, 61]. Critical assessment of the variety of blocking groups employed in azole lithiations has appeared elsewhere, and will not be discussed in detail here [24-26, 62].

Whereas pyrazoles are quite readily N-nitrated by nitric acid in acetic anhydride or with nitronium fluoroborate, imidazoles are usually far too basic, and give the nitrate salts instead. A nitro substituent, however, decreases the basicity sufficiently to allow N-nitration to occur. The N-nitro compounds are subject to thermal rearrangement to the 2- and 4-nitro isomers, pointing to an alternative route to such compounds [63].

7. AROMATIC SUBSTITUTION APPROACHES TO SYNTHESIS

1-Trimethylsilylimidazoles [54]

In general, the imidazole compound is treated with excess hexamethyldisilazane under reflux on an oil bath or heating mantle for several hours. The product is isolated by fractional distillation. For example 1-trimethylsilylimidazole is made from imidazole (13.6 g, 0.20 mol) and hexamethyldisilazane (24.2 g, 0.15 mol) heated for 10 h. The product (85% yield) has a b_{12} of 91°C.

N,N-Dimethylimidazole-1-sulfonamide [64]

N,N-dimethylsulfamoyl chloride (4.3 g, 30 mmol) and imidazole (4.1 g, 60 mmol) in chloroform (50 ml) are allowed to react at 25°C (30 min). The mixture is then taken up in water and extracted with benzene. The organic layer is washed with water, dried and evaporated to a colourless liquid, which is distilled to give a solid (3.3 g, 63%), $b_{0.6}$ 83–87°C, m.p. 45–48°C.

1-Benzenesulfonylimidazole [65]

Imidazole (6.8 g, 0.1 mol) is dissolved in DMSO (50 ml), and 1 eq. of powdered potassium hydroxide is added. When this has dissolved, the solution is cooled to 0°C, and benzenesulfonyl chloride (17.7 g, 0.1 mol) is added during 30 min. The reaction mixture is then poured into water (100 ml) and extracted with chloroform (3 × 50 ml). The organic extracts are washed with water, dried and evaporated to give a yellow oil, which usually solidifies on evacuation (12.4 g, 60%). Recrystallization from chloroform–hexane gives the pure product (8.2 g, 40%), m.p. 78°C.

1-Methylsulfonylimidazole (Begtrup method) [3]

Imidazole sodium salt (10 mmol), dichloromethane (20 ml) and methanesulfonyl chloride (10 mmol) are mixed with stirring and cooling in an ice bath under nitrogen. After stirring at 35°C (24 h) (as for 1-benzoylimidazole, see Section 7.1.4) the mixture is pressure filtered and extracted with dichloromethane (3 × 5 ml) under nitrogen. The solvent is then removed, and the crude product (82%) is recrystallized from toluene, m.p. 85–87°C.

2,4,5-Tribromo-1-vinylimidazole [60]

1,2-Dibromoethane (30 ml, 65.4 g, 348 mmol) and triethylamine (30 ml) are added to a stirred solution of 2,4,5-tribromoimidazole (10.0 g, 32.8 mmol) in triethylamine (20 ml) at ambient temperature, and the resulting mixture is heated under reflux until all of the starting material has been consumed (1 h), as shown by TLC analysis (at this stage the products are 1-bromoethyl- and

1-vinyl-2,4,5-tribromoimidazoles). Aqueous sodium carbonate (10%, 30 ml) and tetrabutylammonium bromide (2.6 g, 8.06 mmol) are added to the mixture, which is refluxed (2 h) when TLC shows the presence of only one product. Water (200 ml) is added to the cooled mixture, and the product is extracted with ether (4 × 30 ml). The combined extracts are washed with 10% aqueous HCl (2 × 50 ml) and dried (MgSO$_4$), and the solvents are removed under reduced pressure. Distillation of the residue using a Kugelrohr apparatus gives the product as a white solid (8.32 g, 77%), b$_{0.05}$ 108–110°C, m.p. 31–32°C, ν_{max} 1642 cm^{-1}.

Devinylation method: 4-bromoimidazole-5-carbonitrile [60]

Ozone is bubbled through a stirred solution of 4-bromo-1-vinylimidazole-5-carbonitrile (0.5 g, 2.53 mmol) in methanol (30 ml) at −78°C until TLC analysis shows complete reaction (30 min). Dimethyl sulfide (0.2 ml, 0.17 g, 2.72 mmol) is added, and the mixture is allowed to warm up to room temperature and stirred (1 h). Distillation of the solvent under reduced pressure gives the product (0.35 g, 80%) as a white solid after recrystallization from aqueous ethanol, m.p. 142–144°C.

1,4-Dinitroimidazole [63]

A suspension of 4-nitroimidazole (0.5 g, 4.42 mmol) in glacial acetic acid (18.2 ml) is treated dropwise with nitric acid (4.25 ml, d 1.5). The temperature rises to 50°C as the suspension dissolves. To this solution is added, with cooling, acetic anhydride (12.1 ml), and the mixture is allowed to stand (1 h). Monitoring by ^{1}H NMR shows that the conversion is essentially complete after a few minutes (δ(CDCl$_3$): 8.20, d, $J = 1.5$ Hz, H-2; 8.33, d, $J = 1.5$ Hz, H-5). The mixture is poured on to ice and extracted with dichloromethane. The extracts are washed with aqueous sodium bicarbonate, dried (MgSO$_4$) and evaporated *in vacuo* to give 1,4-dinitroimidazole as the sole product (0.6 g, 85%). A sample sublimed at 76°C/0.015 mmHg had an m.p. of 92°C.

A variety of sulfo groups can be attached to N-1 of imidazole: *p*-tosyl, mesyl, phenylsulfonyl and dimethylaminosulfonyl [64–66].

REFERENCES

1. M. Begtrup, *Bull. Soc. Chim. Belg.* **97**, 573 (1988).
2. P. A. Benjes and M. R. Grimmett, *Adv. Detailed Reaction Mechanisms* **3**, 199 (1994).
3. M. R. Haque and M. Rasmussen, *Tetrahedron* **50**, 5535 (1994).
4. M. Begtrup and P. Larsen, *Acta Chem. Scand.* **44**, 1050 (1990).
5. M. Begtrup, *J. Chem. Educ.* **64**, 974 (1987).
6. C. G. Begg, M. R. Grimmett and P. D. Wethey, *Aust. J. Chem.* **26**, 2435 (1973).
7. B. Iddon, N. Khan and B. L. Lim, *J. Chem. Soc., Perkin Trans. 1* 1437 (1987).

8. R. M. Claramunt, J. Elguero and R. Garceràn, *Heterocycles* **23**, 2895 (1985).
9. E. Diez-Barra, A. De la Hoz, A. Loupy and A. Sanchez-Migallon, *Heterocycles* **38**, 1367 (1994).
10. R. J. Gallo, M. Makosza, H. J. M. Dou and P. Hassanaly, *Adv. Heterocycl. Chem.* **36**, 175 (1984).
11. M. Hayashi, Y. Hisanaga, K. Yamauchi and M. Kinoshita, *Synth. Commun.* **10**, 791 (1980).
12. M. Searcey, P. L. Pye and J. B. Lee, *Chem. Ind. (London)* 569 (1989).
13. R. W. Middleton, H. Monney and J. Parrick, *Synthesis* 740 (1984).
14. Y. N. Bulychev, M. N. Preobrazhenskaya, A. l. Chernyshev and S. E. Esipov, *Chem. Heterocycl. Compd. (Engl. Transl.)* **24**, 751 (1988).
15. M. Prhavc and J. Kobe, *Tetrahedron Lett.* **31**, 1925 (1990).
16. M. Lissel, S. Schmidt and B. Neumann, *Synthesis* 382 (1986).
17. K. Schofield, M. R. Grimmett and B. R. T. Keene, *Heteroaromatic Nitrogen Compounds. The Azoles*. Cambridge University Press, Cambridge, 1976, p. 95.
18. J. C. Hodges, *Synthesis* 20 (1987).
19. S. Klutchko, J. C. Hodges, C. J. Blankley and N. L. Colbry, *J. Heterocycl. Chem.* **28**, 97 (1991).
20. C. J. Chivikas and J. C. Hodges, *J. Org. Chem.* **52**, 3591 (1987).
21. P. Benjes and M. R. Grimmett, unpublished.
22. D. R. Buckle and C. J. M. Rockell, *J. Chem. Soc., Perkin Trans. 1* 627 (1982).
23. Y. Oikawa, T. Tanaka, K. Horita, T. Yoshioka and O. Yonemitsu, *Tetrahedron Lett.* **25**, 5393 (1984).
24. A. R. Katritzky, G. W. Rewcastle and W. Q. Fan, *J. Org. Chem.* **53**, 5685 (1988).
25. D. J. Chadwick and R. I. Ngochindo, *J. Chem. Soc., Perkin Trans. 1* 481 (1984).
26. B. Iddon and R. Ngochindo, *Heterocycles* **38**, 2487 (1994).
27. J. Winter and J. Rétey, *Synthesis* 245 (1994).
28. T. P. Demuth, D. C. Lever, L. M. Gorgos, C. M. Hogan and J. Chu, *J. Org. Chem.* **57**, 2963 (1992).
29. T. S. Manoharan and R. S. Brown, *J. Org. Chem.* **53**, 1107 (1988).
30. K. L. Kirk, *J. Heterocycl. Chem.* **22**, 57 (1985).
31. J. P. Whitten, D. P. Matthews and J. R. McCarthy, *J. Org. Chem.* **51**, 1891 (1986).
32. B. K. M. Chan, N. H. Chang and M. R. Grimmett, *Aust. J. Chem.* **30**, 2005 (1977).
33. A. R. Katritzky and K. Akutagawa, *J. Org. Chem.* **54**, 2949 (1989).
34. J. K. F. Wilshire, *Aust. J. Chem.* **19**, 1935 (1966).
35. L. M. Sitkina and A. M. Simonov, *Khim. Geterotsiki. Soedin.* **2**, 143 (1966); *Chem. Abstr.* **65**, 13 686 (1966).
36. K. Schofield, M. R. Grimmett and B. R. T. Keene, *Heteroaromatic Nitrogen Compounds. The Azoles*. Cambridge University Press, Cambridge, 1976, p. 292.
37. A. M. Khan and J. B. Polya, *J. Chem. Soc. (C)* 85 (1970).
38. P. Lopez-Alvarado, C. Avendaño and J. C. Menéndez, *Tetrahedron Lett.* **33**, 659 (1992).
39. P. Lopez-Alvarado, C. Avendaño and J. C. Menéndez, *J. Org. Chem.* **60**, 5678 (1995).
40. H. C. Bell, J. R. Kalman, J. T. Pinhey and S. Sternhell, *Aust. J. Chem.* **32**, 1521 (1979).
41. M. L. Cerrada, J. Elguero, J. de la Fuente, C. Pardo and M. Ramos, *Synth. Commun.* **23**, 1947 (1993).
42. K. H. R. Lim and M. R. Grimmett, unpublished.
43. H. A. Staab, *Angew. Chem., Int. Ed. Engl.* **1**, 351 (1962).
44. J. C. Cass, A. R. Katritzky, R. L. Harlow and S. H. Simonsen, *J. Chem. Soc., Chem. Commun.* 48 (1976).
45. A. K. S. B. Rao, C. G. Rao and B. B. Singh, *J. Chem. Res. (S)* 196 (1992).
46. H. A. Staab, M. Luking and F. H. Dürr, *Chem. Ber.* **95**, 1275 (1962).
47. S. Murata, *Chem. Lett.* 1819 (1983).

48. A. Bhattacharya, J. M. Williams, J. S. Amoto, U. H. Dolling and E. J. J. Grabowski, *Synth. Commun.* **20**, 2683 (1990).
49. R. D. Kimbrough, *J. Org. Chem.* **29**, 1242 (1964).
50. M. Daneshtalab and T. Kato, *Heterocycles* **24**, 419 (1986).
51. J. H. Boyer, *J. Am. Chem. Soc.* **74**, 6274 (1952).
52. G. S. Reddy, L. Mandell and J. H. Goldstein, *J. Chem. Soc.* 1414 (1963).
53. J. H. Jones, D. L. Rathbone and P. B. Wyatt, *Synthesis* 1110 (1987).
54. L. Birkofer, P. Richter and A. Ritter, *Chem. Ber.* **93**, 2804 (1960).
55. N. J. Curtis and R. S. Brown, *J. Org. Chem.* **45**, 4038 (1980).
56. R. A. Olofson and R. V. Kendall, *J. Org. Chem.* **35**, 2246 (1970).
57. L. Maat, H. C. Beyerman and A. Noordam, *Tetrahedron*, **35**, 273 (1979).
58. V. Sheludyakov, N. A. Viktorov, G. V. Ryasin and V. F. Mironov, *J. Gen. Chem. USSR (Engl. Transl.)* **42**, 354 (1972).
59. P. Jutzi and W. Sakriss, *Chem. Ber.* **106**, 2815 (1973).
60. D. Hartley and B. Iddon, personal communication.
61. C. van der Steft, P. S. Hofman and A. B. H. Funcke, *Eur. J. Med. Chem. Chim. Therap.* **13**, 251 (1978).
62. A. R. Katritzky, P. Lue and K. Akutagawa, *Tetrahedron* **45**, 4253 (1989).
63. M. R. Grimmett, S. T. Hua, K. C. Chang, S. A. Foley and J. Simpson, *Aust. J. Chem.* **42**, 1281 (1989).
64. J. F. King and T. M. Lee, *Can. J. Chem.* **59**, 356 (1981).
65. R. J. Sundberg, *J. Heterocycl. Chem.* **14**, 517 (1977).
66. F. Effenberger, M. Roos, R. Ahmad and A. Krebs, *Chem. Ber.* **124**, 1639 (1991).

7.2 SYNTHESIS OF C-SUBSTITUTED IMIDAZOLES AND BENZIMIDAZOLES BY ELECTROPHILIC SUBSTITUTION

Unsubstituted imidazole reacts with most electrophiles preferentially at C-4(5), but in neutral or basic media it is often difficult to prevent multiple substitution, e.g. halogenation. Deuterium exchange is unusual in that 2-exchange is more facile than 4- or 5-exchange, and diazo coupling occurs with about equal facility in both 2- and 4-positions. Acidic reagents almost always involve the protonated imidazole (imidazolium) as the reactive species. In this form the reactivity is greatly decreased, and 4(5)-substitution is observed, e.g. nitration and sulfonation. Friedel–Crafts reactions will not occur at all. Under severe conditions, disubstitution may be possible.

When the imidazole or benzimidazole carries an *N*-substituent, reactions which would have involved the conjugate base species either have to take an alternative pathway or may not take place, e.g. diazo coupling and halogenation. Preferential electrophilic attack now occurs at C-5. Exceptions to the usual orientation of substitution are observed in Regel acylations at C-2 (probably via initial *N*-acylation), and some hydroxymethylations.

In benzimidazole, electrophiles preferentially substitute in the fused benzene ring in the 5(6)-position; a powerful electron donor at C-5 will direct subsequent attack to C-4; electron-withdrawing groups lead to subsequent 4- or 6-substitution. Attack at C-2 is virtually unknown.

7.2.1 Conventional electrophilic substitutions

7.2.1.1 Imidazoles

In both imidazoles and their N-substituted derivatives, halogenation occurs preferentially in the 4(5)-positions; there is a slight preference for 5-substitution in 1-substituted substrates. Although the 2-position is much less reactive, it is difficult to prevent substitution at that site. Indeed, polyhalogenation is so facile that it is seldom feasible to make monohalogenated imidazoles directly. Both sodium hypochlorite and NCS convert imidazole into its 4,5-dichloro derivative contaminated by the 2,4,5-trichloro product. Even very mild conditions are unlikely to promote monochlorination, and bromination and iodination are similar. Mechanisms can vary, however, from substrate to substrate. It is likely that C-2 halogenations are the result of addition–elimination [1].

Monohalogeno compounds are more likely to be accessible when imidazoles with an electron-withdrawing substituent are halogenated, by nucleophilic methods (see Section 7.3.1), and via organolithium derivatives (see Section 7.2.2). Alternatively, it may be possible to polyhalogenate, then selectively "reduce" one or more of the halogen substituents.

2,4,5-Tribromo-1-methylimidazole [2]

To a mixture of 1-methylimidazole (24.6 g, 0.3 mol) and sodium acetate (102.0 g, 1.2 mol) in glacial acetic acid (500 ml) a solution of bromine (146.0 g, 0.9 mol) in glacial acetic acid (50 ml) is added dropwise with vigorous stirring and cooling to keep the temperature below 60°C. After the addition is complete the mixture is stirred at ambient temperature (2 h), poured on to ice (1 l), stirred and filtered. Crystallization from acetic acid–water (85:15) gives the product as white crystals (48.7 g, 51%), m.p. 88–89°C.

4-Bromoimidazole [3]

A solution of NBS (11.44 g, 64 mmol) in DMF (100 ml) is added dropwise over 1.5 h at 20°C to a stirred solution of imidazole (4.00 g, 59 mmol) in DMF (100 ml). After 48 h the solution is concentrated to dryness under reduced pressure, and the residue is dissolved in ethyl acetate and percolated through silica gel, eluting with more ethyl acetate to give a solid which is triturated with hot chloroform. On cooling, 4-bromoimidazole (3.15 g, 41%) separates as white cubes, m.p. 125–126°C. Small quantities of 4,5-dibromo- and 2,4,5-tribromoimidazoles are also formed.

7.2. SYNTHESIS OF C-SUBSTITUTED IMIDAZOLES AND BENZIMIDAZOLES

2-Ethyl-4-bromoimidazole [4]

Bromine (10.24 ml, 200 mmol) is added dropwise to a stirred, ice-cooled solution of 2-ethylimidazole (8.54 g, 89 mmol) in ethanol (150 ml). After stirring at room temperature (~3 h), 5 M aqueous NaOH (45 ml) is added to give a solution of pH around 6.5. A solution of sodium sulfite (120 g) in water (1 l) is then added, followed by further ethanol (100 ml) and water (100 ml) to give on warming to about 40°C a yellow solution which is refluxed (18 h). The solution is then concentrated to half of its original volume and extracted with chloroform (3 × 250 ml). The combined extracts are concentrated *in vacuo* to give the title compound as a white solid (6.1 g, 38%). ^{1}H NMR: δ 1.19, t, CH_3; 2.78, q, CH_2; 6.95, s, H-5.

4-Chloro-5-iodoimidazole [5, 6]

4-Iodoimidazole (4.00 g, 8.76 mmol) is dissolved in water (20 ml) containing NaOH (0.92 g, 23 mmol) cooled to 0–5°C in an ice bath. Freshly prepared sodium hypochlorite solution (0.60 mol l^{-1}; 38.3 ml, 23 mmol) is added portionwise with continuous stirring so that the temperature does not exceed 5°C. After complete addition, the solution is allowed to warm up to room temperature, and stirring is continued (1 h). Following decolorization with activated charcoal, the pH of the solution is adjusted to ~3 with concentrated HCl to precipitate a solid. After filtration, washing and recrystallization from water the product (3.21 g, 67%) has an m.p. of 163–164°C.

4-Chloroimidazole [5, 6]

4-Chloro-5-iodoimidazole (2.00 g, 8.76 mmol) is heated under reflux (4 h) in 5% aqueous sodium sulfite solution (70 ml). After cooling, the pH is adjusted to ~2 with concentrated HCl, and then evaporated to dryness under reduced pressure. The solid residue is extracted with boiling ethanol (4 × 100 ml), and the extracts are combined and again evaporated to dryness. The residue is taken up in a small volume of water, and sodium carbonate is added until precipitation occurs. The filtered product is recrystallized from water to give 4-chloroimidazole of greater than 95% purity as colourless needles (0.70 g, 78%). Vacuum sublimation gives the analytically pure product, m.p. 115–117°C.

Mononitration and monosulfonation of imidazoles are much more common. These reactions involve strongly acidic conditions, hence it is the deactivated imidazolium cation which is reacting [7]. Under certain conditions, however, it is possible to get disubstitution, but despite some reports to the contrary, C-2 nitration will not occur readily, if at all. This means that preparation of the pharmacologically important 2-nitroimidazoles such as azomycin (2-nitroimidazole) has to be accomplished indirectly by oxidation of 2-amino analogues or via diazonium salts (see Section 8.2). Dinitrogen tetroxide has been reported to convert 4-substituted imidazoles with electron-withdrawing substituents into a mixture of 5- and 2-nitro derivatives [8], and Katritzky has made 2,4,5-trinitronimidazole from 2,4,5-triiodoimidazole, albeit in only 9% yield [9].

4-Nitroimidazole [5]

Imidazole (15.0 g, 0.22 mol) is added slowly to concentrated nitric acid (30 ml, d 1.42) maintained at 0–5°C in an ice bath. Concentrated sulfuric acid (25 ml) is then added slowly to the solution with continuous stirring. The nitrating mixture is then heated at 100°C (2 h), before cooling and pouring into ice–water to precipitate a solid. This is filtered, washed with water until the filtrate is non-acidic, dried and recrystallized from ethanol to give off-white needles (16.2 g, 65%), m.p. 307–308°C.

Diazo coupling in N-unsubstituted imidazoles occurs with equal ease at either C-2 or C-4(5) (or both) in reactions which have been shown to involve reaction of the imidazole anion with the diazonium ion [10]. The intensely coloured azo dyes which are formed have long been used for identification of imidazoles, especially in qualitative chromatography [7]. The azo groups can be reduced to amino or hydrazino groups, providing a useful alternative approach, especially to 2-aminoimidazoles (see Section 8.3).

Direct hydroxymethylation of imidazoles and 1-substituted imidazoles is a well-known reaction. It takes place quite readily with formaldehyde or paraformaldehyde at elevated temperatures. In sealed tube reactions, imidazole is converted mainly into 2-hydroxymethylimidazole [11, 12] along with smaller amounts of product hydroxymethylated at other ring positions. In solution (e.g. DMSO) or neat, N-unsubstituted imidazoles are hydroxymethylated by formalin in the 4- or 5-positions, while 1-substituted imidazoles give mostly 2-hydroxymethyl products. For example, 4-bromo- and 4-methylimidazoles are converted into the 5-hydroxymethyl derivatives; 1-methyl, 1-benzyl, 1-aryl and 1,5-dimethyl substrates give 2-hydroxymethyl products. In contrast, 1,4-dimethylimidazole forms the 5-hydroxymethyl derivative under similar conditions, and imidazoles bearing strongly electron-withdrawing substituents may fail to react [13]

1-Benzyl-2-hydroxymethylimidazole [14]

A mixture of aqueous formaldehyde (d 1.08, 25 ml) and 1-benzylimidazole (15.8 g, 0.1 mol) is heated in a sealed glass tube at 150°C (3 h). After cooling, the mixture is treated with water (200 ml) and extracted with dichloromethane (3 × 30 ml). The organic extracts are washed with water (2 × 50 ml), dried

7.2. SYNTHESIS OF C-SUBSTITUTED IMIDAZOLES AND BENZIMIDAZOLES

(Na_2SO_4) and evaporated to give the product as an oil (17.86 g, 95%). The hydrochloride salt has an m.p. of 161–163°C.

Although Friedel–Crafts alkylations and acylations will not take place with imidazoles (Lewis acid catalysts deactivate the azole), it is possible to introduce some acyl and aroyl groups at C-2 by what is essentially an electrophilic substitution reaction. The conditions used are modified Schotten–Baumann conditions in which an acyl or aroyl halide reacts with a 1-substituted imidazole in acetonitrile solution in the presence of triethylamine [15–17].

2-Benzoyl-1-methylimidazole [17]
To a solution of 1-methylimidazole (5 g, 61 mmol) in acetonitrile (60 ml) is added sequentially benzoyl chloride (7.1 ml, 8.6 g, 61 mmol) and triethylamine (8.5 ml, 6.2 g, 61 mmol). After 15 h at 25°C the triethylamine hydrochloride is filtered off, the acetonitrile is rotary evaporated, and the residue is boiled (30 min) with aqueous sodium carbonate. Continuous extraction with chloroform of the alkaline solution yields, after removal of the dried solvent, an oil which is distilled to give a pale yellowish liquid (4.2 g, 37%), b_1 149°C. Similarly prepared are ethyl 1-methylimidazole-2-carboxylate ($b_{0.05}$ 60°C), 2-benzoyl-1-phenylimidazole (61%, b_2 210°C) and 2-benzoyl-1-benzylimidazole (68%, m.p. 66°C).

7.2.1.2 Benzimidazoles

As mentioned above, most initial electrophilic substitutions in benzimidazoles take place at C-5, followed by attack at C-6. Electron donors in the 5-position direct attach to C-4; electron-withdrawing groups usually lead to C-4 or C-6 substitution. Unless one is wanting to make simple compounds such as 5-nitro or 5-sulfonic acid derivatives of benzimidazole, it is probably more satisfactory to use a ring-synthetic procedure based on a suitably substituted *o*-phenylenediamine. Halogenations are also inclined to be multiple halogenations, e.g. both benzimidazole and 2-methylbenzimidazole very readily form the 4,5,6-trichloro derivatives. Many bromination regimes are equally non-specific, although NBS supported on silica gel is reported to give 2-bromobenzimidazole in the first instance [18]. If the benzimidazole can be induced to react as its anion, 2-iodination can also be observed [19].

7.2.2 Electrophilic substitutions which involve metallic derivatives

Carbanions generated by the formation of lithium derivatives or Grignard reagents react readily with a wide variety of electrophiles. These include the usual alkylating agents, aldehydes and ketones, acid chlorides, carbon dioxide, disulfides, organic nitriles and halogen sources. In addition there are the quenching agents which introduce a formyl group (DMF or methyl formate), cyano group (tosyl cyanide or cyanogen), nitro group (propyl nitrate, N_2O_4 or tetranitromethane), trimethylsilyl group (trimethylsilyl chloride) or tributylstannyl group (tributylstannyl chloride) [20].

The carbanions can be generated by metal–hydrogen or metal–halogen exchange, and conditions have now been refined to permit regiospecific metallation at the 2-, 4- and/or 5-positions of imidazole. The differing reactivities of anions at these sites can be used as a powerful synthetic tool. The reactions have been reviewed [20, 21].

7.2.2.1 Imidazoles

Provided that imidazole is suitably *N*-protected, metallation at ring carbons occurs in the order C-2 > C-5 > C-4. The range of protecting groups has been surveyed in Section 7.1. *N*-Protected imidazoles are monolithiated in the 2-position, commonly at −78°C, with n-butyllithium frequently the reagent of choice. If, however, the imidazole carries groups susceptible to nucleophilic attack, lithium diisopropylamide (LDA) is usually more successful. Reaction conditions can be selected for specific 2-lithiation even when the imidazole has substituents capable of reacting with the reagent. If the 2-position is already substituted, 5-lithiation occurs readily. With 2-fluoro-1-tritylimidazole, t-butyllithium reacts to form the 4-lithio derivative, presumably because of steric hindrance. One has to be aware of the possibility of lateral metallation of substituent groups such as methyl, benzyl and phenyl, e.g. 1,2-dimethylimidazole can be lithiated at the 2-methyl group or at C-5 depending on the reagent and reaction conditions. The nature of the eventual product is dependent on the relative softness or hardness of the quenching electrophile. Conditions have now been determined to allow regiospecific lithiation at C-2 or C-5, or lateral lithiation. With 2 mol of metallating reagent, 2,5-dilithiation is observed, allowing synthesis of 1,2,5-trisubstituted imidazoles. Since the 5-anion is more reactive than the 2-anion it is possible to carry out selective, sequential quenching with two different electrophiles.

Metal–halogen exchange opens up further possibilities. At −50 to −110°C, 2-bromo-, 2-iodo- (and, rarely, 2-chloro-) imidazoles give the imidazol-2-yllithium derivatives, but any free annular NH group is lithiated first. Excess (at least two molar equivalents) metallating agent will give the *N,C*-dianion. Bromine atoms in other positions are less reactive (2-Br > 5-Br > 4-Br; the adjacent lone pair effect mitigates against anion formation in the 4-position), but this reactivity sequence is synthetically useful, e.g. in the conversion of 2,4,5-tribromoimidazole into 4-bromoimidazole using four molar equivalents of n-butyllithium followed by addition of methanol [22]. In *N*-substituted

7.2. SYNTHESIS OF C-SUBSTITUTED IMIDAZOLES AND BENZIMIDAZOLES

4,5-dibromoimidazoles, exchange occurs initially at C-5, but transmetallation with the 2-position can take place, especially if the reaction is allowed to warm up [23].

It is generally much more difficult to exchange at the 4-position, but an alternative approach to the generation of imidazol-4-yl anions in 2-unsubstituted imidazoles adds ethylmagnesium bromide to an *N*-protected 4-iodoimidazole. Quenching with a variety of aldehydes and ketones gives good yields of imidazolyl-4-carbinols uncontaminated by 2-substituted products [24].

A selection of syntheses illustrates the potential of these procedures. For example, the synthesis of 1-methylimidazole-5-carbaldehyde (**1**) (Scheme 7.2.1) illustrates the uses of blocking groups and regiospecific metallation [25]. In this particular sequence around 10–15% of the regioisomer is formed at the quenching stage, resulting in a small amount of 1-methylimidazole-4-carbaldehyde as a contaminant.

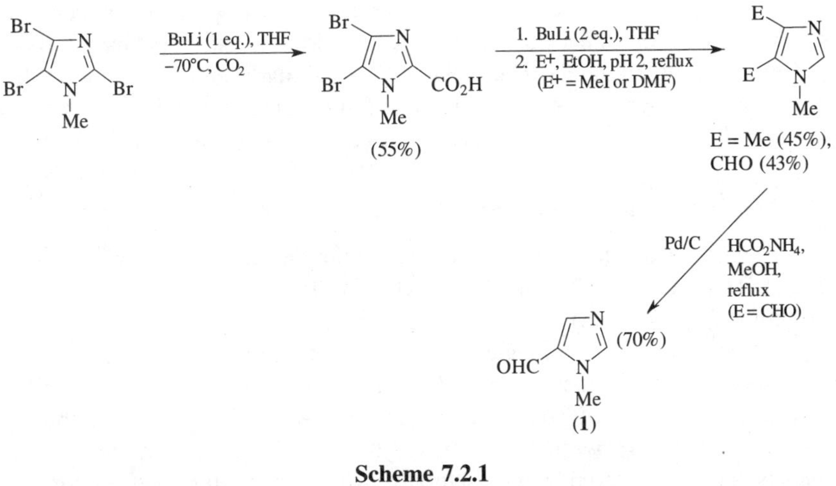

Scheme 7.2.1

Procedures used in the following synthetic examples need dry apparatus and solvents, and often require specialized techniques (see Wakefield's monograph in this series [26]).

4,5-Dibromo-1-methylimidazole-2-carboxylic acid [25]

To a stirred solution of 2,4,5-tribromo-1-methylimidazole (15 g, 47 mmol) in THF at −70°C under argon is added n-butyllithium (1.1 eq.). After 30 min dry carbon dioxide (dry ice evaporated through a sulfuric acid tower) is bubbled into the solution (1 h) before allowing the mixture to come to room temperature. The lithium carboxylate is precipitated by addition of hexane and filtered under argon. Conversion to the free acid is achieved by dissolving the solid in 50% aqueous ethanol, acidifying to pH 2 and refrigeration. The product crystallizes as white needles (7.40 g, 55%), m.p. 112–113°C. The material is dried under vacuum over P_2O_5, and can be stored in a vacuum dessicator for several months without significant decarboxylation. It does, however, decarboxylate in chloroform solution, presumably because of traces of HCl.

4-Bromo-1,5-dimethylimidazole [25]

Two equivalents of n-butyllithium are added to a THF solution of the above product at −70°C under argon to give a fine suspension. After 10 min methyl iodide (1.4 eq.) is added, and the mixture is allowed to warm up to 0°C. A small amount of 2 M HCl is added to quench the reaction, and the mixture is concentrated on a rotary evaporator. The residue is dissolved in ethanol, and HCl is added until the pH reaches 2. The mixture is then refluxed (1 h) and concentrated *in vacuo* to give a residue which is purified by column chromatography on silica gel eluted with dichloromethane containing 10% methanol and 1% ammonia. The yield is 45%. Similarly prepared using DMF as a quenching agent is 4-bromo-1-methylimidazole-5-carbaldehyde (43%), m.p. 88–90°C. Catalytic transfer dehalogenation removes the final bromine.

N,N-Dimethyl-2,5-dimethylimidazole-1-sulfonamide [27]

To a stirred solution of *N,N*-dimethylimidazole-1-sulfonamide (see Section 7.1.5) (1.33 g, 7.6 mmol) in DME (30 ml) at −5°C is added n-butyllithium (16.7 mmol) in hexane (12.12 ml) over 0.25 h. The solution is then cooled to −78°C and quenched by the addition of iodomethane (1.3 ml, 20.9 mmol). Stirring is then continued (12 h) at room temperature, after which the solution is extracted with 2 M aqueous HCl (3 × 10 ml). The combined aqueous extracts are washed with diethyl ether (2 × 50 ml), basified with aqueous sodium hydroxide and saturated with NaCl. Repeated extraction with chloroform (4 × 20 ml), drying and rotary evaporation gives the product (0.81 g, 53%), which is purified further by distillation, $b_{0.26}$ 65°C.

2-Nitroimidazole [28]

In a 250 ml flask a stirred solution of 1-tritylimidazole (3.1 g, 10 mmol) in dry THF (100 ml) is cooled to 0°C before addition of n-butyllithium (8.5 ml, 11 mmol of 1.3 M solution in hexane) over 2 min. The initial colourless solution is stirred at ambient temperature (2 h), by which time a dark red colour has developed and the lithium salt has separated as a white precipitate. A

7.2. SYNTHESIS OF C-SUBSTITUTED IMIDAZOLES AND BENZIMIDAZOLES

solution of n-propyl nitrate (1.47 g, 1.4 ml, 14 mmol) in dry THF (15 ml) is added dropwise over 5 min. Stirring is continued (30–60 min) at ambient temperature (injection of neat n-propyl nitrate from a syringe in a single portion is equally effective). The red colour is quickly discharged, and the solid gradually dissolves with formation of a dark brown solution. Stirring is continued 30 min after the final solution has been achieved. The solution is then cooled to 0°C, diluted with methanol (100 ml) and concentrated HCl (10 ml), and stirred overnight at room temperature to remove the trityl group and to hydrolyse remaining traces of nitrate ester (hydrolysis can also be effected by reflux for several hours). The solvent is evaporated, and the residue triturated with 50% aqueous ethanol (10 ml) and filtered. The filtrate is evaporated to dryness, and the residual orange-brown solid is chromatographed on alumina (150 g, grade III). Eluting the column with chloroform (500 ml) removes triphenylcarbinol and other components. Further elution with 5% methanol in chloroform gives 2-nitroimidazole (0.33 g, 30%), m.p. 283–285°C.

The 2-position can also be blocked by a silyl group to allow the preparation of 4(5)-substituted imidazoles (**2**) (Scheme 7.2.2).

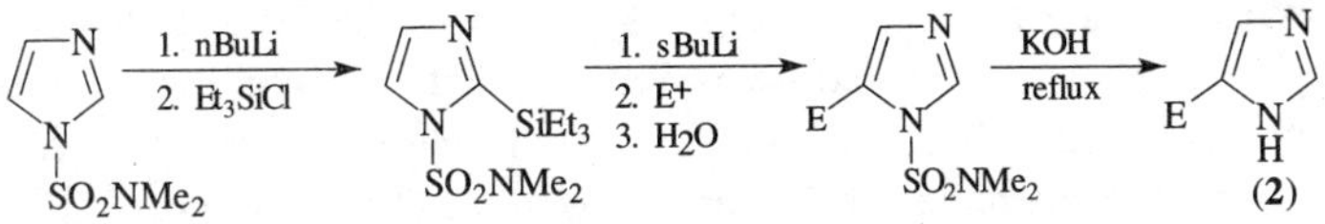

Scheme 7.2.2

General method: 4(5)-substituted imidazoles (**2**) [29, 30]

To a solution of the 1-sulfonamide (see Section 7.1) (1.0 g, 5.71 mmol) in dry THF (30 ml) at $-78°C$ is added n-butyllithium (6.28 mmol) in hexane. The reaction mixture is stirred at $-78°C$ (30 min) to generate the 2-lithio derivative. To this is added triethylchlorosilane (1.92 ml, 11.42 mmol), and the mixture is stirred at 20°C (16 h), after which time the solvent and any excess of chlorosilane is removed by evaporation under reduced pressure and gentle heating. THF (30 ml) is added to the residual oil, and the solution is cooled to $-78°C$ before addition of s-butyllithium (11.42 mmol) in cyclohexane solution and further stirring to generate the 5-lithio derivative. The electrophile, e.g. dimethyldisulfide (1.62 ml, 18 mmol), is added and the mixture is allowed to warm up to 20°C. Stirring is continued (12 h), the solvents are removed by rotary evaporation and the residue is stirred (30 min) with 2 M aqueous HCl (50 ml). The solution is washed with light petroleum (2 × 10 ml), basified to pH 11 with aqueous KOH solution (40% w/w), extracted with diethyl ether (6 × 30 ml), dried (MgSO$_4$) and the solvent is evaporated. The crude product is distilled under reduced pressure to give 5-methylthio-N,N-dimethylimidazole-1-sulfonamide as a clear oil (1.16 g, 92%), $b_{0.3}$ 170°C. Similarly prepared by addition of iodomethane to the 5-lithio species is the 5-methyl derivative (96%), and by addition of chlorotrimethylsilane, the 5-trimethylsilyl derivative (88%). The protecting group is removed by refluxing (12 h) with aqueous KOH (2% w/w, 150 ml), removal of the water, trituration with THF (200 ml), drying (MgSO$_4$) and evaporation. Products (2) prepared in this way are (E, yield given): PhCH$_2$, 64%; Ph$_2$COH, 78%; CO$_2$Et, 74%; Cl, 72%; SMe, 90% (based on the immediate precursor); Me, 92% (based on the immediate precursor); CHO, 80.5%. The imidazole-4-carbaldehyde is isolated by extraction with methanol, evaporation to dryness, extraction of the brownish residue with hot ethyl acetate, evaporation to dryness and, finally, recrystallization from methanol [30].

General procedure: 1-protected 4-imidazolylcarbinols [24]

A 3 M solution of ethylmagnesium bromide (1 eq.) in diethyl ether is added to a 0.25 M solution of N-protected 4-iodoimidazole (suitable N-protecting groups are trityl, SEM and dimethylsulfamoyl) (2 mmol) in dry dichloromethane at room temperature. After 30 min the aldehyde or ketone (1.1 eq.) is added, and the mixture is left to stand overnight. Half-saturated ammonium chloride solution is added, and the aqueous phase is extracted twice with dichloromethane. The combined organic extracts are dried (MgSO$_4$) and concentrated *in vacuo*. Flash chromatography gives the pure products. Made in this way are the following 1-tritylimidazoles: 4-(1'-hydroxyethyl), 83%; 4-(1'-hydroxybenzyl), 79%; 4-(1'-hydroxy-2'-propenyl), 60%; 4-bis(p-chlorophenyl)hydroxymethyl, 69%.

7.2.2.2 Benzimidazoles

1-Protected benzimidazoles are smoothly lithiated in the 2-position by LDA, n-butyllithium and t-butyllithium. Particularly useful are metallations of benzimidazoles protected by formaldehyde as hemiaminals, or as the 1-(N,N-dialkylamino)methyl derivatives. Reaction with a variety of electrophilic reagents, followed by acid-catalysed deprotection under mild conditions, leaves the 2-substituted benzimidazoles [31, 32].

General method: 2-substituted benzimidazoles [32]

1-Hydroxymethylbenzimidazole (see Section 7.1.2), made from benzimidazole (1.57 g, 13.3 mmol) and either paraformaldehyde or aqueous formaldehyde, is suspended in dry THF (60 ml) in a flask which has been evacuated and flushed (three times) with dry argon. The mixture is allowed to warm up to make a homogeneous solution. This is cooled to −78°C to give a precipitate, to which freshly prepared LDA (2.0 eq. from diisopropylamine (26.5 mmol, 3.73 ml) and n-butyllithium (26.5 mmol, 10.6 ml of 2.5 M hexane solution)) or n-butyllithium (26.5 mmol, 10.6 ml of 2.5 M hexane solution) or t-butyllithium (26.5 mmol, 15.6 ml of 1.7 M pentane solution) are slowly added at −78°C. The cooling bath is removed to allow the solution to warm up to −20°C, and then aged at that temperature (30–60 min) with efficient stirring to give a homogeneous yellow-orange solution which is again chilled to −78°C before quenching with the appropriate electrophile (1.0–1.1 eq.). After allowing the mixture to stand (2 h) at −78°C it is then slowly warmed to −20°C over 6 h. Deprotection is accomplished by quenching at 0°C with aqueous ammonium chloride (20 ml), dilution with ether (100 ml) and careful extraction (four times) with 2 M HCl. The aqueous extracts are combined and made basic with ammonia and ammonium carbonate with efficient stirring at 0°C. The precipitate which forms is filtered and dried (first at the filter pump and then under vacuum for 24 h) to give a solid which is washed with the appropriate solvent and recrystallized. If no precipitate forms, the aqueous layer is extracted with ether or ethyl acetate, the solvent is dried and evaporated, and the crude product remaining is recrystallized, e.g. (2-benimidazolyl)phenylcarbinol (72%), (2-benzimidazolyl)-*p*-chlorophenylcarbinol (55%), (2-benzimidazolyl)diphenylcarbinol (46%), (2-benzimidazolyl)cyclohexanol (50%), 2-methylthiobenzimidazole (58%) and 2-phenylthiobenzimidazole (46%).

REFERENCES

1. M. R. Grimmett, *Adv. Heterocycl. Chem.* **57**, 291 (1993).
2. J. F. O'Connell, J. Parquette, W. E. Yelle, W. Wang and H. Rapoport, *Synthesis* 767 (1988).
3. B. D. Palmer and W. A. Denny, *J. Chem. Soc., Perkin Trans. 1* 95 (1989).
4. S. P. Watson, *Synth. Commun.* **22**, 2971 (1992).
5. P. Benjes and M. R. Grimmett, unpublished.
6. A. W. Lutz and S. DeLorenzo, *J. Heterocycl. Chem.* **4**, 399 (1970).
7. M. R. Grimmett, *Adv. Heterocycl. Chem.* **12**, 103 (1970).
8. S. S. Novikov, L. V. Khmel'nitskii, O. V. Lebedev, V. V. Sevost'yanova and L. V. Epishina, *Chem. Heterocycl. Compd. USSR (Engl. Transl.)* **6**, 669 (1970).
9. A. R. Katritzky, D. J. Cundy and J. Chen, *J. Energetic Mat.* **11**, 345 (1993).
10. L. M. Anderson, A. R. Butler, C. Glidewell, D. Hart and N. Isaacs, *J. Chem. Soc., Perkin Trans. 2*, 2055 (1989).
11. P. W. Alley, *J. Org. Chem.* **40**, 1837 (1975).
12. C. Rufer, K. Schwarz and E. Winterfeldt, *Liebigs Ann. Chem.* 1465 (1975).
13. M. R. Grimmett, in *Comprehensive Heterocyclic Chemistry* (ed. A. R. Katritzky and C. W. Rees). Pergamon Press, Oxford, 1984, Vol. 5 (ed. K. T. Potts).
14. H. Galons, I. Bergerat, C. C. Farnoux and M. Miocque, *Synthesis* 1103 (1982).

15. E. Regel and K. H. Buechel, German Patent 1,926,206 (1971); *Chem. Abstr.* **74**, 31 754 (1971).
16. E. Regel and K. H. Buechel, German Patent 1,956,711 (1971); *Chem. Abstr.* **75**, 49 086 (1971).
17. C. G. Begg, M. R. Grimmett and Lee Yu-Man, *Aust. J. Chem.* **26**, 415 (1973).
18. A. G. Mistry, K. Smith and M. R. Bye, *Tetrahedron Lett.* **27**, 1051 (1986).
19. M. Moreno-Manas, J. Bassa, N. Llado and R. Pleixats, *J. Heterocycl. Chem.* **27**, 673 (1990).
20. B. Iddon and R. I. Ngochindo, *Heterocycles* **38**, 2487 (1994).
21. B. Iddon, *Heterocycles* **23**, 417 (1985).
22. K. Stensio, K. Wahlberg and R. Wahren, *Acta Chem. Scand.* **27**, 2179 (1973).
23. M. P. Groziak and L. Wei, *J. Org. Chem.* **56**, 4296 (1991).
24. R. M. Turner, S. D. Lindell and S. V. Ley, *J. Org. Chem.* **56**, 5739 (1991).
25. G. Shapiro and B. Gomez-Lor, *J. Org. Chem.* **59**, 5524 (1994).
26. B. J. Wakefield, *Organolithium Methods*. Academic Press, London, 1988.
27. D. J. Chadwick and R. I. Ngochindo, *J. Chem. Soc., Perkin Trans. 1.* 481 (1984).
28. D. P. Davis, K. L. Kirk and L. A. Cohen, *J. Heterocycl. Chem.* **19**, 253 (2982).
29. A. J. Carpenter and D. J. Chadwick, *Tetrahedron* **42**, 2351 (1986).
30. J. Winter and J. Retey, *Synthesis* 245 (1994).
31. A. R. Katritzky, G. W. Rewcastle and W.-Q. Fan, *J. Org. Chem.* **53**, 5685 (1988).
32. A. R. Katritzky and K. Akutagawa, *J. Org. Chem.* **54**, 2949 (1989).

7.3 OTHER SUBSTITUTION METHODS

7.3.1 Nucleophilic substitution approaches

Nucleophilic substitutions are rather less synthetically useful than electrophilic methods except under special circumstances. Imidazoles are usually resistant to nucleophiles unless the molecules are activated by an electron-withdrawing group, or by quaternization. Benzimidazoles are rather more susceptible to nucleophilic attack, especially in the 2-position. Indeed, both imidazole and its condensed analogue are most likely to react at this site. A nitro group at C-4 of imidazole, though, can render a bromine or iodine at C-5 more reactive than one at C-2. In 1-substituted imidazoles, 5-halogens are replaced more readily than those at C-4, a function of the better delocalization in the developing Meisenheimer complex. Nucleophilic displacements of hydrogen are unknown or rare in imidazole chemistry; usually a halide, mesyl, tosyl or nitro group is displaced. In benzimidazoles, though, direct amination becomes possible at C-2 with sodamide, e.g. 1-benzylbenzimidazole can be converted into 2-amino-1-benzylbenzimidazole [1]. 2-Aminobenzimidazoles are, however, not difficult to make by ring-synthetic methods (see Section 8.3). The usual interconversions of 2-benzimidazolones and 2-chlorobenzimidazoles may have occasional application.

Perhaps more useful are the transformations of diazonium salts. Imidazole-2-diazonium fluoroborate can be converted into the 2-fluoro- and 2-nitroimidazoles, the former by heating or photolysis, the latter with sodium

7.3. OTHER SUBSTITUTION METHODS

nitrite. A number of 4-fluoroimidazoles can be made similarly, albeit in only moderate yields [2–5]. The use of diazonium salts in imidazole synthesis is limited both by the difficulty of making (and storing) the amino precursors (see Section 8.3), and by the high stability and concomitant unreactivity of some 4-diazoimidazoles [6].

Passing reference should be made here to the *cine* substitution reactions of 1,4-dinitroimidazoles, which give 5(4)-alkoxy-4(5)-nitroimidazoles when treated with alkoxides; similar reactions take place with secondary amines. Such reactions may prove to have preparative value [7, 8]. Nucleophilic attack by primary amines tends to cause ring opening in these compounds, and this may be followed by ring closure to give rearranged products. For example, the treatment of 2-methyl-1,4-dinitroimidazole with *p*-toluidine in aqueous methanol gives a high yield of 2-methyl-4-nitro-1-(*p*-tolyl)imidazole (76%); aniline converts 4-nitro-1-(*o*-nitrobenzenesulfonyl)imidazole in aqueous methanol into 4-nitro-1-phenylimidazole [9]. This last reaction seems to provide a mild approach to the synthesis of 1-arylimidazoles provided that there is sufficient electron withdrawal in the hetero ring.

4-Nitro-1-phenylimidazole [9]

A suspension of 4-nitro-1-(*o*-nitrobenzenesulfonyl)imidazole (made from 4-nitroimidazole, *o*-nitrobenzenesulfonyl chloride, and triethylamine) (0.745 g, 2.5 mmol) and aniline (1.0 g, 11 mmol) in aqueous methanol (1:1; 20 ml) is stirred and heated at 70°C (2 h). The mixture is then steam distilled, and the hot undistilled residue is filtered to give the above product (0.41 g, 87%), m.p. 185–187°C. Cooling of the filtrate results in separation of *o*-nitrobenzenesulfonamide (0.42 g, 86%).

2-Amino-1-benzylimidazole [1]

Benzimidazole (4.8 g, 25 mmol), sodium hydroxide (1.6 g, 40 mmol) and a saturated aqueous solution of *N*-benzyl-*N*-phenyldimethylammonium chloride (10 g, 38 mmol) are heated together (1 h) on a steam bath to give, on cooling, 1-benzylbenzimidazole (7.6 g, 90%), m.p. 116–117°C. This is heated with sodamide in *N*,*N*-dimethylaniline (1 h) at 90°C and finally at 120°C until evolution of hydrogen has ceased. The product (3.5 g, 42%) has an m.p. of 194–195°C.

Ethyl 2,4-difluoroimidazole 5-carboxylate [5]

To a solution of 2-amino-4-fluoroimidazole-5-carboxylate (2.5 g, 14.5 mmol) in chilled 50% fluoroboric acid (100 ml) at −20 to −10°C is added a solution of sodium nitrite (1.2 g, 17.4 mmol) in water (3 ml). The solution is irradiated

(3 h) at −60 to −40°C, at which time the diazonium chromophore (324 nm) has disappeared. The mixture is neutralized to pH 5–6 (at −20 to −10°C) with cold 1 M NaOH, and then extracted with ethyl acetate (3 × 20 ml). The combined organic extracts are dried (Na_2SO_4) and evaporated, and the residue is again extracted with chloroform (2 × 100 ml). The extracts are evaporated to give an oil which is chromatographed on a short silica gel column eluted with ethyl acetate–chloroform (1:3). Rotary evaporation of the solvents gives the pure product (1.35 g, 53%), m.p. 107–109°C.

The irradiation light source is a Hanovia 450 W medium-pressure Hg vapour lamp placed in a quartz immersion well. The reaction solution is contained in a 150 ml quartz semicircular flask mounted as closely as possible to the immersion well. No light filter is used. The entire apparatus is immersed in a large Dewar flask charged with dry ice–isopropanol with the bath temperature maintained between −60 and −40°C during irradiation.

7.3.2 Radical substitution methods

Although a number of such radical reactions are known, few promise much synthetic potential. Examples include the 2-phenylation of imidazole and benzimidazole by benzoyl peroxide, but both products are more readily obtained by other routes. Homolytic alkylations of imidazole and benzimidazole also occur at C-2, but usually give indifferent yields [10]. A potentially useful reaction is the synthesis of 2- and 4-trifluoromethylimidazoles from imidazoles and photochemically generated trifluoromethyl radicals. 1-Substituted imidazoles are largely substituted at C-5 in these reactions; benzimidazole reacts initially at the 4-position [11–14].

Trifluoromethylation of 4-methylimidazole [11]

4-Methylimidazole (8.21 g, 0.1 mol) and trifluoromethyl iodide (9.8 g, 0.05 mol) are dissolved in methanol (40 ml) and irradiated (7 days) at room temperature in a quartz tube (2 × 20 cm) fitted with a Teflon stopper. The remaining air space is filled with dry argon, and the tube is mounted 5 cm from a 15 W low-pressure Hg lamp provided with air cooling. There is no noticeable increase in temperature or pressure in the tube. The reaction mixture is then evaporated to dryness under reduced pressure, and the residue is suspended in water (50 ml) and extracted with ethyl acetate (3 × 50 ml). The combined organic extracts are dried ($MgSO_4$) and again rotary evaporated to dryness. The residue is chromatographed on a column of silica gel (200 ml), eluting with diethyl ether. Recrystallization of the products from the column gives 4-methyl-2-trifluoromethylimidazole (1.61 g, 21.5%), m.p. 107–108°C, as colourless plates from benzene, 5-methyl-4-trifluoromethylimidazole (4.7 g,

63.4%), m.p. 178-179°C, as plates from benzene-ethanol and 5-methyl-2,4-bis(trifluoromethyl)imidazole (0.32 g, 2.9%), m.p. 174-176°C, as needles from ether.

REFERENCES

1. R. Rastogi and S. Sharma, *Synthesis* 861 (1983).
2. K. L. Kirk and L. A. Cohen, *J. Am. Chem. Soc.* **95**, 4619 (1973).
3. K. L. Kirk and L. A. Cohen, *J. Org. Chem.* **38**, 3647 (1973).
4. K. L. Kirk, *J. Org. Chem.* **43**, 4381 (1978).
5. K. Takahashi, K. L. Kirk and L. A. Cohen, *J. Org. Chem.* **49**, 1951 (1984).
6. G. Cirrincione, A. M. Almericao, E. Aiello and G. Dattolo, *Adv. Heterocycl. Chem.* **48**, 65 (1990).
7. J. Suwinski, *Pol. J. Chem.* **58**, 311 (1984).
8. J. Suwinski, E. Salwinska and M. Bialecki, *Pol. J. Chem.* **65**, 1071 (1991).
9. J. Suwinski, J. Pawlus, E. Salwinska and K. Swierczek, *Heterocycles* **37**, 1511 (1994).
10. C. G. Begg, M. R. Grimmett and Y. M. Lee, *Aust. J. Chem.* **26**, 415 (1973).
11. H. Kimoto, S. Fujii and L. A. Cohen, *J. Org. Chem.* **47**, 2867 (1982).
12. H. Kimoto, S. Fujii and L. A. Cohen, *J. Org. Chem.* **49**, 1060 (1984).
13. Q.-Y. Chen and Z.-T. Li, *J. Chem. Soc., Perkin Trans. 1* 645 (1993).
14. V. M. Labroo, R. B. Labroo and L. A. Cohen, *Tetrahedron Lett.* **31**, 5705 (1990).

–8–
Synthesis of Specifically Substituted Imidazoles and Benzimidazoles

This chapter focuses on methods of preparation of imidazoles and benzimidazoles with a particular substituent at a specific ring position. The survey is largely confined to the preparation of compounds with substituent groups difficult to introduce by ring-synthetic methods. Accordingly, there are only a few references to benzimidazoles with substituents in the homocyclic ring, and these relate to groups best introduced by direct substitution in the preformed benzazole, or by modification of groups already present. There is naturally considerable overlap with the topics covered in Chapter 7, and in most instances ring-synthetic procedures are merely listed with a cross-reference to earlier chapters. The content should be studied in concert with the index.

8.1 HALOGEN GROUPS

As discussed earlier (see Section 7.2), electrophilic monochlorination, -bromination and -iodination of imidazoles and 1-substituted imidazoles is difficult because of the great propensity of the molecules to polyhalogenate. Methods have, however, been developed to achieve selective halogen introduction on all ring sites. Fluorination is a special case (see below).

8.1.1 Fluoro derivatives

2-Fluoroimidazole and -benzimidazole can be made from the corresponding diazonium salts. This necessitates the prior synthesis of the 2-amino derivatives, a relatively simple process for benzimidazoles [1], but less so for imidazoles. The Balz–Schiemann method gives the best yields for 2-fluoroimidazoles, although those yields can be as low as 30–40%. Working in fluoroboric acid will often improve matters. Both 4- and 5-fluoroimidazoles have been made similarly; they are somewhat more stable than 2-fluoroimidazoles, but their preparation is made difficult by the problems of making and handling the 4(5)-aminoimidazole precursors, and the high stabilities of the derived diazonium salts [2–5]. (See also Section 7.4.)

Although some nucleophilic fluorinations have been reported in which potassium, caesium, silver or xenon fluorides react with suitably activated bromo- or iodoimidazoles, such procedures appear to be limited in their application [6, 7]. Electrophilic fluorinations, too, are confined to reactions of metallic derivatives, e.g. both 2- and 4-fluoroimidazoles have been made from the lithium derivatives. Reaction of 2-lithio-1-methylimidazole with perchloryl fluoride gives a greater than 50% yield of 2-fluoro-1-methylimidazole [8]. Rather more convenient appear to be the reactions of caesium fluoroxysulfate or fluorine at −78°C with the appropriate 2- and 5-trimethylstannyl derivatives which are readily available, although methods do not yet appear to be optimized. Mercury derivatives react similarly, but they are more difficult to prepare and purify [9, 34, 35].

2-Fluoro-1-methylimidazole [2]

A solution of sodium nitrite (3.04 g, 44 mmol) in water (10 ml) is slowly added to a stirred solution of 2-amino-1-methylimidazole [33] (6.06 g, 40 mmol) in 8.5 M tetrafluoroboric acid (250 ml) cooled to −20°C. Sodium tetrafluoroborate is then added until saturation has been achieved. Nitrogen is bubbled through the reaction mixture, which is now irradiated at −30°C until the α-naphthol test for diazonium salts is negative (∼5 h). The resulting solution is neutralized first with cold sodium hydroxide solution, then sodium bicarbonate, before filtering under vacuum. The filtrate is continuously extracted with dichloromethane (24 h), the extracts are dried (Na_2SO_4) and the solvent is removed *in vacuo*. Vacuum distillation of the residual oil gives the product (1.90 g, 48%), b_{24} 66–69°C.

1-Methyl-2-trimethylstannylimidazole [34]

To a solution of 1-methylimidazole (3.28 g, 40 mmol) in freshly distilled THF (50 ml) cooled to −10°C is added a solution of n-butyllithium (40 mmol) in hexane (25 ml) over 1 h. The solution is kept at −10°C and stirred (1 h) before adding a solution of trimethyltin chloride (40 ml) in diethyl ether (50 ml) over 1 h via a syringe. The mixture is allowed to warm to room temperature, filtered using conventional Schlenk apparatus, and the solvents are removed under reduced pressure. The product is isolated by distillation (10.79 g, 73%), $b_{0.5}$ 84–86°C.

8.1.2 Chloro derivatives (see also Sections 7.2.1.1 and 7.2.2.1)

Clean monochlorination of all possible ring positions in imidazole is most likely to be successful using the lithio precursors, e.g. 1-benzyl-2-lithioimidazole reacts with hexachloroethane to give 1-benzyl-2-chloroimidazole [10, 11]; quenching the 2-lithio derivative of 1-tritylimidazole with chlorine, followed by deprotection gives 2-chloroimidazole in 39% yield

8.1. HALOGEN GROUPS

(see Section 8.1.4) [3]. There are many other examples, including those in which C-2 of imidazole is protected by a triethylsilyl, t-butyldimethylsilyl or dimethylcarboxamido group. The anion is then formed at C-5, opening the way to the preparation of 4(5)-chloroimidazole (Scheme 8.1.1) [12, 13]. Conversions of 2-imidazolinones into 2-chloroimidazoles with phosphoryl chloride (ideally with copper(I) chloride catalyst) may have occasional application [14], but other nucleophilic chlorinations depend on sufficient electron withdrawal being present in the ring, e.g. 5-chloro-4-nitroimidazole can be made by heating the bromo analogue with concentrated hydrochloric acid [15]. Halodenitrations are also possible if there is more than one nitro group in the ring [5].

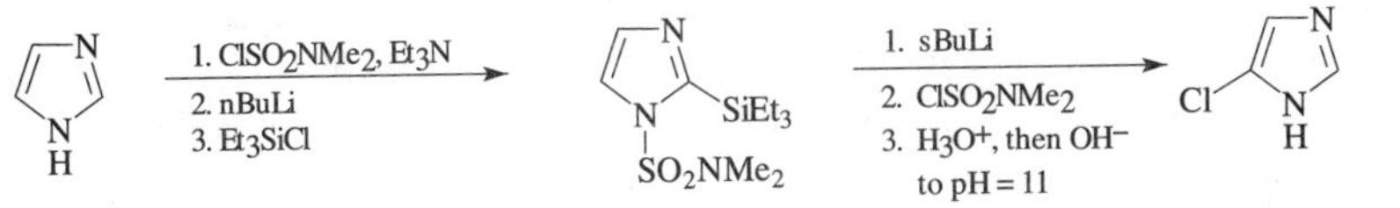

Scheme 8.1.1

Both benzimidazole and 2-methylbenzimidazole undergo polychlorination in the fused benzene ring (the former forms 4,5,6-trichlorobenzimidazole), but the reactions are difficult to control. If 2-chlorobenzimidazole is required, the most convenient routes are from 2-benzimidazolone or from benzimidazole 1-oxide treated with phosphoryl chloride [14, 15]. Ring-synthetic approaches to chloroimidazoles include cyclizations of oxamides with PCl_5 to give 1-substituted 5-chloroimidazoles (see Section 2.1.1). There is also a specific synthesis of 4,5-dichloroimidazoles from cyanogen and aldehydes in the presence of HCl (see Section 3.1.1).

4(5)-Chloroimidazole [12]

To a solution of N,N-dimethylimidazole-1-sulfonamide [17] (see Section 7.2.6) (1.0 g, 5.71 mmol) in THF (30 ml) at −78°C is added n-butyllithium (6.28 mmol) in hexane. The reaction mixture is stirred at −78°C (30 min). To this is added triethylchlorosilane (1.92 ml, 11.42 mmol), and the mixture is stirred at 20°C (16 h), before excess chlorosilane and solvent are removed under reduced pressure and gentle heating. THF (50 ml) is added to the resulting oil, the solution is cooled to −78°C, s-butyllithium (11.42 mmol) in cyclohexane solution is added, and the mixture is stirred at −78°C (30 min) to generate the 5-lithio derivative. To this is added dimethylsulfamoyl chloride (1.84 ml, 17.14 mmol), and the mixture is stirred at −78°C (30 min), then at 20°C (2 h). Solvents are removed, and the residue is stirred with 2 M aqueous HCl (100 ml) at 20°C (1 h). The resulting solution is washed with light petroleum (2 × 10 ml), and basified to pH 11 with aqueous potassium hydroxide (40% w/w). Extraction of the aqueous layer with ethyl acetate (6 × 30 ml), drying (MgSO$_4$) and rotary evaporation gives the crude product as a tan solid. Recrystallization from cyclohexane-dichloromethane gives pure 4-chloroimidazole (0.42 g, 72%), m.p. 120-121°C. (See also the alternative route in Section 7.2.1.)

8.1.3 Bromo derivatives

Imidazoles and benzimidazoles polybrominate with great ease with most brominating agents (see Section 7.2.1), and even imidazoles with electron-withdrawing substituents can be brominated at vacant sites [16]. Attempts to make use of the relative positional reactivities (0.45:0.2 (4(5)-:2-substitution of imidazole); 2.3:1.7:0.8 (5-:4-:2-substitution for 1-methylimidazole) are seldom successful for synthetic purposes, although Denny has successfully monobrominated imidazole in the 4-position in an acceptable yield by careful addition of NBS in DMF (see Section 7.3.1). The conventional route to 4-bromoimidazole has been by "reduction" of 2,4,5-tribromoimidazole with sulfite or triphenylphosphite [5]. The more recent sequential use of butyllithium and water provides a logical alternative route to 4-bromoimidazole (the anions form most readily at C-2 and C-5). It is therefore possible to replace the bromine atoms by hydrogen in a stepwise manner in the order 2 > 5 > 4 by employing one, two or three molar equivalents of butyllithium [18-20]. For the preparation of 2-iodoimidazole see Section 8.1.4.

4-Bromo-2-ethylimidazole [21]

Bromine (10.24 ml, 200 mmol) is added dropwise to a stirred, ice-cooled solution of 2-ethylimidazole (8.54 g, 89 mmol) in ethanol (150 ml). After stirring at room temperature (~3 h), 5 M aqueous sodium hydroxide (45 ml) is added, to give a solution with a pH of about 6.5. To this is added a solution of sodium sulfite (120 g) in water (1 l), followed by further ethanol (100 ml) and water (100 ml), to give on warming to about 40°C a yellow solution which is refluxed (18 h). The resulting solution is concentrated to half its volume, and then extracted with chloroform (3 × 250 ml). The combined extracts are dried (MgSO$_4$), filtered and rotary evaporated to give a white solid (6.1 g, 38%). ^{1}H NMR: 1.19, t, 3H, CH$_3$; 2.78, q, 2H, CH$_2$; 6.95, s, 1H, H-5.

1-Benzyl-4-bromoimidazole [20, 22]

n-Butyllithium (1.78 M in hexane, 68.2 ml, 120 mmol) is added dropwise to a stirred solution of 1-benzyl-2,4,5-tribromoimidazole (20.0 g, 50.63 mmol) in dry diethyl ether (175 ml) at −78°C under dry, oxygen-free nitrogen. The

8.1. HALOGEN GROUPS

resulting mixture is stirred at $-78°C$ (1 h), water (50 ml) is then added, and stirring is continued for a few minutes before separation of the ethereal and aqueous layers. The latter is extracted several times with ether, and the organic layers are combined, dried ($MgSO_4$) and evaporated to give an oil which is chromatographed on alumina using gradient elution with light petroleum containing increasing proportions of ethyl acetate. The product (8.5 g, 71%), after removal of the solvent, is recrystallized from ethyl acetate-light petroleum, m.p. 88-90°C.

The difficulty of monobrominating benzimidazole and its 1-substituted derivatives mirrors the state of affairs with the uncondensed imidazoles. Electrophilic bromination occurs at first in the 5-position, then at C-7, but excess brominating agent often substitutes all available positions on the fused benzene ring [23]. It has been found, though, that NBS supported on silica gel forms the 2-bromobenzimidazole (67%) in the first instance [32]. The same compound can also be made from 2-benzimidazolone, and it should be readily available via the 2-anion formed by reaction of an N-protected benzimidazole with LDA, n-butyllithium or t-butyllithium. Hydroxymethyl and N-(dialkylamino)methyl protecting groups would appear to be the best choices [24, 25].

8.1.4 Iodo derivatives

Many earlier references to iodoimidazoles need to be read with caution because of the perpetuation of an historical error which represented 4,5-diiodoimidazole as its 2,4-isomer. The matter has now been clarified, but a number of publications prior to 1980 contain misleading data [5, 26].

Iodination of imidazole with iodine and potassium iodide in aqueous sodium hydroxide gives mainly 2,4,5-triiodoimidazole, but minor changes to the reaction conditions can give 4,5-diiodoimidazole (not 2,4-diiodoimidazole — see above) as the major product [27]. The 4-monoiodoimidazole becomes available by reductive methods (e.g. with sodium sulfite) in up to 70% yield [26-28]. 1-Methylimidazole is much more difficult to triiodinate, but under a variety of reaction conditions mixtures of mono-, di- and triiodo products are formed, with the 4,5-diiodo species usually predominating. It can be formed almost exclusively (up to 95%) [29]. The 2- and 5-iodo derivatives can be made from the lithio derivatives [3, 11], but 2-iodo- (or 2-bromo-) imidazole is not readily available from 2-lithio-1-phenylsulfonylimidazole [30]. When 1-benzyl-2-lithioimidazole is quenched with 2-nitroiodobenzene the 2-iodo product is formed in only 35% yield [11], while lithiation of 1,2-dimethylimidazole and quenching with finely divided iodine gives a poor yield (~16%) of 2-iodo-1,2-dimethylimidazole. It is possible that some lateral metallation is taking place [22]. Considerable work has been done on conversions of 1-tritylimidazole into 2-halogenoimidazoles. When quenched with NIS or iodine the 2-lithio-1-tritylimidazole gives a 40% yield of the 2-iodo product; with NBS the bromo analogue is formed in 35% yield; NCS gives only about 5% of the 2-chloro product, but t-butyl hypochlorite raises the yield to 39%. Since the trityl group can be removed almost quantitatively, this approach appears to be a useful one [3].

2-Iodoimidazole [3]

To a solution of 1-tritylimidazole (see Section 7.1.1) (0.62 g, 2 mmol) in THF (freshly distilled from $LiAlH_4$, 25 ml) at 0°C in a nitrogen atmosphere is added n-butyllithium (1.6 M in hexane, 1.5 ml). The solution, which gradually turns red, is stirred at room temperature (1.5 h), then cooled to 0°C before the dropwise addition of finely divided iodine (0.508 g, 2 mmol) (similar results are obtained using NIS) in THF (5 ml) over a period of 5 min. After an additional 10 min at 0°C, the mixture is poured into water (25 ml), the solution is concentrated by rotary evaporation and extracted with ether, and the organic layers are dried (Na_2SO_4) and concentrated. Chromatography on silica gel eluted with ether–petroleum ether (1:1) gives 1-trityl-2-iodoimidazole (0.35 g, 40%) which, on recrystallization from ethyl acetate–cyclohexane has an m.p. of 170–172°C. Similarly prepared are (2-halogen substituent, quenching agent, yield, m.p. given): bromo, NBS, 35%, 208–209°C; chloro, t-butyl hypochlorite, 39%, 208–210°C. Removal of the trityl group in each instance is achieved by the following method. 2-Iodo-1-tritylimidazole (0.350 g, 0.8 mmol) is heated under reflux (30 min) in 5% methanolic acetic acid (5 ml). After evaporation of the solvent, water is added to the residue, the solution is chilled and filtered, and the filtrate is evaporated to give 2-iodoimidazole (0.155 g, 99%), m.p. 190–192°C. Similarly prepared are the 2-bromo- (>99%, m.p. 197–198°C) and 2-chloro- (98%, m.p. 166–167°C) imidazoles.

4,5-Diiodoimidazole [31]

A solution of iodine (0.15 g, 59 mmol) in 10% aqueous potassium iodide (20 g, 120 mmol in 200 ml) is added dropwise to a stirred solution of imidazole (2.30 g, 34 mmol) in 2 M sodium hydroxide (200 ml) at room temperature. The mixture is stirred overnight. Addition of 25% aqueous acetic acid is continued until the mixture is neutral. The white precipitate which forms is filtered, washed with water and air dried before recrystallization from ethanol as colourless crystals (6.42 g, 42%), m.p. 197–198°C.

2,4,5-Triiodoimidazole [31]

An aqueous solution of iodine (20.32 g, 80 mmol) and potassium iodide (26.56 g, 160 mmol) in water (200 ml) is added dropwise to a stirred solution of imidazole (1.36 g, 20 mmol) in 2 M sodium hydroxide (200 ml) at room temperature. The mixture is stirred overnight, and 25% aqueous acetic acid is added to bring the pH close to 7, when a creamy precipitate forms. This is filtered, washed with water and air dried before recrystallization from ethanol as colourless crystals (8.15 g, 91%), m.p. 191°C.

REFERENCES

1. K. L. Kirk and L. A. Cohen, *J. Org. Chem.* **38**, 3647 (1973).
2. F. Fabra, C. Galvez, A. Gonzalez, P. Viladoms and J. Vilarassa, *J. Heterocycl. Chem.* **15**, 1227 (1978).
3. K. L. Kirk, *J. Org. Chem.* **43**, 4381 (1978).
4. K. Takahashi, K. L. Kirk and L. A. Cohen, *J. Org. Chem.* **49**, 1951 (1984).
5. M. R. Grimmett, *Adv. Heterocycl. Chem.* **57**, 291 (1993).
6. K. L. Kirk and L. A. Cohen, *J. Am. Chem. Soc.* **95**, 4619 (1973).
7. L. N. Nikolenko, N. S. Tolmucheva and L. D. Shustov, *J. Gen. Chem. USSR (Engl. Transl.)* **49**, 1251 (1979).
8. P. Bouchet, C. Coquelet and J. Elguero, *Bull. Soc. Chim. Fr.* 171 (1977).
9. M. R. Bryce, R. D. Chambers, S. T. Mullins and A. Parkin. *Bull. Soc. Chim. Fr.* 930 (1986).
10. D. K. Anderson, J. A. Sikorski, D. B. Reitz and L. T. Pilla, *J. Heterocycl. Chem.* **23**, 1257 (1986).
11. M. Moreno-Manas, J. Bassa, N. Llado and R. Pleixats, *J. Heterocycl. Chem.* **27**, 673 (1990).
12. A. J. Carpenter and D. J. Chadwick, *Tetrahedron* **42**, 2351 (1986).
13. R. I. Ngochindo, *J. Chem. Soc., Perkin Trans. 1* 1645 (1990).
14. C. P. Whittle, *Aust. J. Chem.* **33**, 1545 (1980).
15. T. J. Benson and B. Robinson, *J. Chem. Soc., Perkin Trans. 1* 211 (1992).
16. M. R. Grimmett, in *Comprehensive Heterocyclic Chemistry* (ed. A. R. Katritzky and C. W. Rees). Pergamon Press, Oxford, 1984, Vol. 5 (ed. K. T. Potts).
17. D. J. Chadwick and R. Ngochindo, *J. Chem. Soc., Perkin Trans. 1* 481 (1984).
18. B. Iddon, *Heterocycles* **23**, 417 (1985).
19. B. Iddon and N. Khan, *J. Chem. Soc., Perkin Trans. 1* 1445 (1987).
20. B. Iddon and N. Khan, *J. Chem. Soc., Perkin Trans. 1* 1453 (1987).
21. S. P. Watson, *Synth. Commun.* **22**, 2971 (1992).
22. B. Iddon and B. L. Lim, *J. Chem. Soc., Perkin Trans. 1* 271 (1983).
23. M. R. Grimmett, *Adv. Heterocycl. Chem.* **59**, 245 (1994).
24. A. R. Katritzky, G. W. Rewcastle and W. Q. Fang, *J. Org. Chem.* **53**, 5685 (1988).
25. A. R. Katritzky and K. Akutagawa, *J. Org. Chem.* **54**, 2949 (1989).
26. J. P. Dickens, R. L. Dyer, B. J. Hamill, T. A. Harrow, R. H. Bible, R. M. Finnegan, K. Henrick and P. G. Owsten, *J. Org. Chem.* **46**, 1781 (1981).
27. R. M. Turner, S. V. Ley and S. D. Lindell, *J. Org. Chem.* **56**, 5739 (1991).
28. B. Iddon and B. L. Lim, *J. Chem. Soc., Perkin Trans. 1* 735 (1983).
29. M. El Borai, A. H. Moustafa, M. Anwar and F. I. Adbel Hay, *Pol. J. Chem.* **55**, 1659 (1981); *Chem. Abstr.* **99**, 5551 (1983).

30. R. J. Sundberg, *J. Heterocyclic Chem.* **14**, 517 (1977).
31. A. R. Katritzky, D. J. Cundy and J. Chen, *J. Energetic Mater.* **11**, 345 (1993).
32. A. G. Mistry, K. Smith and M. R. Bye, *Tetrahedron Lett.* **27**, 1051 (1986).
33. B. T. Storey, W. W. Sullivan and C. L. Moyer, *J. Org. Chem.* **29**, 3118 (1964).
34. K. C. Molloy, P. C. Waterfield and M. F. Mahon, *J. Organomet. Chem.* **365**, 61 (1989).
35. K. Gaare, T. Repstad, T. Benneche and K. Undheim, *Acta Chem. Scand.* **47**, 57 (1993).

8.2 NITROGEN GROUPS

8.2.1 Nitro derivatives

4(5)-Nitroimidazoles are readily made by nitration of imidazole or 1-substituted imidazoles in concentrated sulfuric acid (see Section 7.2.1). It is much more difficult to make 2-nitroimidazoles since direct nitration is seldom observed in the 2-position. Although electrophilic nitrodehalogenation reactions, too, occur mainly at C-4(5) [1], Katritzky has recently selectively nitrodeiodinated 2,4,5-triiodoimidazole to prepare 2,4(5)-dinitro-5(4)-iodo- and 2,4,5-trinitroimidazoles, albeit in poor yield [2]. Other routes to 2-nitroimidazoles include those which react a diazonium fluoroborate with the nitrite ion, and methods which oxidize 2-amino derivatives, themselves often only available by laborious sequences. The most appealing routes to 2-nitroimidazoles are the methods which make the 2-lithio derivative and treat it with a source of nitronium ion (e.g. n-propyl nitrate or N_2O_4) [3–5] (see Section 7.2.2).

Biological oxidation of a 2-aminoimidazole gives poor yields (<38%), and none at all with 1-alkyl-2-aminoimidazoles. Nor will oxidation with peroxytrifluoroacetic acid work. It is, however, satisfactory for the oxidation of 4-aminoimidazoles (which are usually rather unstable compounds). The most common way of making 2-nitroimidazoles is from the diazonium fluoroborates subjected to the Gattermann reaction (see Section 7.3). Yields vary from 20 to 50% [6, 7], and again are dependent on the availability of the 2-aminoimidazoles (see Section 8.2.2).

1-Alkyl-5- and 1-alkyl-4-nitroimidazoles can be made with a good degree of regiospecificity by alkylation of 4(5)-nitroimidazole in neutral and basic media, respectively (see Section 7.1.1 and Table 7.1.1). See also the *cine* nucleophilic substitutions of 1,4-dinitroimidazoles (Section 7.3.1).

8.2.2 Amino derivatives

Attention here focuses on the synthesis of primary amino derivatives, especially 2-aminoimidazoles and -benzimidazoles. 4(5)-aminoimidazoles are often quite unstable compounds which may be difficult to isolate [8]. The

8.2. NITROGEN GROUPS

4(5)-amino-5(4)-cyanoimidazoles are, however, now well known and are of special interest because they can be converted into purines.

2-Aminobenzimidazoles are readily made from o-phenylenediamines cyclized in the presence of cyanogen chloride (or bromide) or cyanamide (see Section 3.1.2). Yields are high, and the general method can be adapted to give 2-substituted amino and 1,2-diamino derivatives. This general approach, then, would normally be the first to be considered. Alternatives utilize the susceptibility of 2-bromobenzimidazole to nucleophilic displacement by an amine in the presence of copper(I) bromide (see Section 7.3.1), direct nucleophilic amination of benzimidazoles by sodamide (see Section 7.4), the oxidative cyclization of an o-aminoarylthiourea, or reduction of o-cyanoaminonitrobenzenes (see Section. 2.1.1).

In theory, 2-aminoimidazoles should also be accessible by ring synthesis, by reduction of 2-nitro- or 2-azoimidazoles, or by nucleophilic substitution methods. Practically, none of these are straightforward. One route which gives quite good yields is an adaptation of the Marckwald synthesis in which the cyanate, thiocyanate or isothiocyanate is replaced by cyanamide (see Section 4.1). Guanidines will react with suitably α-functionalized carbonyl compounds to give up to 90% yields of 2-aminoimidazoles, e.g. reaction of benzoin with diguanylhydrazine (see Section 4.3). DAMN reacts with isocyanide dichlorides or cyanogen chloride to give 2-aminoimidazoles (see Section 3.1.1), and 2-acylaminoimidazoles can be made from 3-amino-1,2,4-oxadiazoles (see Section 2.2.1).

N-Unsubstituted 4-aminoimidazoles can be made in 60–80% yields by cyclization of α-amidinonitriles, or more directly by heating an imidate with an α-aminonitrile. The products are esters of 4-aminoimidazole-2-carboxylic acids. An alternative route cyclizes suitably substituted guanidines in acidic conditions. The guanidines are made, for example, from aminoacetaldehyde dialkyl acetals and cyanamide (see Section 2.2).

If a 1-substituted 4-aminoimidazole is required, the reaction between an orthoformic ester and benzyloxycarbonyl or tosyl derivatives of aminoacetamidine may be appropriate (see Section 3.1.1).

The alternative 5-amino-1-substituted imidazoles are usually made by reactions between primary amines and alkyl N-cyanoalkylimidates (see Section 3.2). Other possibilities include ring closure of formylglycineamidines by heating them alone or with phosphoryl chloride (yields are usually low) (see Section 2.1.1), cyclization of α-cyanoalkylcyanamides (to form 5-amino-2-bromoimidazoles) (see Section 2.2), alkylation of arylaminomethylene cyanamides or cyanoimidothiocarbanates with α-halogenocarbonyl compounds (gives 4-acyl-5-aminoimidazoles) (see Section 2.3), and cyclizations of DAMN with amidrazones (to 1,5-diaminoimidazoles) (see Section 2.2.1).

The synthetically important 4-amino-5-cyanoimidazoles are mainly derived from reactions of the commercially available reagent DAMN. With nitrilium triflates, amidinium salts are formed in the first instance, but in the presence of bases these cyclize to give a variety of products. Careful control of the base can lead to poor to moderate yields of 2-substituted 4-amino-5-cyanoimidazoles. With formamidine, DAMN gives the parent 5-amino-4-cyanoimidazole (in the N-unsubstituted species it is likely that the 4-cyano tautomer predominates). Better yields are possible if DAMN is refluxed in dry dioxan with triethyl orthoformate to form the imidate, then treating this with ammonia to form the formamidine intermediate, which is cyclized as before (see Section 2.2.1). Base-catalysed cyclization of the product which is formed when N-cyanoacetamido esters and glycine esters (or α-aminonitriles) react in the presence of triethylamine is also reported to give 40–90% yields. N-Cyanoiminodithiocarbamic esters react with α-cyanoammonium salts to form isothiourea intermediates, which also cyclize in basic media (see Section 2.3), and low yields (10–60%) of 5-amino-4-cyanoimidazoles are also obtained when imidates react with aminomalonodinitrile (see Section 4.1).

REFERENCES

1. K. C. Chang, M. R. Grimmett, D. D. Ward and R. T. Weavers, *Aust. J. Chem.* **32**, 1727 (1979).
2. A. R. Katritzky, D. J. Cundy and J. Chen, *J. Energetic Mater.* **11**, 345 (1993).
3. D. P. Davis, K. L. Kirk and L. A. Cohen, *J. Heterocycl. Chem.* **19**, 253 (1982).
4. B. D. Palmer and W. A. Denny, *J. Chem. Soc., Perkin Trans. 1* 95 (1989).
5. B. Iddon and N. Khan, *J. Chem. Soc., Perkin Trans. 1* 1453 (1987).
6. B. Cavalleri, in *Nitroimidazoles: Chemistry, Pharmacology and Clinical Application* (ed. A. Breccia, B. Cavalleri and G. E. Adams), *NATO Adv. Study Inst. Ser. A* **42**. Plenum Press, New York, 1982.
7. J. H. Boyer, *Organic Nitro Chemistry*. VCH, Florida, 1986, Vol. 1, p. 79
8. K. Schofield, M. R. Grimmett and B. R. T. Keene, *Heteroaromatic Nitrogen Compounds: The Azoles*. Cambridge University Press, Cambridge, 1976, p. 208.

8.3 ACYL GROUPS

8.3.1 Carboxylic acid derivatives

This section is limited to a description of the synthesis of imidazoles substituted by carboxyl or carboxylate at the 2-, 4- and 5-positions, and benzimidazoles substituted at C-2.

In addition to the many ring-synthetic procedures which give carboxyl-substituted products, there is also a variety of other methods which can be applied to the preformed rings. Oxidation of an alkyl group is an obvious

8.3. ACYL GROUPS

possibility provided that other groups in the ring are unaffected by strong oxidizing agents [1].

Direct introduction of a carboxyl group can be achieved via the Grignard reagent or lithiated imidazole (see Section 7.2.2). This approach is especially valid for making 2-carboxylic acids, but 4- and/or 5-carboxylation is possible when C-2 is blocked (see Scheme 7.2.2), and if the differential reactivities of anions at these positions are utilized by careful control of reaction conditions. An example has been included in Section 7.2.2, and many others are listed in Iddon and Ngochindo's recent review [2].

Oxidation of substituents such as hydroxyalkyl or aldehyde can also give carboxylic acid groups, and once the carboxyl group has been introduced it can easily be converted into the usual acid derivatives, or reduced to an alcohol or aldehyde [3]; it is usually preferable to reduce the carboxyl group all the way to hydroxymethyl, then reoxidize this to the aldehyde (see Section 8.3.2).

Usually decarboxylation is accomplished by heating the acids above their melting points, often in the presence of a copper–chromium catalyst. Imidazole-4,5-dicarboxylic acid can be monodecarboxylated by heating its monoanilide; imidazole- and benzimidazole-2-carboxylic acids decarboxylate very readily indeed, so readily that the carboxyl function makes a useful blocking group in metallation procedures (see Scheme 7.2.1) [3–5]. A potentially useful method of preparation of imidazole-4-carboxylic acid derivatives heats the 4,5-dicarboxylic acid (2) with acetic anhydride to form (1), which is essentially an azolide and very prone to nucleophilic attack which cleaves the nitrogen–carbonyl bond (Scheme 8.3.1). With methanol the methyl ester (3) is formed; with hydrazines the 4-hydrazides (4) result [6].

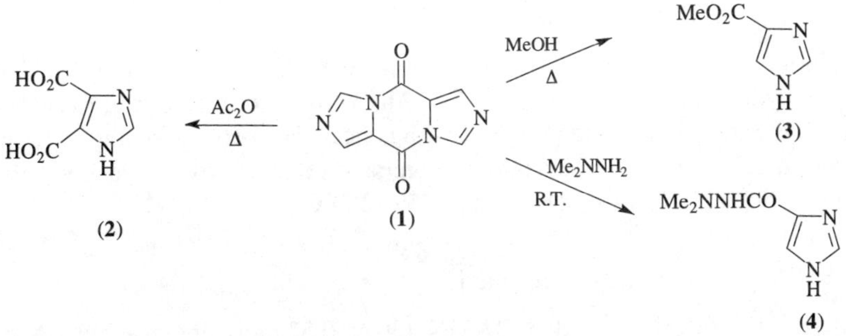

Scheme 8.3.1

Diimidazo[3,4-a;3',4'-d]piperazin-2,4-dione (1) [6]

A mixture containing an excess of acetic anhydride and imidazole-4,5-dicarboxylic acid (2) (5.0 g, 32 mmol) is magnetically stirred and heated under reflux (36 h). The solution is then evaporated *in vacuo*, and the residue is washed with tetrachloromethane and then sublimed to give (**1**) (3.1 g, 51%), m.p. 254–255°C, after recrystallization from benzene.

4-(2',2'-Dimethyl)imidazolecarbohydrazide (**4**) [6]

1,1-Dimethylhydrazine (1.8 g, 30 mmol) is added to (**1**) (0.5 g, 2.7 mmol) and left at room temperature (24 h). The semisolid product is washed with tetrachloromethane and dried under vacuum to give the crude dimethylhydrazide (0.39 g, 94%). Recrystallization from benzene–methanol gives the pure product, m.p. 277–278°C.

The earliest method of making imidazole-4,5-dicarboxylic acids combined tartaric acid dinitrate with an aldehyde in the presence of ammonium ions (see Chapter 5). While this method may still have some validity, the parent compound (**2**) is also formed on oxidation of benzimidazole (see Section 6.1.2.4), and a range of 1-substituted derivatives can be prepared starting from 4,5-dicyanoimidazole (now commercially available or readily made by heating DAMN and triethyl orthoformate in anisole). The dinitrile can be alkylated and then hydrolysed to give 1-alkylimidazole-4,5-dicarboxylic acids in high yields. Alternatively, the sequence from DAMN can be performed in diglyme, a solvent which facilitates large-scale preparation of the dicarboxylic acids in comparable yields without isolation of the 4,5-dicyanoimidazole [7]. The 1-substituted dicarboxylic esters can also be made by modification of the thiocyanate method [8].

1-Methylimidazole-4,5-dicarboxylic acid [7]

A solution of 1-methyl-4,5-dicyanoimidazole (105.0 g, 0.796 mol) in 6 M sodium hydroxide (750 ml) is refluxed (2 h) with vigorous stirring. The hot solution is carefully acidified to pH 2 with concentrated HCl. Cooling to room temperature, filtration, and drying the precipate at 100°C overnight gives the dicarboxylic acid (108.8 g, 81%), m.p. 259–260°C.

Alternative precedure using diglyme [7]

To diglyme (100 ml) is added DAMN (30 g, 0.32 mol) and triethyl orthoformate (54 ml, 0.32 mol). The homogeneous mixture is heated to a bath temperature of 135°C with stirring until distillation of ethanol ceases. Sodium methoxide (1.0 g, 20 mmol) is added slowly, and heating is continued till no more distillate is collected. The diglyme is then removed by distillation *in vacuo* (b_{21} 135°C), removing final traces of diglyme at 0.13 mbar. To the residue is added water (250 ml) followed by careful addition of sodium

8.3. ACYL GROUPS

bicarbonate (42 g, 0.50 mol). The solution is heated to 65°C with stirring, dimethyl sulfate (38 ml, 0.41 mol) is added dropwise (1 h) and the solution is stirred at 65°C (1 h). Cooling is followed by extraction with 5% methanol in ethyl acetate (4 × 250 ml). The combined organic extracts are washed with brine, dried (Na_2SO_4) and evaporated. The residue is refluxed (2 h) with 6 M sodium hydroxide (300 ml). Acidification to pH 2 with concentrated HCl precipatates a light brown solid which is purified by suspension in water (250 ml), then adding solid sodium bicarbonate until the solid dissolves. The solution is then carefully acidified with 4 M HCl and allowed to crystallize at 0°C to give 1-methylimidazole-4,5-dicarboxylic acid (25.5 g, 54%), m.p. 254-256°C.

Once this dicarboxylic acid has been made it should be possible to remove selectively one of the carboxyl groups, e.g. treatment with N,N-dimethylacetamide at 180°C for 4 h is reported to give the 1-methyl-4-carboxylic acid, whereas heating in acetic anhydride at 100°C (4 h) leads to specific 4-decarboxylation to give the 1-methyl-5-carboxylic acid [9]. It is apparent, however, that the specificity is not quite this good, with mixtures of isomeric products being obtained, especially with N,N-dimethylacetamide. Separation of the isomeric products is best accomplished by converting the products into their methyl esters which can be efficiently separated by fractional sublimation. The 1-methyl-5-carboxylate ester is the more volatile [7].

1-Methylimidazole-5-carboxylic acid [7]

A mixture of the dicarboxylic acid (21.0 g, 0.12 mol) and acetic anhydride (600 ml, dried over 4 Å sieves) is heated at 100°C (4 h) with vigorous stirring. The mixture is evaporated *in vacuo*, and the residue is triturated with acetone, and azeotropically dried with toluene to afford the product (14.6 g, 95%), m.p. 245-248°C (dec).

1-Methylimidazole-4-carboxylic acid [7]

A mixture of the dicarboxylic acid (0.850 g, 5 mmol) and N,N-dimethylacetamide (25 ml) is heated at 180°C (3 h). The mixture is evaporated, and the residue is triturated with benzene, followed by recrystallization from ethanol to give the product (0.470 g, 75%), m.p. 245-246°C. A substantial amount of the other isomer is also formed.

Specific preparation of each of the above isomers should theoretically be possible by methylation of imidazole-4-carboxylic acid in neutral and basic media. Methylation yields, however, are unlikely to be high, and the separation of mixtures will be necessary. Some methyl ester formation is also to be expected.

Ring-synthetic procedures which lead to carboxyl-substituted imidazoles are usually multistep and seldom give high yields. Nevertheless, they are frequently the only viable routes to products with a particular substitution pattern. A summary of the methods available follows.

Imidazole-2-carboxylates can be made by amidine cyclization (see Section 2.2.1 and Table 2.2.2), by reaction of an aminocarbonyl compound with thioxamate (see Section 4.1 and Scheme 4.1.6), and from 1-cyano- or 1-carbethoxy-substituted 4-amino-2-azabutadienes (see Section 3.2 and Scheme 3.2.3). An improved amidine cyclization treats trichloroacetonitrile with aminoacetaldehyde dimethyl acetal to give the amidine (5), which cyclizes with trifluoroacetic acid at room temperature to give 2-trichloromethylimidazole (Scheme 8.3.2). This is not purified, but converted immediately into ethyl imidazole-2-carboxylate or imidazole-2-carboxylic acid in high yields [10].

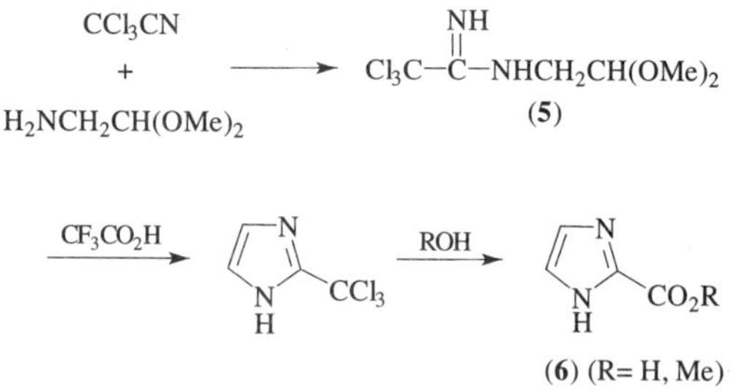

$$CCl_3CN + H_2NCH_2CH(OMe)_2 \longrightarrow Cl_3C-\underset{\underset{}{\overset{NH}{\|}}}{C}-NHCH_2CH(OMe)_2$$
(5)

$$\xrightarrow{CF_3CO_2H} \underset{\underset{H}{}}{\text{imidazole-CCl}_3} \xrightarrow{ROH} \underset{\underset{H}{}}{\text{imidazole-CO}_2R}$$
(6) (R = H, Me)

Scheme 8.3.2

N-(2,2-Dimethoxyethyl)trichloroacetamidine (5) [10]

Aminoacetaldehyde dimethyl acetal (10.9 ml, 100 mmol) is added dropwise to a stirred solution of trichloroacetonitrile (14.4 g, 100 mmol) in THF (25 ml) at −35 to −40°C under an argon atmosphere. The cooling bath is removed to allow the temperature to rise to ambient, before diluting the mixture with ethyl acetate, washing with water, drying (Na$_2$SO$_4$) and concentrating *in vacuo* to give an oil (∼24.5 g, 98%) which spontaneously crystallizes, m.p. 44–45°C.

Ethyl imidazole-2-carboxylate (6) *(R = Et)* [10]

The amidine (5) (2.00 g, 8.0 mmol) is added at 0°C to trifluoroacetic acid (2 ml). The solution is left to stand at room temperature (24 h). Absolute ethanol (20 ml) is added, and the solution is heated under reflux (4 h) before removal of the solvent under reduced pressure. Ethyl acetate is added to the

8.3. ACYL GROUPS

residue, and the solution is filtered through a pad of Act.1 neutral alumina. Evaporation of the filtrate *in vacuo* gives the ester (**6**) (R = Et) as a solid (0.87 g, 73%). Crystallization from ethyl acetate gives the pure product, m.p. 175–177°C. Similarly prepared by addition of water to the trifluoroacetic acid solution of the 2-trichloromethylimidazole is imidazole-2-carboxylic acid (**6**) (R = H) (94%), m.p. 166–167°C.

Benzimidazole-2-carboxylates can be made by cyclization of 2,4-dinitrophenylaminoalkenes (see Section 2.1.2.2 and Scheme 2.1.24), and from the reaction of oxalic acid with an *o*-diaminobenzene (see Section 3.1.2). See also Section 7.2.2 for methods which utilize 2-lithiobenzimidazole.

Imidazole-4-carboxylates have been made from amidines derived from α-amino acids (see Section 2.2.1 and Table 2.2.1), by Claisen rearrangement of the adduct formed when an arylamidoxime reacts with a propiolate ester (see Section 2.2.1 and Scheme 2.2.6), from α-aminocarbonyls with cyanates or thiocyanates (see Section 4.1 and Table 4.1.1), from α-oximino-β-dicarbonyl compounds heated with an alkylamine (see Section 4.1 and Scheme 4.1.7), and by anionic cycloaddition of an alkyl isocyanoacetate to diethoxyacetonitrile (see Section 4.2 and Scheme 4.2.11; see also Scheme 4.2.12). A further useful approach is to use an appropriate tricarbonyl compound with an aldehyde and a source of ammonia (see Chapter 5 and Scheme 5.1.1). Irradiation of 1-alkenyltetrazoles bearing an ester substituent may have applications (see Section 6.1.2.3).

Regiochemical synthesis of *1-substituted imidazole-4-carboxylates* can be achieved by treatment of a (Z)-β-dimethylamino-α-isocyanoacrylate with an alkyl or acyl halide (see Section 2.1.1 and Scheme 2.1.8), by cyclization of 3-alkylamino-2-aminopropanoic acids with triethyl orthoformate followed by dehydrogenation of the initially formed imidazoline (see Section 3.1.1 and Scheme 3.1.2), by condensation of 3-arylamino-2-nitro-2-enones with ortho esters in the presence of reducing agents (see Section 3.1.1 and Scheme 3.1.4), by reaction of an alkyl *N*-cyanoalkylimidate with a primary amine (see Section 3.2 and Scheme 3.2.1), the poor-yielding acid-catalysed cyclization of a 2-azabutadiene with a primary amine (see Section 3.2 and Scheme 3.2.3), the cyclocondensation of an isothiourea with the enolate form of ethyl isocyanoacetate (see Section 4.2 and Scheme 4.2.5), and from the interaction of α-aminonitrile, primary amine and triethyl orthoformate (see Chapter 5, Scheme 5.1.5, and Tables 5.1.1 and 5.1.2).

The isomeric *1-substituted imidazole-5-carboxylates* are made by cyclization of 3-amino-2-alkylaminopropanoic acids with triethyl orthoformate followed by active manganese dioxide oxidation of the imidazoline product (see Section 3.1.1), or from *N*-substituted glycine esters, which are formylated, converted into the enolates and then condensed with potassium thiocyanate (see Section 4.1 and Scheme 4.1.2). Alternatively, one can use the reaction of a carboxy substituted α-aminocarbonyl compound with an isothiocyanate (see Section 4.1, Scheme 4.1.1 and Table 4.1.1). The sulfur group at C-2 can be removed either oxidatively or by Raney nickel treatment. One can also use Gold's salt condensed with sarcosine methyl ester (see Section 4.2).

Methyl 1-methylimidazole-5-carboxylate [11]

Methyl *N*-methylglycinate hydrochloride (76 g, 0.54 mol) is suspended in t-butylmethyl ether (1 l). To the suspension is added dimethyl oxalate (13 g, 0.11 mol) and a 30% solution of sodium methoxide in methanol (291 g, 1.62 mol). Nitrogen is bubbled through while Gold's salt [15] (freshly prepared from cyanuric chloride (46.1 g, 0.25 mol) and DMF (120.5 g, 1.65 mol)) (116 g, 0.71 mol) in t-butylmethyl ether (250 ml) is introduced. The temperature rises to 32°C before gradually dropping. Stirring is continued (1 h) while nitrogen continually flushes out the liberated dimethylamine. TLC (t-butylmethyl ether–methanol, 7:3) of the orange solution should indicate at this stage that only minor amounts of the amide are present, and the ester is the major product. Celite (25 g) is added to the reaction mixture, and the suspension is filtered through a 3 cm layer of silica gel. The residue is washed twice with t-butylmethyl ether–methanol (7:3), the combined filtrates are concentrated and the residue is vacuum distilled. Once all of the DMF has been collected, the product distils, $b_{1.2}$ 92–96°C (69 g, 84%). Sublimation *in vacuo* gives the pure compound, m.p. 55–56°C. Similarly prepared is ethyl 1-benzylimidazole-5-carboxylate (56%), m.p. 64–65°C.

8.3.2 Aldehyde and ketone derivatives

Direct formylation of imidazole is not possible. Nor can 2- or 4(5)-formylimidazoles be made readily by ring-synthetic methods. Rather, it is necessary to make them by oxidation of an hydroxymethyl group, by reduction of carboxyl, via lithio derivatives, or from the dichloromethyl precursors. Since hydroxymethylimidazoles are often quite readily made, their oxidation is an appealing approach. The most satisfactory oxidizing agent appears to be activated manganese dioxide, which has been used to convert 2- and 4(5)-hydroxymethylimidazoles into the corresponding aldehydes [12]. The reagent (as manganese(IV) oxide) is available commercially, or it can be made quite readily [13].

1-Benzylimidazole-5-carbaldehyde [12]

A stirred solution of 1-benzyl-5-hydroxymethylimidazole (0.79 g, 4.2 mmol) and activated manganese dioxide (3.6 g, 41 mmol) in dioxan (25 ml) is refluxed

(6 h). The manganese oxide is filtered, and the filtrate is evaporated to dryness under reduced pressure. The residue is chromatographed on a silica gel column eluted with ethyl acetate–methanol (9:1) to give an oil (0.49 g, 63%), v_{max} 1665 cm^{-1}. Similarly prepared are 1-benzyl-4-carbaldehyde (65%), 1-(o-nitrobenzyl)-4-carbaldehyde (48%), 1-(o-nitrobenzyl)-5-carbaldehyde (24%), 1-(p-nitrobenzyl)-4-carbaldehyde (38%), 1-(p-nitrobenzyl)-5-carbaldehyde (10%). 1-(o-nitrophenyl)-4-carbaldehyde (76%) and 1-(p-nitrophenyl)-5-carbaldehyde (87%).

Reductive procedures are seldom sufficiently specific to give high yields of pure aldehydes, and so it is customary to reduce an ester function with lithium aluminium hydride to hydroxymethyl and then reoxidize [11]. A range of 1-substituted 4,5-dicarbethoxyimidazoles have been reduced to the 4,5-bis(hydroxymethyl) derivatives in 50–88% yields; oxidation to the bis-aldehydes is reported with 35–45% efficiency [8].

Approaches to imidazolecarbaldehydes via lithioimidazoles (usually quenched with DMF) have already been discussed (see Section 7.2.2 and Schemes 7.2.1 and 7.2.2), and the literature now abounds with examples which specifically produce 2- [14, 15], 4- [16–19] and 5-formyl isomers [5, 14, 15] by careful choice of blocking agents and reaction conditions.

A process analogous to that used for making ethyl imidazole-2-carboxylate (see Section 8.3) gives the 2-carbaldehyde in high yield. Reaction of the imidate derived from dichloroacetonitrile and methanolic methoxide with aminoacetaldehyde dimethyl acetal gives an amidine analogous to (5) (Scheme 8.3.2). Refluxing this with trifluoroacetic acid induces cyclization, presumably to 2-(dichloromethyl) imidazole and ultimately to imidazole-2-carbaldehyde [10].

Imidazole-2-carbaldehyde [10]

A solution of *N*-(2,2-dimethoxyethyl)dichloroacetamidine (analogous to (5) (Scheme 8.3.2), made in 88% yield, m.p. 85–89°C) (5.00 g, 23.2 mmol) in 95–97% formic acid (10 ml) is heated at 70–80°C in an oil bath (20 h). The solvent is removed under reduced pressure, benzene is added to the residue, and the mixture is again evaporated to dryness (repeated three times). The final residue is dissolved in water (9 ml), and solid sodium bicarbonate is added to raise the pH to 8, when the aldehyde precipitates. The mixture is cooled overnight in a refrigerator, and the product is collected by filtration and dried *in vacuo* (2.21 g, 99%). Sublimation at 80–90°C/2 mmHg gives the pure aldehyde, m.p. 204–205°C.

Other synthetic approaches include ring synthesis of 1,3-disubstituted imidazolium-2-carbaldehydes (see Section 3.1.1) and the conversion of suitably substituted isoxazoles into 5-acylimidazoles (see Section 6.1.2, Scheme 6.13, and Tables 6.1.1 and 6.1.2). Methods for regiochemical synthesis of 4- and 5-carbaldehydes have been discussed in some detail earlier (see Section 6.1.2; see also Scheme 2.1.3 and Table 2.1.12).

The introduction of ketone groups conjugated with the imidazole ring is now quite a routine matter starting from lithiated imidazoles and quenching them with acid halides, nitriles, 1-acylpyrrolidine, or with aldehydes followed by oxidation (which may be spontaneous) [2] (see Section 7.2.2). Direct acylation at the 2-position is possible with 1-substituted imidazoles under Regel conditions (see Section 7.2.1) although alkanoyl halides do not appear to react as well as aroyl halides. Friedel–Crafts acylations will not occur. Pyrolysis of 1-acetylimidazole is known to give a mixture of 2- and 4-acetyl isomers, but this is unlikely to be synthetically useful [20].

2-Acylamino-4-acylimidazoles have been made from 3-amino-1,2,4-oxadiazoles and 1,3-dicarbonyl reagents (see Section 2.2.1 and Scheme 2.2.5). 4(5)-Acylimidazoles can be derived from 4-acylaminoisoxazoles (see Section 6.1.2 and Scheme 6.1.3). (See also the discussion in Section 2.2.1 on 4-acylimidazole synthesis.) 5-Acyl-1-arylimidazoles can be made from α-oxoketene-S,N-acetals and nitrosoaromatics (see Section 3.2 and Scheme 3.2.5), and 4-acyl-imidazoles by nitration of 1,3-dicarbonyl compounds in their enolic forms, reduction to N-alkenylformamides and subsequent cyclization (see Section 3.2 and Scheme 3.2.4). Examples have also been isolated from reactions of 2-oximino-1,2,3-tricarbonyls and amines (see Section 4.1 and Scheme 4.1.7), from compounds such as 3-chloro-4,4-dimethoxy-2-butanone and 3,4-disubstituted 3-buten-2-ones (see Section 4.3 and Scheme 4.3.5), and by ultraviolet irradiation of 1-alkenyltetrazoles which bear an acyl group conjugated with the exocyclic double bond (see Section 6.1.2.3).

2-n-Heptanoyl-1-methylimidazole [21]

To a stirred solution of 1-methylimidazole (3.28 g, 40 mmol) in THF (80 ml) at −78°C is added n-butyllithium (25.6 ml, 1.6 M solution in hexane, 40 mmol), and the mixture is stirred (10 min). 1-n-Heptanoylpyrrolidine (7.32 g, 40 mmol) (prepared by treating either 1-n-heptanoylimidazole or n-heptanoyl chloride with pyrrolidine) is added to the mixture, which is then stirred at ambient temperature (30 min). After addition of diethyl ether (80 ml) and 10% aqueous HCl (80 ml) to the mixture, the aqueous phase is separated, washed with ether and made alkaline by addition of solid potassium carbonate. The oil which separates is extracted with ethyl acetate, the solution is dried (Na_2SO_4), and the solvent is evaporated again to an oil which is distilled in a Kugelrohr apparatus, b_3 118–120°C (7.76 g, 100%), ν_{max} 1776 cm^{-1}. Similarly prepared are the following 1-methyl-2-acyl(aroyl)imidazoles (2-substituent, b.p. or m.p., yield given): COPh, b_3 140–145°C, 100%; $C_6H_{11}CO$, b_3 112–114°C, 100%; PhCH=CH–CO, m.p. 120–121°C, 62%; 4'-pyridyl-CO, m.p. 66–68°C, 84%.

REFERENCES

1. K. Schofield, M. R. Grimmett and B. R. T. Keene, *Heteroaromatic Nitrogen Compounds: The Azoles*. Cambridge University Press, Cambridge, 1976, p. 204.
2. B. Iddon and R. Ngochindo, *Heterocycles*, **38**, 2487 (1994).
3. M. R. Grimmett, in *Comprehensive Heterocyclic Chemistry* (ed. A. R. Katritzky and C. W. Rees). Pergamon Press, Oxford, 1984, Vol. 5 (ed. K. T. Potts).
4. F. Seng and K. Ley, *Synthesis* 703 (1975).
5. G. Shapiro and B. Gomez-Lor, *J. Org. Chem.* **59**, 5524 (1994).
6. S. Kasina and J. Nematollahi, *Synthesis* 162 (1975).
7. J. F. O'Connell, J. Parquette, W. E. Yelle, W. Wang and H. Rapoport, *Synthesis* 767 (1988).
8. H. Schubert and W. D. Rudorf, *Z. Chem.* **11**, 175 (1971).
9. K. Takahashi and K. Mitsuhashi, *Bull. Chem. Soc. Jpn.* **53**, 557 (1980).
10. E. Galeazzi, A. Guzman, J. L. Nava, J. Liu, M. L. Maddox and J. M. Muchowski, *J. Org. Chem.* **60**, 1090 (1995).
11. R. Kirchlechner, M. Casutt, U. Heywang and M. W. Schwarz, *Synthesis* 247 (1994).
12. I. Antonini, B. Cristalli, P. Franchetti, M. Grifantini and S. Martelli, *Synthesis* 47 (1983).
13. L. F. Fieser and M. Fieser, *Reagents for Organic Synthesis*. Wiley, 1967, p. 637.
14. M. P. Groziak and L. Wei, *J. Org. Chem.* **56**, 4296 (1991).
15. B. Iddon and N. Khan, *J. Chem. Soc., Perkin Trans. 1* 1445 (1987).
16. G. Shapiro and B. Gomez-Lor, *Heterocycles* **41**, 215 (1995).
17. B. Iddon and N. Khan, *J. Chem. Soc., Perkin Trans. 1* 1453 (1987).
18. K. L. Kirk, *J. Heterocycl. Chem.* **22**, 57 (1985).
19. J. Winter and J. Retey, *Synthesis* 245 (1994).
20. A. Maquestiau, A. Tommasetti, C. Pedregal-Friere, J. Elguero and R. Flammang, *Bull. Soc. Chim. Belg.* **93**, 1067 (1984).
21. S. Ohta, S. Hayakawa, H. Moriwaki, S. Tsuboi and M. Okamoto, *Heterocycles* **23**, 1759 (1985).
22. H. Gold, *Argew. Chem.* **72**, 956 (1960).

8.4 THIOL AND OTHER SULFUR FUNCTIONS

Imidazole- and benzimidazole-2-thiols usually exist largely as the thione tautomers. The thiol (thione) group is susceptible to alkylation (especially in alkaline media), and can be oxidized to sulfide, disulfide and sulfonic acid. This oxidation can often be carried out quite selectively by careful choice of oxidizing agent. The sulfur function can be removed with nitric acid, iron(III) chloride, hydrogen peroxide or, most commonly, Raney nickel. Alkyl- and arylthio groups can be oxidized to sulfoxide or sulfone.

Benzimidazole-2-thiones are most commonly made from o-arylenediamines in reaction with thiophosgene, carbon disulfide and other CS sources (see Section 3.1.2), by direct introduction of sulfur at elevated temperatures to the benzimidazole (see below), or by the interaction of alcohols and aryl-1,2-diisothiocyanates (see Section 2.1.1 and Scheme 2.1.18).

In the uncondensed imidazoles the standard method reacts an α-aminocarbonyl compound with a thiocyanate (see Section 4.1 and Table 4.1.1). If a 2-alkylthioimidazole is required directly, one can combine an N-alkyl- or N-arylcarbonimidodithioate in refluxing acetic acid with the aminocarbonyl substrate (see Section 4.1 and Scheme 4.1.3). Alternatively, reaction between thiourea and a two-carbon synthon (α-hydroxy-, α-halogeno-, α-dicarbonyl) leads to imidazoline-2-thiones (see Section 4.3). In sulfuric acid, 3-butynylthiourea cyclizes to 4,5-dimethylimidazolin-2-thione (see Section 2.2.1). 1-Substituted 2-methylthioimidazoles can be made, albeit in rather poor yields, from appropriately substituted 2-azabutadienes (see Section 3.2 and Scheme 3.2.3), and 2-arylthioimidazoles are available in moderate yields from benzyl isocyanides and arylsulfenyl chlorides (see Section 4.2 and Scheme 4.2.12). Ring transformations of 5-amino-2-alkylaminothiazoles and 2-acylamino-5-aminothiazoles may have occasional applications (see Section 6.1.2.7). The ease with which a thiol group or imidazole or benzimidazole can be alkylated, in comparison with the annular nitrogens, usually makes it more convenient to prepare alkylthioimidazoles from the thiols (or thiones).

Imidazole-4- and -5-thiols are less well known than the 2-isomers, and they are not always easy to isolate in the free thiol form — it may be preferable to isolate them "masked" as disulfides or sulfides [1]. A simple, but somewhat limited method of making the 4- or 5-thiols is by nucleophilic displacement of a halogen (e.g. using ammonium sulfide), but the success of this approach is dependent on the presence of strong electron withdrawal elsewhere in the ring [2]. Alternative ring-synthetic methods include cyclization of thionamides (see Section 2.1.1, Scheme 2.1.12 and Table 2.1.6), exothermic reaction of α-oxothionamides with aldimines to form 1,2,5-trisubstituted imidazole-4-thiols (see Section 4.1 and Scheme 4.1.6) and rearrangement of 5-aminothiazoles in basic medium to imidazole-5-thiols, a reaction of limited synthetic application (see Section 6.1.2.7). The reactions of TOSMIC, especially with alkyl isothiocyanates in the presence of base, give 1-substituted 4-tosylimidazole-5-thiols (see Section 4.2 and Scheme 4.2.1).

4-Arylthioimidazoles can be made in high yields when thiophenates react with N-(1-cyanoalkyl) alkylidene N-oxides. Alkyl and aralkyl thiolates, however, react much less readily (see Section 2.1.1 and Scheme 2.1.10). Imidazole-4-thioethers can be made in a general reaction between nitriles and a range of isocyanides which are susceptible to α-metallation (see Section 4.2, Scheme 4.2.2 and Table 4.2.1). Oxoketene acetals bearing alkylthio substituents react with nitrosoaromatics to give 5-acyl-4-alkylthio-1-arylimidazoles in moderate to good yields (see Section 3.2 and Scheme 3.2.5).

8.4. THIOL AND OTHER SULFUR FUNCTIONS

When cyanoimidothiocarbamates react with α-halogenocarbonyl compounds, the products cyclize in base to form 1-substituted 2-alkylthio-5-aminoimidazoles (see Section 2.3, Scheme 2.3.1 and Table 2.3.1), while similar condensations of N-cyanoiminodithiocarbamates with sarcosine nitrile salts, sarcosine ester salts or methylaminoacetophenone lead to analogous products, often in quite high yields (see Scheme 2.3.2).

By careful control of metallating agent, blocking groups and reaction conditions, it is possible to make 2-alkyl(aryl)thio- and 4(5)-alkyl(aryl)thioimidazoles in high yields, e.g. 1-trityl-2-phenylthio- [3], 4(5)-methylthio- [4] and 1-benzyl-2,4,5-tris(methylthio)-imidazole [5].

Although it is frequently more convenient to make imidazole and benzimidazole sulfones (and sometimes sulfoxides) by direct oxidation of the thioethers, 4-tosyl groups can also be introduced quite conveniently by a ring synthesis based on the reagent TOSMIC (see Section 4.2 and Scheme 4.2.1).

Direct electrophilic sulfonation of imidazole and 2-substituted imidazoles has been traditionally carried out using oleum at around 160°C. Yields of the 4-sulfonic acid can be above 80%, but if a three-fold excess of sulfur trioxide in refluxing 1,2-dichloroethane is used, the acids are obtained under much milder conditions. Care has to be taken, for if only a 1:1 or 1:2 ratio of SO_3:imidazole is used, a charge transfer complex is formed instead [6].

1-Methylbenzimidazole-2-thiol [7]

A mixture of 1-methylbenzimidazole (2.48 g, 19 mmol) and powdered sulfur (0.58 g, 18 mmol) is heated in a bath at 240–260°C (1 h). After cooling, dissolving in 10% aqueous sodium hydroxide (20 ml), treating with activated charcoal, and filtering through a pad of Celite, the filtrate is acidified with concentrated HCl to precipitate the product (2.94 g, 98%), m.p. 188–190°C.

1-Methylbenzimidazole-2-sulfonic acid [7]

To a stirred solution of 1-methylbenzimidazole-2-thiol (2.74 g, 167 mmol) in 50% aqueous potassium hydroxide (20 ml) at 20°C is added dropwise a saturated solution of potassium permanganate (5.28 g, 33 mmol) in water. After the addition is complete, stirring is continued (30 min), the manganese dioxide is filtered off and the filtrate is acidified with hydrochloric acid. The precipitated product is filtered and recrystallized from hot water (3.2 g, 90%), m.p. 326–328°C.

5-Methylsulfinyl-4-nitroimidazole [8]

5-Methylthio-4-nitroimidazole (0.250 g, 1.57 mmol) is dissolved with slight warming in glacial acetic acid (50 ml). Magnesium monoperoxyphthalate hexahydrate (80%, 0.5 mol eq., 0.485 g) is then added, and the reaction mixture is warmed to about 50°C and stirred to effect solution. Immediately, the mixture is poured into ice–water (50 ml) and concentrated under reduced pressure until solid is observed to precipitate. After chilling (3–4 h), the solid is filtered, washed with water and recrystallized from ethanol to give colourless crystals (0.185 g, 67%), m.p. 228–230°C (dec.), ν_{max} 1535, 1480, 1355, 1070, 1040 cm^{-1}.

REFERENCES

1. A. Spaltenstein, T. P. Holler and P. B. Hopkins, *J. Org. Chem.* **52**, 2977 (1987).
2. S. Kulkarni and M. R. Grimmett, *Aust. J. Chem.* **40**, 1415 (1987).
3. A. R. Katritzky, J. J. Slawinski, F. Brunner and S. Gorun, *J. Chem Soc., Perkin Trans. 1* 1139 (1989).
4. A. J. Carpenter and D. J. Chadwick, *Tetrahedron*, **42**, 2351 (1986).
5. B. Iddon and N. Khan, *J. Chem. Soc., Perkin Trans. 1* 1453 (1987).
6. M. R. Grimmett, in *Comprehensive Heterocyclic Chemistry* (ed. A. R. Katritzky and C. W. Rees), Elsevier, Oxford, 1996, Vol. 3.02 (ed. I. Shinkai) p. 77.
7. A. V. El'tsov, K. M. Krivozheiko and M. B. Kolesova, *J. Org. Chem. USSR (Engl. Transl.)* **3**, 1475 (1967).
8. P. Benjes and M. R. Grimmett, unpublished.

Index

Bold page numbers refer to compounds for which full experimental details are provided in the text. *Italicised* page numbers refer to table entries.
Substitution patterns are emphasised by the ordering of the index so that 2-substituted precedes 4- and 5-substituted entries. This means that strict alphabetical order may not always be followed. Where prototropic tautomerism is possible 4(5)-substituted imidazole derivatives are listed as 4-substituted. In cases where the 4(5)-substituent is electron withdrawing the equilibrium commonly favours the 4-tautomer. Esters are listed as carbalkoxy rather than systematically.

acetamidines 6
acetamidoacetone **105**
acetylamidrazones 147
N-(2-acetylaminophenyl)-1H-pyrrol-1-amine **24**
ω-acylamines **96**
α-C-acylaminoanilines, cyclization 36
α-C-acylamino esters, hydrolysis 105, **106**
α-acylaminoketones 9
adenine 13, 49
Akabori method 41
alkylaminoethanal dimethyl acetals **106**
N-alkylation of benzimidazoles 202
N-alkylation of imidazoles 195
amidines 10, 40, 43, 45, 46, 54, 55, 57, 59, 134, 140, 143, 144, 179, 240
amidinoamides 8, 9
α-amidinoketones 179
α-amidoketones 8, 96
amidrazones 52
aminoacetone hydrochloride **105**
α-amino-α-carbethoxy-N-methylnitrones **143**
α-aminocarbonyls 103, 106, 111
(Z)-N-(2-amino-1,2-dicyanovinyl)formamidine **50**
(Z)-N^3-(2-amino-1,2-dicyanovinyl)formamidrazone **52**
α-aminoketones 9, 179, **104**, **106**
3-amino-1,2,4-oxadiazole 45
o-aminophenylhydrazines, acylated 23

N-(2-aminophenyl)-1H-pyrrol-1-amine **24**
aromatization 27, **139**, 143, 147, 168
2-azabutadienes 97, 246
2H-azirines 19, 167
Balz-Schiemann conditions 227

Benzimidazole 71, 194
  2-acetoxy-1-ethyl- 35
  1-acetyl-2-methyl- **25**
  1-acyl- 25, 205
  1-alkoxy- **83**
  1-alkoxy-2-alkyl- 82
  1-alkyl- 54, 187
  1-alkyl-2-amino- **186**
  1-alkyl-2-aryl- *21*
  2-alkyl- 20, *21*, 22, 71, 78, 79, **88**
  2-alkyl-1-hydroxy- 3-oxides 188
  2-alkyl-1-substituted 74
  7-alkyl- 186
  2-alkylamino- 28
  2-alkylthio- 221
  1-allyloxy-2-vinyl- 82
  1-amino- 20, 23
  1-amino-2-methyl- 23
  1-amino-2-phenyl- 23
  2-amino- 27, **28**, 29, 55, 78, 84, *85*, **86**, 186, 187, 222, 227, 234
  2-amino-1-benzyl- *85*, **85**, 222
  2-amino-1-carbalkoxy- 29
  2-amino-1-carboxylic acid 29
  2-amino-5-chloro- *85*

Benzimidazole (*continued*)
2-amino-1,5-dimethyl- *85*
2-amino-4,6-dimethyl- *85*
2-amino-5,6-dimethyl- *85*
2-amino-1-hydroxy- 29
2-amino-1-methyl- *85*
2-amino-5-methyl- *85*
2-amino-5-nitro- 29
2-amino-1-phenyl- *85*
2-amino-1-(2'-pyridyl)- **86**
2-amino-5-substituted 29
2-amino-1,5,6-trimethyl- *85*
2-aminoalkyl- 74
2-aminocarbonyl-1-hydroxy- 3-oxide **189**
2-(*o*-aminophenyl)- *73*
2-(*p*-aminophenyl)- *21*
1-aryl- 21, 54, **204**
2-aryl- *21*, 22, 26, *27*, *73*, 74, 76–78, 84, **88**, 187
2-aryl-1-benzyloxy- 82
2-aryl-1-substituted 74
2-arylamino- 28, 84, 86
2-arylthio- 221
1-benzoyl-2-phenyl- 26, 190
2-benzoyl-5-methyl-1-oxide *31*
1-benzyl- 85, 222
1-benzyl-2-methyl- *21*
2-benzyl- **55**, 79, 80
2-benzyl-5-nitro- 79
2-benzylamino- 28
2,2'-bis- 20, **75**
2-bromo- 215, 231, 235
2-(*o*-bromophenyl)- *73*
2-(*m*-bromophenyl)- *73*
2-(*p*-bromophenyl)- *73*
2-*t*-butyl- 72, 73, **75**, 88
2-butylamino- 28
2-butylthiomethyl- *72*
2-carbalkoxy- 29, 77
2-carbamates 29, 30, 55, 78, 86, 88, 190
2-carbethoxy-1-hydroxy-3-oxide 189
2-carbethoxy-5-methoxy-1-oxide *31*
2-carbethoxy-5-methyl-1-oxide *31*
2-carbethoxy-5-nitro-1-oxide *31*
2-carbethoxy-1-oxide *31*
2-carbethoxy-1-phenyl- 77
5-carbethoxy- *72*
5(6)-carbethoxy-2-(4'-hydroxyphenyl)- **80**
2-carbinols 221
2-carbomethoxy- 77
2-carbomethoxy-5-nitro-1-oxide *31*
2-carbomethoxy-1-oxide *31*
2-[(2'-carbomethoxyphenoxy)methyl]- **79**
2-carboxamido-6-chloro-1-hydroxy-3-oxide 189
2-carboxylates 240
2-carboxylic acid 72, 189, 237
2-(*o*-carboxyphenyl)- *73*
2-chloro- 72, 87, 222, 228
5-chloro-2-cyano-1-oxide *31*
5-chloro-1-oxide *31*
5-chloro-2-(3'-pyridyl)- *74*
5-chloro-2-(4'-pyridyl)- *74*
5-chloro-2-trichloromethyl- 79
7-chloro-2-(3'-pyridyl)- *74*
2-(*p*-chlorobenzyl)- 79
2-chloromethyl-1-ethoxyethyl- *72*
2-chloronitromethyl- 84
2-(*o*-chlorophenyl)- *73*
2-(*m*-chlorophenyl)- *73*
2-(*p*-chlorophenyl)- *73*
2-(*p*-chlorophenyl)methoxy- 221
2-cinnamyl- *72*
2-cyano- 32
2-cyano-5-fluoro-1-oxide *31*
2-cyano-6-fluoro-1-oxide *31*
2-cyano-1-hydroxy- 190
2-cyano-5-methoxy-1-oxide *31*
2-cyano-5-methyl-1-oxide *31*
2-cyano-4-nitro-1-oxide *31*
2-cyano-1-oxide *31*, **32**, 33
1-cyanobenzoyl- 190
1,2-cycloalkyl- **34**, 35
2-cyclohexylaminocarbonyl-1-hydroxy-3-oxide 189
5,6-diacyl- 186
1,2-dialkyl- *21*, 88
1,2-diamino- 87, 235
1,2-diamino-5-trifluoromethyl- **87**
2-dibenzylamino-1-isopropyl-5-methoxy- 187
5,6-dichloro-2-trichloromethyl- 79
2-(3',4'-dichlorophenyl)- *73*
2-diethylamino- 87
4,7-dimethoxy- *72*
4,7-dimethoxy-2-ethyl- *72*
4,7-dimethoxy-2-hydroxymethyl- *72*
4,7-dimethoxy-2-methyl- *72*
4,7-dimethoxy-2-propyl- *72*
5,6-dimethoxy- *72*
5,6-dimethoxy-2-methyl- *72*

INDEX

1,2-dimethyl- *21*, 35
1,2-dimethyl-3-oxide **21**, 30
5,6-dimethyl- 72, 82
5,6-dimethyl-2-(4'-pyridyl)- *74*
5,6-dimethyl-2-trichloromethyl- 79
2-dimethylamino- 87
2-dimethylamino-1-propyl- 187
2-(*p*-dimethylaminophenyl)- *27*
4,7-dinitro- *72*
4,7-dinitro-2-methyl- *72*
5,7-dinitro- 75
2-(2',4'-dinitrobenzyl)- **79**
2-diphenylmethoxy- 221
2-diphenylmethyl- 88
1,2-disubstituted 20, 34, 35, 78, 80
1,4-disubstituted 202
1,5-disubstituted 202
1,6-disubstituted 202
1,7-disubstituted 202
2,4-di(trifluoromethyl)- *72*
2,5-di(trifluoromethyl)- *72*
1-ethyl-2-morpholino- 187
2-ethyl- 72, **76**,
2-ethyl-1-hydroxy-3-oxide 189
2-ethylaminocarbonyl-1-hydroxy-3-oxide 189
2-fluoro- 227
4-fluoro- *72*
5-fluoro-1-oxide *31*
2-(*o*-fluorophenyl)- *73*
2-(*m*-fluorophenyl)- *73*
2-(*p*-fluorophenyl)- *73*
2-(2'-furyl)- 80, 88
2-guanidino- 86
1-hetaryl- 24
2-hetaryl- *74*, 75, 80, **88**
1-hydroxy- 188, 189
1-hydroxy-2-carboxylic acid 188, 189
1-hydroxy-2-(*p*-dimethylaminophenyl)-3-oxide 89
1-hydroxy-2-(*o*-hydroxyphenyl)-3-oxide 89
1-hydroxy-2-(*p*-hydroxyphenyl)-3-oxide 89
1-hydroxy-2-methyl-3-oxide 189
1-hydroxy-2-methylaminocarbonyl-3-oxide 189
1-hydroxy-2-(*m*-nitrophenyl)-3-oxide 89
1-hydroxy-3-oxides **89, 189**
1-hydroxy-2-phenyl-3-oxide 89, 189

1-hydroxy-2-phenylaminocarbonyl-3-oxide 189
1-hydroxy-2-styryl-3-oxide 89
1-hydroxy-2-(2'-thienyl)-3-oxide 89
2-hydroxy-1-oxides 33
4-hydroxy- **77**, 188
4-hydroxy-1-methyl- 179, 187, 188
1-hydroxymethyl- **203**, 220
2-(*o*-hydroxyphenyl)- *73*
2-(*o*-iodophenyl)- *73*
2-(*m*-iodophenyl)- *73*
2-(*p*-iodophenyl)- *73*
2-iodo- 215
1-isopropyl-5-methoxy-2-morpholino- 187
1-isopropyl-2-pyrrolidino- 187
2-isopropyl- 80, 88
2-lithio- 203, 241
5-methoxy-1-methyl- **74**
5-methoxy-2-(2'-pyridyl)- *74*
5-methoxy-2-(3'-pyridyl)- *74*
5-methoxy-2-(4'-pyridyl)- *74*
2-(*p*-methoxyphenyl)- 187
1-methyl- 35
1-methyl-4-nitro- **80**, 202
1-methyl-7-nitro- 202
1-methyl-3-oxide 30, *34*
1-methyl-2-phenyl- 35
1-methyl-2-sulfonic acid **247**
1-methyl-4,5,6,7-tetrachloro- 35
1-methyl-2-thiol **247**
1-methyl-2-trichloromethyl- 79
1-methyl-2-trimethylsilyl- 77
2-methyl- 20, *21, 23, 25, 72*, 77, 215
2-methyl-5,7-dinitro-1-phenyl- **75**
2-methyl-4-nitro- **82**
2-methyl-1-(*o*-nitrobenzyl)- *21*
4-methyl-2-phenyl- 187
5-methyl-1-oxide *31*
5-methyl-2-trichloromethyl- 79
2-methylaminocarbonyl- 189
2-methylsulfonylamino- **85**
2-methylthio- 221
2-morpholino-1-propyl- 187
2-(1'-naphthyl)- *74*
4-nitro- 80, 82
4-nitro-2-phenyl- 187
5-nitro- 28, 215
5-nitro-1-oxide *31*, 32
5-nitro-2-phenyl- 187

Benzimidazole (*continued*)
  2-nitromethyl- 84
  1-(*p*-nitrophenyl)- 203, 204
  2-(*o*-nitrophenyl)- *73*
  2-(*p*-nitrophenyl)- *27, 73*, 88, 187
  2-(*m*-nitrophenyl)- *27*
  1-oxides 21, 30, *31*, 32, 33, 89, 167, 189, 229
  3-oxides 21, 30, 33, *34*, 89, 188, 189
  5-phenoxy- 186
  1-phenyl- 75
  1-phenyl-3-oxide 30
  2-phenyl- *21*, 24, 26, *27*, 55, *73*, 88, 187, 224
  2-phenyl-1-oxide 26, *31*, 33
  2-phenyl-1-(*p*-tolyl)- 204
  2-phenylamino- **28, 84**, 87
  2-phenylmethoxy- 221
  2-phenylsulfonyl- 84
  2-phenylthio- 221
  2-propyl- 88
  2-(2'-pyridyl)- **75**, 80
  2-(3'-pyridyl)- *74*
  1-(1'-pyrryl)-2-methyl- 24
  1-SEM- 201
  1-substituted 74, 80
  2-subsituted 23, 71, 80. 84, **220**
  sulfones 247
  2-sulfonic acid 247
  5-sulfonic acid 215
  2-sulfonylamino- 84
  tetrahydro- 15
  2-(4'-thiazolyl)- *21*
  2-(2'-thienyl)- 88
  2-thiol (thione) 245, 247
  2-(*o*-thiophenyl)- *73*
  1-(*p*-tolyl)- 204
  2-(*o*-tolyl)- *73*
  2-(*m*-tolyl)- *73*
  2-(*p*-tolyl)- *27, 73*, 88
  2-(*o*-tolylamino)- 28
  2-(*p*-tolylamino)- 87
  4,5,6-trichloro- 215, 229
  2-trichloromethyl- **79**
  2-trifluoromethyl- **72**, *72*
  5-trifluoromethyl- 87
  2-(*m*-trifluoromethylphenyl)amino- 87
  2-(1',2',2'-trimethylpropyl)- *21*
  2-trimethylsilyl- 77
  2-vinyl- 82
  2-vinyl-1-oxide 30, *31*

Benzimidazolin-2-one 20, 23, 25, 27, 28, 32, 78, 80, **81**, 83, 222, 229, 231
  1-acetyl-3-methyl- 34
  1-alkyl- 22
  1-aryl- 55
  5-chloro- 81
  6-chloro-1-phenyl- 55
  1,3-dihydroxy- 188
  1-methoxy- 55
  5-methoxy- 81
  5-methyl- 81
  5-nitro- 28

Benzimidazolin-2-thione 25, 26–28, 81, 86, 245
  5-chloro-1-trifluoroacetyl- 82
  5-fluoro-1-trifluoroacetyl- **81**
  5-methoxy-1-trifluoroacetyl- 82
  5-methyl-1-trifluoroacetyl- 82

3-benzoylaminobutanone 9
α-benzoylaniline methyl ester **105**
4-(*N*-benzylamino)isoxazole **175**
N-benzyl-*N*-(4-isoxazolyl)formamide **175**
betmip 147, 148
Bredereck, formamide synthesis 129, 151, 157, 167
3-bromo-4-ethoxy-3-buten-2-one 7, **141**
carbodiimides 86, 147
N-chloroamidines 138
3-chloro-4,4-dimethoxy-2-butanone 7, 140, **142**
N-(1-cyanoalkyl)alkylidene *N*-oxides 15
N-cyanomethyl-*o*-nitroaniline **32**
diaminomaleonitrile (DAMN) 11, 12–14, 48, 66, 113, 236, 238
diaminomaleonitrile, monoacyl 6
diaminomaleonitrile monoamide 14
N-(2,2-dimethoxyethyl)dichloroacetamidine **44**
N-(2,2-dimethoxyethyl)trichloroacetamidine **240**
dimethyl *N*-aryldithiocarbonimidates 84
N,*N*-dimethyl-(1-butyl-3,3-bis-[dimethylamino]-2-aza-3-propenyliden)ammonium perchlorate **60**
enamines 138
enaminones 6, 45, 101, 172
epoxides, ring opening **171**

INDEX 253

N-ethoxycarbonyl-*N*-alkyl-*o*-nitroanilines 22
N-ethoxycarbonyl-*N*-alkyl-*o*-
   phenylenediamines 22
N-ethoxycarbonylmethyl-*N*'-cyano-*N*-
   phenylformamidine 59
ethyl *N*-cyanomethylimidate 11
(Z)-*N*-(2-amino-1,2-dicyanovinyl)formimidate
   49
ethyl 2, 4 -dinitrophenylacetimidate 78
Gattermann reaction 234

General reactions
   α-aminoketone synthesis 104
   aromatization of 5, 5 -disubstituted
      2-imidazolines 139
   cyclization of enaminones 45
   deprotonation of *NH*-imidazoles 196
   devinylation 209
   hydrolysis of α-*C*-acylamino esters 106
   hydrolysis of 2-carbethoxy groups 32
   hydrolysis of 2-cyano groups 32
   isoxazole-imidazole interconversion 174
   α-ketoaldehydes 152
   oxidative cyclization of
      o-acylaminoanilines 36
   photolysis of 1-vinyltetrazole 171
   ring opening of epoxides 171

Gold's salt 133
guanidine 45, 47, 54, 55, 59, 110, 144, 146,
   147, 235
guanine 49
L-homohistidine 137
hydantoin 53
imidates 41, 66, 78, 95, 113, 236

Imidazole 194
   1-acetamido- 147
   2-acetamido- 145, **146**
   2-acetamido-4-*t*-butyl- 146
   2-acetamido-4,5-dimethyl- 146
   2-acetamido-4,5-diphenyl- 146
   2-acetamido-4-ethyl- 146
   2-acetamido-4-methyl- 146
   2-acetamido-5-methyl-4-phenyl- 146
   2-acetamido-4-phenyl- 146
   4-acetamido- 45
   4-acetamido-1-acetyl-2-phenyl- *47*
   4-acetamido-5-methyl-2-phenyl- *47*
   4-acetamido-2-phenyl- *47*

   1-acetyl- 205, 244
   2-acetyl- 244
   2-acetyl-4-methyl- 151
   2-acetyl-4-tetrahydroxybutyl- 152
   4-acetyl- 7, 116, *173*, **176**, 244
   4-acetyl-2-*t*-butyl- *7, 173*
   4-acetyl-2-(*p*-chlorophenyl)-5-methyl-
      116
   4-acetyl-2-dimethylaminomethyl- *7, 173*
   4-acetyl-2-ethyl- *7, 173*
   4-acetyl-2-hexyl- 142
   4-acetyl-2-hydroxymethyl- 141
   4-acetyl-2-isopropyl- *173*
   4-acetyl-2-methoxymethyl- *7, 173*
   4-acetyl-2-methyl- 7, 140, **141**, 142, *173*,
      177
   4-acetyl-5-methyl-2-(*p*-nitrophenyl)- 116
   4-acetyl-5-methyl-2-phenyl- 116
   4-acetyl-5-methyl-2-(*p*-tolyl)- 116
   4-acetyl-5-methyl-2-vinyl- **116**
   4-acetyl-2-phenyl- 7, **142**, 171, *173*
   4-acetyl-2-(4'-pyridyl)- 141
   4-acetyl-2-substituted 7
   4-acetyl-2-trifluoromethyl- *7, 173*
   5-acetyl-1-benzyl-2-methyl- *173*
   5-acetyl-1-benzyl-2-methyl-4-carbaldehyde
      173
   5-acetyl-1,2-dimethyl- *173*, **177**
   5-acetyl-1,2-diphenyl-4-methylthio- 101
   1-(*o*-acetylphenyl)- 204
   1-(*m*-acetylphenyl)- 204
   1-(*p*-acetylphenyl)- 204
   1-acyl- 6, 205
   1-acyl-4-substituted 205, 206
   2-acyl- 14, 156, 215
   2-acyl-1-methyl-4-carboxylates 14
   4-acyl 6, 7, 46, 47, 65, 66, 100, **116**,
      141, 167, 171-174, 220, 244
   4-acyl-2-acetamido- 45
   4-acyl-2-acylamino- 244
   4-acyl-5-alkyl(aryl)-2-amido- 45
   4-acyl-5-amino- 170, 235
   4-acyl-2-aryl- 46
   4-acyl-5-substituted 6
   4-acyl-1,2,5-trisubstituted 6
   5-acyl- 6, 7, 58, 60, 101, 174, 244
   5-acyl-4-alkylthio-1-aryl- 246
   5-acyl-4-amino- 58
   5-acyl-1-aryl- 244
   5-acyl-4-thio- **101**

Imidazole (*continued*)
  5-acyl-1,2,4-trisubstituted 6
  1-acylamino-2-amino-4-aryl- 180
  4-acylamino- 69
  1-alkoxy 53
  2-alkoxy- 11, 65, 181
  5-alkoxy-4-nitro- 223
  1-alkoxyalkyl- 200
  1-alkyl- 11, 195
  1-alkyl-2-amino- 234
  1-alkyl-5-amino- 11
  1-alkyl-4-aryl- **126**
  1-alkyl-5-chloro- 5
  1-alkyl-2,4-diamino- 61
  1-alkyl-4,5-diaryl- 148
  1-alkyl-4,5-dicarboxlates 238
  1-alkyl-4,5-dicarboxylic acid 238
  1-alkyl-4,5-dicyano- 51, 66
  1-alkyl-4,5-dicyano-2-substituted 11
  1-alkyl(aryl)-2-methylthio- **113**
  1-alkyl-4-nitro- 234
  1-alkyl-5-nitro- 234
  2-alkyl- 44, 63, 65, 88, 157, 169
  2-alkyl-4,5-diaryl- *152*, 168
  2-alkyl-4,5-dicyano- 67
  2-alkyl-1-methyl-4-carboxylates 14
  4-alkyl- 14, **126**, *130*, 151, 157, 170, 178, 220
  4-alkyl-2-aryl- 170
  4-alkyl-2-carboxylates 111
  5-alkyl- 139
  5-alkyl-4-aryl- *158*
  4,4'-alkylene-bis- 137
  2-alkylthio- 246, 247
  2-alkylthio-5-amino- 247
  4-alkylthio- 15, 101, *124*, 220, 247
  1-allyl-4-amino-5-carbethoxy-2-methylthio- 58
  1-allyl-2-(*p*-chlorophenyl)-5-methyl-4-phenyl- *10*
  2-allyloxy-4,5-dicyano- *12*
  4-amidino- 45
  1-amido-4-aryl-2-benzimidoylamino- 46
  2-amido- 46
  1-amino- 9, **10**, 52, 97, 98, 178
  2-amino- 11, 45, *47,* 67, 68, **110**, 135, 136, 143–145, 214, 227, 234, 235
  2-amino-1-aryl- 45, 143
  2-amino-4-aryl-1-benzylideneamino- 137
  2-amino-1-benzyl- **223**
  2-amino-5-carbethoxy-1-methyl- **110**
  2-amino-5-cyano- 143
  2-amino-4,5-di(*p*-chlorophenyl)- 145
  2-amino-4,5-dicyano- *67,* 68
  2-amino-4,5-dimethyl- 145
  2-amino-1-dimethylaminoethyl- *47*
  2-amino-4,5-diphenyl- 145
  2-amino-4,5-disubstituted- **145**
  2-amino-5-ethyl-4-methyl- 145
  2-amino-4-methoxy- 71
  4-amino- 15, 43, 45, 53, *54,* 57, 58, 60, 160, 227, 234, 235
  4-amino-5-benzoyl-2-methylthio-1-phenyl- *58*
  4-amino-5-benzoyl-1-phenyl- *58*
  4-amino-1-benzyl-2-bromo- *58*
  4-amino-2-bromo- 53, 54
  4-amino-2-bromo-1-butyl- *54*
  4-amino-2-bromo-5-(*o*-chlorophenyl)-1-phenyl- *54*
  4-amino-2-bromo-5-(*m*-chlorophenyl)-1-phenyl- *54*
  4-amino-2-bromo-5-(*p*-chlorophenyl)-1-phenyl- *54*
  4-amino-2-bromo-1-cyanomethyl- *54*
  4-amino-2-bromo-5-(2',4'-dichlorophenyl)-1-phenyl- *54*
  4-amino-2-bromo-1,5-diphenyl- *54*
  4-amino-2-bromo-1-ethyl- *47*
  4-amino-2-bromo-1-methyl- *54*
  4-amino-2-bromo-1-phenyl- *54*
  4-amino-2-carbethoxy- 44, *47*
  4-amino-5-carbethoxy- 143
  4-amino-5-carbethoxy-1-(*p*-chlorophenyl)- *58*
  4-amino-5-carbethoxy-1-methyl-2-methylthio- *58*
  4-amino-5-carbethoxy-2-methylthio-1-phenyl- *58*
  4-amino-5-carbethoxy-1-phenyl- *58*
  4-amino-2-carboxylate 44, 235
  4-amino-5-cyano- 57–60, 236
  4-amino-5-cyano-1,2-dimethyl- *58*
  4-amino-5-cyano-2-methyl- *58*
  4-amino-1-substituted 69, 235
  5-amino- 11, 13, 15, *47*, 48, 49, *50,* 52, 66, 69, 95, 127, 160, 170
  5-amino-1-(*p*-aminophenyl)-4-cyano- *50*
  5-amino-1-(*m*-aminophenyl)-4-cyano- *50*
  5-amino-1-aryl-4-cyano- **51**, 113
  5-amino-2-bromo- 235
  5-amino-1-aryl-4-(cyanoformimidoyl)- **51**

INDEX

Imidazole (*continued*)
5-amino-1-benzyl-4-carbethoxy- **95**
5-amino-1-benzyl-4-cyano- 51
5-amino-1-(*p*-benzyloxyphenyl)-4-cyano- 50
5-amino-1-*t*-butyl-4-carbethoxy- 95
5-amino-2-carbethoxy-4-carboxamido- *47*
5-amino-4-carbethoxy- 127
5-amino-4-carbethoxy-1-(*p*-methoxybenzyl)- 95
5-amino-4-carbethoxy-2-phenyl- **59**
5-amino-4-carboxamido- 48
5-amino-4-carboxamido-1-ethylaminoethyl- 161
5-amino-4-carboxamido-1-hydroxy- 45
5-amino-4-carboxamido-2-hydroxy- *47*
5-amino-2-carboxylate 44
5-amino-4-carboxylate 127
5-amino-1-(*p*-chlorophenyl)-4-cyano- 113
5-amino-1-(*p*-chlorophenyl)-4-cyano- 50, 113
5-amino-1-(5'-chloro-2'-pyridyl)-4-cyano- 113
5-amino-4-cyano- 13, 15, 48, 49, 113, 144, 236
5-amino-4-cyano-1-(*p*-cyanophenyl)- 50
5-amino-4-cyano-1-(2'-hydroxy-3'-nonyl)- **161**
5-amino-4-(cyanoformimidoyl)- 49, **50**
5-amino-4-cyano-1-(2',4'-dichlorophenyl)- 113
5-amino-4-cyano-1-(2',4'-dimethoxyphenyl)- *50*
5-amino-4-cyano-1-(3',4'-dimethoxyphenyl)- *50*
5-amino-4-cyano-1-(2',4'-dimethylphenyl)- *50*
5-amino-4-cyano-1-ethylaminoethyl- 161
5-amino-4-cyano-1-(*p*-fluorophenyl)- *50*
5-amino-4-cyano-1-(*p*-methoxyphenyl)- *50*
5-amino-4-cyano-1-(*p*-nitrophenyl)- *50*
5-amino-4-cyano-1-phenyl- *50*
5-amino-4-cyano-1-(2'-pyridyl)- 113
5-amino-4-cyano-1-(3'-pyridyl)- 113
5-amino-4-cyano-1-(2'-pyrimidinyl)- 113
5-amino-4-cyano-1-(*p*-tolyl)- *50*
5-amino-2,4-dicarbethoxy- *47*
5-amino-1,4-disubstituted 95
5-amino nucleosides 95

5-amino-1-substituted *47*, 235
1-aroyl- 206
2-aroyl- 14, 117, 215, 245
4-aroyl- *173*
5-aroyl- 60, 101
1-aryl- 50, 66, 101, 113, **204**, 223
1-aryl-5-chloro- *4, 5*
1-aryl-2,5-diphenyl- 96
1-aryl-2-hydroxymethyl- 214
2-aryl- 43, 63, 65, 157, 168, 170
2-aryl-4-alkyl- 170
2-aryl-4-carboxylates 47
2-aryl-4,5-dicarboxylic acids 160
2-aryl-4,5-dichloro- 155
2-aryl-5-trifluoromethyl-4-phenyl- 153
4-aryl- 46, **126**, *130*, 131, 151, 156, *158*, 178
4-aryl-5-trifluoromethyl- *17*
5-aryl- 139
2-arylamino- 45
N-arylation 203, 204
2-aryloxymethyl- *47*
2-arylthio- 133, 246, 247
4-arylthio- **15**, *124*, 246, 247
2-azo- 214
4-azo- 214
2,2'-azo- 136
1-benzenesulfonyl- **208**
1-benzoyl- 206
1-benzoyl-4-phenyl- **206**
2-benzoyl-1-benzyl- 215
2-benzoyl-1-methyl- **215**, 245
2-benzoyl-1-phenyl- 215
2-benzoyl-4-phenyl- 117, 156
4-benzoyl-2-methyl-5-phenyl- *173*
5-benzoyl-4-benzylthio-1,2-diphenyl- 101
5-benzoyl-1,2-diphenyl-4-ethylthio- 101
5-benzoyl-1,2-diphenyl-4-methylthio- 101
5-benzoyl-2-methyl-4-methylthio-1-phenyl- 101
5-benzoyl-4-methylthio-2-phenyl-1-(*p*-tolyl)- 101
5-benzoyl-1-phenyl-4-phenylthio- 101
1-benzyl- 5, 95, 196, **197**, 223
1-benzyl-4-bromo- **230**
1-benzyl-4-carbaldehyde **176**, 243
1-benzyl-5-carbaldehyde *173*, **176**, **242**
1-benzyl-5-carbethoxy- 242
1-benzyl-4-carbomethoxy-5-phenyl- 100

Imidazole (*continued*)
1-benzyl-4-carbomethoxy-5-(3'-pentyl)- 100
1-benzyl-2-chloro- 228
1-benzyl-5-chloro- **5**
1-benzyl-5-chloro-2-phenyl- *4*
1-benzyl-4-(*p*-chlorophenyl)-5-phenyl- 149
1-benzyl-4,5-dicarbaldehyde *173*
1-benzyl-2,4-dimethyl-5-carbaldehyde *173*
1-benzyl-4,5-dimethyl-3-oxide **115**
1-benzyl-4,5-dimethyl-2-phenyl- *10*
1-benzyl-2,5-diphenyl- 139
1-benzyl-2-hydroxymethyl- **214**
1-benzyl-5-hydroxymethyl-4-methyl-3-oxide 115
1-benzyl-4-isopropyl-5-methyl-2-phenyl- *10*
1-benzyl-2-lithio- 228, 231
1-benzyl-4-methyl-5-carbaldehyde *173*
1-benzyl-5-methyl-3-oxide 115
1-benzyl-2-methylthio- 113
1-benzyl-2,4,5-tribromo- **196**
1-benzyl-2,4,5-trimethyl- *10*
1-benzyl-2,4,5-tris(methylthio)- 247
2-benzyl-4-hydroxymethyl- 157
2-benzyl-4,5-diphenyl- 168
4-benzyl- 220
4-benzyl-1-*t*-butyl- 112
4-benzyl-5-ethyl- 179
4-benzyl-5-methyl- 172
4-benzyl-1-phenethyl- *126*
4-benzyl-2-phenyl- 42
5-benzyl-4-butyl- **129**, *130*
5-benzyl-4-(*p*-tolyl)thio- *124*
4-benzylamido-5-methyl-2-(*p*-nitrophenyl)- 116
1-benzyloxymethyl- *126*
4-benzylthio-1-methyl-5-[(2-tetrahydropyranyl)oxy]methyl- **18**
4-bis-(*p*-chlorophenyl)hydroxymethyl-1-trityl- 220
2,4-bis-(dimethylamino)-1-methyl- **61**
2,4-bis-(trifluoromethyl)-5-methyl- **225**
2,5-bis-(*m*-bromophenyl)-4-methyl- 137
2,5-bis-(*m*-nitrophenyl)-4-methyl- 137
2,5-bis-(*p*-nitrophenyl)-4-methyl- 137
2,5-bis-(trifluoromethyl)-4-phenyl- *152*
4,5-bis-(*p*-fluorophenyl)- **159**
4,5-bis-(hydroxymethyl)- 243
2-bromo- *54*, 216, 232
4-bromo- **212**, 213, 216, 230
4-bromo-5-cyano- 209
4-bromo-1,5-dimethyl- **218**
4-bromo-2-ethyl- **213, 230**
4-bromo-5-hydroxymethyl- 214
4-bromo-1-methyl-5-carbaldehyde 218
1-(*p*-bromophenyl)-2,5-diphenyl- 139
1-(*p*-bromophenyl)-2-methylthio- 112
2-(*p*-bromophenyl)-4,5-bis-(*p*-methoxyphenyl)- *152*, **154**
2-(*o*-bromophenyl)-4,5-dicyano- *12*
2-(*m*-bromophenyl)-4,5-dicyano- *12*
2-(*p*-bromophenyl)-4,5-dicyano- *12*
2-(*p*-bromophenyl)-1,5-diphenyl- 139
2-(*p*-bromophenyl)-4,5-diphenyl- *152*
2-(*p*-bromophenyl)-4-(2'-thienyl)- 97
4-butanoyl-2-phenyl- 171
1-butyl- 154
1-butyl-5-chloro-2-propyl- *4*
1-butyl-2,5-diisopropyl-4-thiol *19*
1-butyl-5-isopropyl-2-phenyl-4-thiol *19*
2-butyl-5-chloro-1-pentyl- *4*
4-butyl- 129, *130*, 131, 132
4-butyl-1,5-dipropyl- 131
4-butyl-1-ethyl- *126*
4-butyl-2-methyl- 157
4-butyl-5-methyl- *130*
4-butyl-5-phenyl-1-propyl- 132
4-butyl-5-propyl- 172
1-*t*-butyl- 17, *126*, 154
1-*t*-butyl-4-ethyl- 112
1-*t*-butyl-4-isopropyl- 112
1-*t*-butyl-4-isopropyl-5-trifluoromethyl- **17**
1-*t*-butyl-4-methyl- 126
1-*t*-butyl-5-methyl- **120**
1-*t*-butyl-5-methyl-4-phenyl- **121**
1-*t*-butyl-5-phenyl- 99
1-*t*-butyl-4-(*p*-tolyl)- 112
2-*t*-butyl-4,5-diphenyl- 168
4-*t*-butyl- 157
4-*t*-butyl-1-isopropyl- 112
4-*t*-butyl-1-methyl- 112
4-*t*-butyl-2-methyl- 157
5-*t*-butyl-1-methoxyethyl-2-thiol **108**
5-*t*-butyl-1-methyl- 99
5-*t*-butyl-4-(*p*-tolyl)thio- *124*
2-*t*-butylamino-4,5-dicyano- **67**, *67*
2-carbaldehyde 9, 43, **44**, *47*, **243**

INDEX

Imidazole (*continued*)
4-carbaldehyde   114, 128, *173*, **175**, 176, 217, 220, 243
5-carbaldehyde   *173*, 176, 217, 218, 243
4-carbalkoxy-   171, 241
4-carbalkoxy-5-aryl-2-arylthio-   133
5-carbalkoxy-   108
4-carbamoyl-   *47*, 181
4-carbinols   220
2-carbethoxy-   14, 43, *47*, 115, **240**, 243
2-carbethoxy-4-carbomethoxy-   **144**
2-carbethoxy-1-methyl-   215
2-carbethoxy-4-methyl-   115
4-carbethoxy-   14, 59, 95, 112, 116, 127, 133, 143, 144, *152*, 153, 160, 220
4-carbethoxy-2,5-diphenyl-   *152*
4-carbethoxy-2,5-dipropyl-   *152*
4-carbethoxy-5-methyl-2-phenyl-   116
4-carbethoxy-1-phenyl-5-phenylamino-   **127**
4-carbethoxy-2-phenyl-   *42*, *47*
4-carbethoxy-2-phenyl-5-propyl-   *152*
4-carbethoxy-5-phenyl-   143, *152*
4-carbethoxy-5-phenyl-2-propyl-   *152*
5-carbethoxy-   *58*, 110, 143
5-carbethoxy-2,4-difluoro-   **223**
5-carbethoxy-2,4-diphenyl-   *152*, **155**
4-carbethoxyethyl-2-phenyl-   *42*
4-carbethoxymethyl-2-phenyl-   *42*
4-carbinols   217, 220
2-carbomethoxy-   98, 240
4-carbomethoxy-   *10*, 95, 98, **109**, 116, 132, 144, 171, 237
4-carbomethoxy-2-(*p*-chlorophenyl)-   *47*
4-carbomethoxy-5-diethoxymethyl-   **132**
4-carbomethoxy-1,5-diphenyl-   100
4-carbomethoxy-5-hydroxy-   95
4-carbomethoxy-1-methyl-   **65, 169**
4-carbomethoxy-1-methyl-5-phenyl-   100
4-carbomethoxy-2-methyl-   171
4-carbomethoxy-5-methyl-2-phenyl-   116
4-carbomethoxy-5-methyl-2-phenyl-1-phthalimido-   *10*
4-carbomethoxy-1-phenethyl-5-phenyl-   **100**
4-carbomethoxy-2-phenyl-   *47*, 171
4-carbomethoxy-5-phenyl-1-(3'-picolyl)-   100
5-carbomethoxy-1-methyl-   **242**
5-carbomethoxy-4-phenyl-2-thiol   **106**
1,1'-carbonyldi-   80

2-carboxamido-   46, 114, 167, 168
2-carboxamido-5-ethyl-4-methylphenylamino-   167
2-carboxamido-5-methyl-4-methylphenylamino-   167
2-carboxamido-4-methylphenylamino-5-phenyl-   167
4-carboxamido-   14, *47*, 48, 116, 160, 181
5-carboxamido-4-cyano-   **68**
2-carboxylates   111, 115, 240
4-carboxylates   100, 108, 109, 116, 241
5-carboxylates   108, 241
2-carboxylic acid   44, *47*, 217, 237, 240
4-carboxylic acid   *47*, 64, 237, 239
5-carboxylic acid   237, 239
(R)-1α-carboxy-γ-methylthiopropyl-4,5-dimethyl-3-oxide   **160**
2-chloro-   216, 229, 232
4-chloro-   183, **213**, 220, **229**
4-chloro-1-cyano-   183
4-chloro-1-cyano-2,5-dibutyl-   183
4-chloro-1-cyano-2,5-diethyl-   183
4-chloro-1-cyano-2,5-diisobutyl-   183
4-chloro-1-cyano-2,5-dimethyl-   183
4-chloro-1-cyano-2,5-diphenyl-   183
4-chloro-1-cyano-2,5-dipropyl-   183
4-chloro-5-iodo-   **212**
5-chloro-   3, 4, 5, 229
5-chloro-1-benzyl-   *4*
5-chloro-1-butyl-   *4*
5-chloro-2-(*o*-chlorophenyl)-1-methyl-   *4*
5-chloro-2-(*p*-chlorophenyl)-1-methyl-   *4*
5-chloro-1,2-disubstituted   3, 4
5-chloro-1,2-diphenyl-   5
5-chloro-1-ethyl-   *4*
5-chloro-1-ethyl-2-methyl-   *4*
5-chloro-1-ethyl-2-phenyl-   *5*
5-chloro-2-ethyl-1-propyl-   *4*
5-chloro-1-isobutyl-   *4*
5-chloro-1-isobutyl-2-isopropyl-   *4*
5-chloro-1-isopropyl-   *4*
5-chloro-1-methoxyethyl-2-phenyl-   *4*
5-chloro-1-methyl-   *4*
5-chloro-1-methyl-2-phenyl-   **4**, *4*
5-chloro-1-methyl-2-(3'-pyridyl)-   *4*
5-chloro-4-nitro-   229
5-chloro-1-phenyl-   *4*
5-(*p*-chlorobenzoyl)-1,2-diphenyl-4-methylthio-   101
5-(*p*-chlorobenzoyl)-2-methyl-4-methylthio-1-phenyl-   101

Imidazole (*continued*)
5-(*p*-chlorobenzoyl)-4-methylthio-1-phenyl- 101
1-(2-chloroethyl)-4-isopropyl-5-methyl-2-phenyl- *10*
1-(*o*-chlorophenyl)-2-methylthio- 113
1-(*p*-chlorophenyl)-2-methylthio- 113
2-(*o*-chlorophenyl)-4,5-dicyano- *12*
2-(*m*-chlorophenyl)-4,5-dicyano- *12*
2-(*p*-chlorophenyl)-4,5-dicyano- *12*
2-(*o*-chlorophenyl)-1-(2',6'-dimethylphenyl)-5-fluoro-4-trifluoromethyl- *47*
2-(*p*-chlorophenyl)-4,5-diphenyl- *152*
2-(*p*-chlorophenyl)-5-phenyl-4-(2'-thienyl)- 97
4-(*p*-chlorophenyl)- *158*
5-(*p*-chlorophenyl)-2-(2'-furyl)-4-phenyl- *152*
5-(*p*-chlorophenyl)-4-phenyl-2-(2'-thienyl)- *152*
2-cinnamoyl-1-methyl- 245
1-cyano- 181, 182
1-cyano-2,5-dibutyl- 183
1-cyano-2,5-di-*s*-butyl- 183
1-cyano-2,5-diethyl- 183
1-cyano-2,5-diisopropyl- 183
1-cyano-2,4-dimethoxy- 181
1-cyano-2,5-dimethyl- 183
1-cyano-2,4-diphenyl- 183
1-cyano-2,5-diphenyl- 183
1-cyano-4,5-diphenyl- 183
1-cyano-2,5-dipropyl- 183
1-cyano-2-methyl-4-phenyl- 183
1-cyano-2-methyl-5-phenyl- 183
1-cyano-4-methyl-2-methylthio- 181
1-cyano-4-methyl-2-phenyl- 183
1-cyano-5-methyl-2-phenyl- 183
1-cyano-5-methyl-4-phenyl- 183
2-cyano- 114, 181, 182
2-cyano-1-hydroxy- 182
2-cyano-4-phenyl- 182
4-cyano- 13-15, 48-52, 68, 113, 160, 168, 181, 209
4-cyano-5-amino- 13, 235
4-cyano-5-carboxamide 14, 68
4-cyano-5-hydroxy- 143
4-cyano-1-(2'-hydroxy-3'-nonyl)- **161**
4-cyano-5-methoxy-4-tetrahydroxybutyl- 12
4-cyano-1-methyl- 168
4-cyano-2-phenyl- 181
5-cyano- 58, 60, 143
5-cyano-1,2-diaryl-4-hydroxy- 143
5-cyano-1-methyl- 60, 168
4-cyanoformimidoyl- 50, 51
1-(*o*-cyanophenyl)- 204
1-(*m*-cyanophenyl)- 204
1-(*p*-cyanophenyl)- 204
1-cycloheptyl-4,5-diphenyl- 148
1-cyclohexyl- 154
1-cyclohexyl-5-(*p*-nitrophenyl)-4-tosyl- 122
1-cyclohexyl-5-phenyl- **99**
1-cyclohexyl-4-tosyl-5-thiol 122
2-cyclohexyl-4,5-dicyano- *12*
2-cyclohexyl-4,5-diphenyl- 156
2-cyclohexyl-4-phenyl- 156
5-cyclohexyl-2,4-diphenyl- 156
deprotonation **196**
1,4-diacetyl-2-methyl- **177**
2,4-diacyl- 143, 144
1,2-dialkyl-4,5-dicyano- 68
1,4-dialkyl- **126**, 147
1,5-dialkyl- 197
2,4-dialkyl- 15, 170, 178
2,4(5)-dialkyl-5(4)-arylthio- **15**
2,5-dialkyl-1-cyano- **182**
2,5-dialkyl-4-chloro-1-cyano- **183**
4,5-dialkyl- 135, 151, 157, *158*
2-dialkylamino- 11
1-dialkylaminomethyl- 199
1,2-diamino-4-aryl- 137
1,5-diamino- 52, 94
1,5-diamino-4-cyano- **52**
1,5-diamino-4-cyanoformimidoyl- 52
2,5-diamino- 147, 148
1,2-diaryl- 143
1,5-diaryl- 125
2,4-diaryl- *42*, 170
2,5-diaryl-4-methyl- 137
4,5-diaryl- 130, 148, 151, *152*, 157, 158
4,5-diaryl-1-methyl- **130**
2-diazonium fluoroborate 222
4-diazo- 223
1,4-dibenzyl-5-methyl-2-phenyl- *10*
4,5-dibromo-1-methyl-2-carboxylic acid **217**
4,5-dibutyl- *158*
1,4-di-*t*-butyl- 112
4,5-di-*t*-butyl- 153
4,5-dicarbaldehyde 173, 217

INDEX

Imidazole (*continued*)
  2,4-dicarbalkoxy- 144
  2,4-dicarbethoxy- 14, *47*
  4,5-dicarbethoxy- 243
  2,4-dicarbomethoxy- 171
  2,4-dicarbomethoxy-1-methyl- 144
  4,5-dicarbomethoxy-2-phenyl- 171
  4,5-dicarboxamide 68
  4,5-dicarboxylates 160
  4,5-dicarboxylic acid 160, **172**, 237, 238
  4,5-dichloro- 66, 155, 212, 229
  4,5-dichloro-2-substituted 66
  2-dichloromethyl- 44, 243
  2-(2',4'-dichlorophenyl)-4,5-diphenyl- *153*
  1,5-di(*p*-chlorophenyl)- 121
  1,5-di(*p*-chlorophenyl)-2-methylthio- 125
  4,5-di(*p*-chlorophenyl)-1-methyl- 131
  4,5-dicyano- 11, **12**, *12*, 48, 49, 66–68, 238
  4,5-dicyano-2-*t*-butyl- *12*
  4,5-dicyano-1,2-dimethyl- 52
  4,5-dicyano-2-dimethylamino- *12*
  4,5-dicyano-2-ethoxy- *12*
  4,5-dicyano-1-ethyl- 51, 52
  4,5-dicyano-2-ethyl- *12*
  4,5-dicyano-2-(*o*-fluorophenyl)- *12*
  4,5-dicyano-2-(*m*-fluorophenyl)- *12*
  4,5-dicyano-2-methoxy- *12*
  4,5-dicyano-2-(*p*-methoxyphenyl)- *12*
  4,5-dicyano-1-methyl- **51**
  4,5-dicyano-2-methyl- *12, 67*
  4,5-dicyano-2-methylsulfonamido- 67
  4,5-dicyano-2-phenyl- **12**, *12*
  4,5-dicyano-2-phenylsulfonamido- 67
  4,5-dicyano-2-propyl- *12*
  4,5-dicyano-2-sulfonylamino- **67**
  4,5-dicyano-2-trifluoromethyl- *12*
  1-diethoxymethyl- **206**
  5-diethoxymethyl-4-methoxycarbonyl- 132
  2,5-diethyl-4-phenyl- 179
  4,5-diethyl- *158*
  2,4-difluoro- 223
  2,5-di(*p*-fluorophenyl)-4-phenyl- 162
  4,5-di(*p*-fluorophenyl)- *158*
  2,4-di(2'-furyl)- 156
  4,5-di(2'-furyl)-2-phenyl- 156
  diimidazo[3,4-a;3',4'-d]piperazin-2,4-dione 237
  4,5-diiodo- 231, **232**
  4,5-diisobutyl- *158*
  1,4-diisopropyl- 112
  1,5-diisopropyl- 121
  4,5-diisopropyl- *158*
  2-(2',6'-dimethoxy)benzoyl- 111
  2,4-di(*p*-methoxyphenyl)- 156
  2,5-di(*p*-methoxyphenyl)-4-(*p*-fluorophenyl)- 162
  1,2-dimethyl- 168, 177, 216, 231
  1,2-dimethyl-4,5-diaryl- 128
  1,2-dimethyl-2-iodo- 232
  1,2-dimethyl-4-(*p*-nitrophenyl)-5-phenyl- 128
  1,4-dimethyl- 168, 197
  1,4-dimethyl-2,5-diaryl- 128
  1,4-dimethyl-2,5-diphenyl- 128
  1,4-dimethyl-5-hydroxymethyl- 214
  1,5-dimethyl- 168, 197, 218
  1,5-dimethyl-2,4-diaryl- 128
  1,5-dimethyl-2,4-diphenyl- 9, 128
  1,5-dimethyl-2-hydroxymethyl- 214
  1,5-dimethyl-4-(*p*-methoxyphenyl)-2-phenyl- 128
  1,5-dimethyl-4-methylcarbamoyl-2-phenyl- 181
  1,5-dimethyl-4-(*p*-nitrophenyl)-2-phenyl- 128
  1,5-dimethyl-2-phenyl-4-phenylcarbamoyl- 181
  2,4-dimethyl- 151, 157, 169, 172
  2,4-dimethyl-5-carbaldehyde *173*
  2,5-dimethyl- 114
  2,5-dimethyl-1-dimethylamino- *10*
  2,5-dimethyl-1-(*N,N*-dimethyl)sulfonamido- **218**
  2,5-dimethyl-1-isopropyl- **114**
  2,5-dimethyl-1-phenyl-4-phenylcarbamoyl- 181
  4-(2',2'-dimethyl)imidazolecarbohydrazide **238**
  2,5-dimethyl-4-methylcarbamoyl-1-phenyl- **182**
  4,5-dimethyl- 115, *158*, 159, 160, 172
  4,5-dimethyl-1-dimethylamino-2-phenyl- *10*
  4,5-dimethyl-2-ethyl-1-hydroxy- 3-oxide 115
  1-[(dimethylamino)methyl]- **200**
  1-dimethylamino-2,4,5-trimethyl- *10*
  2-dimethylamino-4,5-di-(*p*-chlorophenyl)- 145

Imidazole (*continued*)
  2-dimethylamino-4,5-di-
    (*p*-methoxyphenyl)- 145
  2-dimethylamino-4,5-diphenyl- 145
  1-(*p*-dimethylaminophenyl)-4,5-diphenyl-
    149
  2-(*p*-dimethylaminophenyl)-4-phenyl-5-
    (2'-thienyl)- *152*
  1-(2',6'-dimethylphenyl)-5-fluoro-2-
    phenyl-4-trifluoromethyl- *47*
  1-(*N*,*N*-dimethylsulfamoyl)- **200**, 207
  1-(*N*,*N*-dimethyl)sulfonamido- **208**, 218,
    219, 229
  4,5-di(2'-naphthyl)-1-methyl- 131
  1,4-dinitro- **209**, 223
  1,4-dinitro-2-methyl- 223
  2,4-dinitro- 234
  2,4-dinitro-5-iodo- 234
  1,5-di(*p*-nitrophenyl)- 121
  1,2-diphenyl- 18, 19, 140
  1,2-diphenyl-5-ethyl- 140
  1,2-diphenyl-5-(*p*-methoxybenzoyl)-4-
    methylthio- 101
  1,2-diphenyl-5-methyl- 140
  1,2-diphenyl-4-methylthio-5-(*p*-toluoyl)-
    101
  1,5-diphenyl- 122
  1,5-diphenyl-2-methyl- 139
  1,5-diphenyl-4-tosyl- **122**
  2,4-diphenyl- *42*, 155, **169**, 170, 172
  2,4-diphenyl-1,5-dipropyl- 132
  2,4-dipheny-1-(*p*-fluorophenyl)-5-methyl-
    139
  2,4-diphenyl-5-(*p*-methoxystyryl)-1-
    methyl- 128
  2,4-diphenyl-5-methyl-1-(*p*-tolyl)- 139
  2,4-diphenyl-5-(1'-naphthyl)- *152*
  2,4-diphenyl-5-propyl- 131
  2,4-diphenyl-5-(3'-pyridyl)- 131
  2,4-diphenyl-5-(2'-thienyl)- *152*
  2,4-diphenyl-5-(*m*-tolyl)- 131
  2,4-diphenyl-5-(*p*-tolyl)- 131
  2,4-diphenyl-5-trifluoromethyl- *152*
  2,5-diphenyl- 128, 137
  2,5-diphenyl-1-(*p*-fluorophenyl)- 139
  2,5-diphenyl-4-(*p*-
    fluorophenyl)- 162
  2,5-diphenyl-1(*p*-methoxyphcnyl)- 139
  2,5-diphenyl-4-(*p*-methoxyphenyl)-1-
    methyl- 128
  2,5-diphenyl-1-methyl-4-(*p*-nitrophenyl)-
    128
  2,5-diphenyl-1-(*p*-nitrophenyl)- 139
  4,5-diphenyl- 54, 148, *152, 158*
  4,5-diphenyl-2-(2'-furyl)- *152*, 156
  4,5-diphenyl-2-(*p*-methoxyphenyl)- *152*
  4,5-diphenyl-1-methyl- 121, 128, **130**
  4,5-diphenyl-2-methyl- *152*, 168
  4,5-diphenyl-1-(*p*-nitrophenyl)- 121
  4,5-diphenyl-1-propyl- 131
  4,5-diphenyl-2-(2'-thienyl)- *152*
  4,5-diphenyl-2-trifluoromethyl- *152*
  4,5-diphenyl-2-vinyl- 168
  4-diphenylmethoxy- 220
  1,5-dipropyl-4-phenyl- 131
  4,5-dipropyl- 135, *158*
  4,5-dipropyl-2-methyl- 179
  1,2-disubstituted 45, 112, 139
  1,4-disubstituted 64, 69, **112, 120**, 125,
    126, 157, 160, 198
  1,5-disubstituted 64, 98, 99, 108, 119,
    120, *173*, 195, 198, 206
  2,4-disubstituted *7*, **41, 42**, *42*, 46, 68,
    71, 112, 115, 141, 156, 169, 181,
    183
  2,5-disubstituted 41
  4,5-disubstituted 100, 120, 124,
    127–129, *130*, 132, 135, 145, 157,
    158, 237
  2,5-dithiol 179
  4,5-di(*p*-tolyl)-1-methyl- 131
  2,4-di(trifluoromethyl)- *152*
  2,5-di(*p*-trifluoromethylphenyl)-4-
    (*p*-fluorophenyl)- 162
  2-ethoxy-1-hydroxy- 70
  2-ethoxy-1-hydroxy-4-methyl-5-phenyl-
    70
  2-ethoxy-1-hydroxy-5-methyl-4-phenyl-
    70
  4-ethoxy- 5
  4-ethoxy-2-phenyl- **5**
  1-(1-ethoxyethyl)-2-methyl- 200
  1-(1-ethoxyethyl)- 200
  1-(1-ethoxyethyl)- 2-phenyl- 200
  1-ethoxymethyl-2-phenyl- **200**
  1-ethyl- *126*
  1-ethyl-2-hydroxy-5-methyl-4-phenyl-
    110
  1-ethyl-5-methyl- 121
  1-ethyl-2-methylthio- 113
  1-ethyl-5-phenyl- **206**

INDEX

Imidazole (*continued*)
- 1-ethyl-4-tosyl-5-thiol 123
- 2-ethyl- 230
- 2-ethyl-1-hydroxy-3-oxide 155
- 2-ethyl-4-methyl-5-phenyl- 161
- 4-ethyl-1-isopropyl- 112
- 4-ethyl-1-methyl-2-phenyl- 9
- 4-ethyl-5-methyl- 172
- 4-ethyl-2-phenyl- *42*
- 4-ethyl-5-propyl- *158*
- 5-ethyl-1-(*p*-fluorophenyl)-2-phenyl- 139
- 5-ethyl-4-phenyl- *158*, 179
- 5-ethyl-4-(*p*-tolyl)thio- *124*
- 4-ethynyl-2-phenyl- 181
- 2-fluoro- 222, 227
- 2-fluoro-1-methyl- **228**
- 2-fluoro-1-trityl- 216
- 4-fluoro- 223, 227
- 5-fluoro- 47, 48, 227
- 5-fluoro-2-phenyl-4-trifluoromethyl-1-(2',4',6'-trimethylphenyl)- *47*
- 5-fluoro-4-trifluoromethyl- 47, 48
- 2-(*p*-fluorophenyl)-4-methyl- **112**
- 2-(*o*-fluorophenyl)-4-(2'-thienyl)- 97
- 4-(*p*-fluorophenyl)-5-phenyl-2-(4'-pyridyl)- **163**
- 2-formyl-4-methyl-1-phenyl-3-oxide 9
- 2-(2'-furyl)-5-(*p*-methoxyphenyl)-4-phenyl- *152*
- 2-(2'-furyl)-4-phenyl-2-(*p*-tolyl)- *152*
- 2-*H*- 15, 68
- 4-*H*- 17, 145
- 4-halogeno- 220
- 4-halogeno-1-methyl- 198
- 5-halogeno-1-methyl- 198
- 2-*n*-heptanoyl-1-methyl- **244**
- 2-hetaryl- *152*
- 4-hetaryl- 131, *152*
- 1-hydroxy- 53, 69, 70, 115, 116, 155, 159
- 1-hydroxy-3-oxide 115, 116, 155
- 1-hydroxy-2-phenyl-3-oxide 155
- 1-hydroxy-2,4,5-trimethyl-3-oxide 115, 155
- 2-hydroxy- (one) *67*, 109, 147
- 2-hydroxy-4,5-dicyano- 67
- 4-hydroxy- 143, 145
- 5-hydroxy- 65, 66, 95, 96
- 5-hydroxy-1-isopropyl-2-methyl-4-propyl- 114
- 4-(1'-hydroxybenzyl)-1-trityl- 220
- 4-(*p*-hydroxybenzyl)-2-phenyl- *42*
- 2-(1'-hydroxyethyl)- 43, 116
- 4-(1'-hydroxyethyl)-5-methyl- 159
- 4-(1'-hydroxyethyl)-1-trityl- 220
- 2-hydroxymethyl- 47, 214, 242
- 2-hydroxymethyl-1-methyl- 47
- 4-hydroxymethyl- 42, 135, **136**, 156, 157, 242
- 4-hydroxymethyl-5-methyl- *158*
- 4-hydroxymethyl-2-phenyl- *42*, 157
- 4-hydroxymethyl-2-(*p*-tolyl)- 157
- 5-hydroxymethyl- 214
- 4-(1'-hydroxy-2'-propenyl)-1-trityl- 220
- 4-(*p*-hydroxyphenyl)-2-phenyl- *42*
- 4-hydroxypropyl- 159
- 2-iodo- 206, 231, **232**
- 2-iodo-1,2-dimethyl- 231
- 4-iodo- 213, 231
- 4-iodo-1-triphenylmethyl- **200**
- 5-iodo- 213, 231
- 4-isobutyl- 157
- 4-isobutyl-2-methyl- 157
- 1-isopropyl- 154
- 1-isopropyl-2,5-dimethyl- **114**
- 1-isopropyl-5-methyl- 121
- 1-isopropyl-5-methyl-4-phenyl- 121
- 2-isopropyl-4-(2'-thienyl)- 99
- 4-isopropyl- 17
- 4-isopropyl-2-methyl- **156**
- 4-isopropyl-5-methyl-2-phenyl-1-phthalimido- *10*
- 5-isopropyl-1-methyl- 121
- 5-isopropyl-4-phenyl- *158*
- 5-isopropyl-4-(*p*-tolyl)thio- *124*
- 2-lithio- 216, 228, 231
- 2-lithio-1-methyl- 228
- 2-lithio-1-phenylsulfonyl- 231
- 2-lithio-1-trityl- 232
- 4-lithio- 216
- 5-lithio- 216
- 2-methoxy-5-methyl-4-phenylcarbamoyl- **182**
- 4-methoxy- 17, 71
- 4-methoxy-2-substituted **70**
- 5-methoxy-4-methyl- **17**
- 4-methoxycarbonyl- 65
- 4-methoxycarbonyl-5-(2'-furyl)- 133
- 4-methoxycarbonyl-5-phenyl- 133
- 4-methoxycarbonyl-5-(2'-pyridyl)- 133
- 4-methoxycarbonyl-5-(3'-pyridyl)- 133
- 4-methoxycarbonyl-5-(4'-pyridyl)- 133

Imidazole (*continued*)
  5-methoxymethyl-4-(*p*-tolyl)thio-   *124*
  1-(*p*-methoxyphenyl)-   *126*
  1-(*p*-methoxyphenyl)-4-methyl-   *126*
  1-(*p*-methoxyphenyl)-2-methylthio-   113
  2-(*p*-methoxyphenyl)-4-phenyl-5-(2'-thienyl)-   *152*
  4-(*p*-methoxyphenyl)-   *158*
  5-(*p*-methoxyphenyl)-4-phenyl-2-(2'-thienyl)-   *152*
  5-(*p*-methoxyphenyl)-4-(*p*-tolyl)thio-   *124*
  1-methyl-   4, 9, 18, 51, 61, 64, 130, 154, 168, **196**, 212, 217, 228
  1-methyl-5-*t*-butyl-   99
  1-methyl-4-carbaldehyde   217
  1-methyl-5-carbaldehyde   217
  1-methyl-4-carboxylic acid   **239**
  1-methyl-5-carboxylic acid   **239**
  1-methyl-4-(*p*-chlorophenyl)-5-trifluoromethyl-   *17*
  1-methyl-5-(*p*-chlorophenyl)-4-trifluoromethyl-   *17*
  1-methyl-4,5-diaryl-   128
  1-methyl-4,5-dicarbaldehyde   217
  1-methyl-4,5-dicarboxylic acid   **238**
  1-methyl-2,5-diphenyl-   68
  1-methyl-4,5-diphenyl-   16
  1-methyl-2,5-diphenyl-4-styryl-   **128**
  1-methyl-2-hydroxymethyl-   214
  1-methyl-5-methoxybenzyl-4-thiol   *19*
  1-methyl-4-(*p*-methoxyphenyl)-5-trifluoromethyl-   *17*
  1-methyl-5-(*p*-methoxyphenyl)-4-trifluoromethyl-   *17*
  1-methyl-2-methylthio-   113
  1-methyl-4-nitro-   198
  1-methyl-5-nitro-   198
  1-methyl-5-(*p*-nitrophenyl)-   99, 121
  1-methyl-4-(*p*-nitrophenyl)-5-trifluoromethyl-   *17*
  1-methyl-5-(*p*-nitrophenyl)-4-trifluoromethyl-   *17*
  1-methyl-2-pentyl-   43
  1-methyl-2-phenyl-   43
  1-methyl-5-phenyl-   99
  1-methyl-2-phenyl-4,5-tetramethylene-   **9**
  1-methyl-4-phenyl-5-trifluoromethyl-   *17*
  1-methyl-5-phenyl-   121
  1-methyl-5-phenyl-4-trifluoromethyl-   *17*
  1-methyl-4-thiol   *19*
  1-methyl-4-(*p*-tolyl)-5-trifluoromethyl-   *17*
  1-methyl-5-(*p*-tolyl)-   121
  1-methyl-5-(*p*-tolyl)-4-trifluoromethyl-   *17*
  1-methyl-2,4,5-triaryl-   128
  1-methyl-2,4,5-tribromo-   217
  1-methyl-2-trimethylstannyl-   **228**
  1-methyl-2,4,5-triphenyl-   9, 128
  1-methyl-4-tosyl-5-thiol   **123**
  2-methyl-   63, 65, 141, 156, 157, 159, **169**, 179
  2-methyl-d_3 **111**
  2-methyl-4-carbaldehyde   *173*
  2-methyl-4,5-dicarboxylic acid   *172*
  2-methyl-4,5-diphenyl-   168
  2-methyl-1-hydroxy-3-oxide   155
  2-methyl-4-methylthio-1-phenyl-5-(*p*-toluoyl)-   101
  2-methyl-4-nitro-1-(*p*-tolyl)-   223
  2-methyl-1-phenyl-   179
  2-methyl-4-propyl-   157
  4-methyl-   17, 112, *130*, 151, 156, 168, 170, 172, 220, 224,
  4-methyl-2-carbaldehyde   *173*
  4-methyl-2,5-diphenyl-   **137**
  4-methyl-5-hydroxymethyl-   214
  4-methyl-1-(*p*-nitrophenyl)-5-phenyl-   121
  4-methyl-2-pentyl-   *42*
  4-methyl-1-phenyl-   *126*
  4-methyl-2-phenyl-   *42*, 170, 172
  4-methyl-5-propyl-   172
  4-methyl-2-trifluoromethyl-   **224**
  5-methyl-   110, 120, 121
  5-methyl-2-methylamino-4-phenyl-   **145**
  5-methyl-2-phenyl-4-carbaldehyde   *173*
  5-methyl-2-phenyl-1-(*p*-tolyl)-   140
  5-methyl-2-phenyl-   *158*, 158
  5-methyl-4-phenyl-   *158*, 158
  5-methyl-4-(*p*-tolyl)thio-   *124*
  5-methyl-4-trifluoromethyl-   **224**
  5-methyl-1,2,4-triphenyl-   139
  2-methylamino-   145
  methylation ratios   *199*
  4-methylcarbamoyl-1,2,5-trimethyl-   181
  5-methylsulfinyl-4-nitro-   **247**
  1-methylsulfonyl-   **208**, 209
  2-methylthio   58, 60, 98, 111–113
  2-methylthio-1-phenyl-   113
  2-methylthio-1-substituted   113, 246
  2-methylthio-1-(*o*-tolyl)-   113

INDEX

Imidazole (*continued*)
2-methylthio-1-(*p*-tolyl)- 113
4-methylthio- 220, 247
4-methylthio-1-phenyl-5-(*p*-toluoyl)- 101
5-methylthiomethyl-4-(*p*-tolyl)thio- *124*
1-nitro- 207, 209, 223
2-nitro- 214, **218**, 234
4-nitro- **214**, 234, 247
4-nitro-1-phenyl- **223**
4-nitro-(1-*o*-nitrobenzenesulfonyl)- 223
5-nitro- 234
1-(*o*-nitrobenzyl)-4-carbaldehyde 243
1-(*p*-nitrobenzyl)-4-carbaldehyde 243
1-(*o*-nitrobenzyl)-5-carbaldehyde 243
1-(*p*-nitrobenzyl)-5-carbaldehyde 243
1-(*o*-nitrophenyl)- 204
1-(*m*-nitrophenyl)- 204
1-(*p*-nitrophenyl)- 204
1-(*o*-nitrophenyl)-4-carbaldehyde 243
1-(*p*-nitrophenyl)-5-carbaldehyde 243
1-(*p*-nitrophenyl)-5-phenyl- 121
1-(*p*-nitrophenyl)-5-phenyl-4-tosyl- 122
2-(*m*-nitrophenyl)-4-phenyl-5-(2'-thienyl)- *152*
4-(*p*-nitrophenyl)- *158*
5-(*p*-nitrophenyl)-1-phenyl- **120**
5-(*p*-nitrophenyl)-1-phenyl-4-tosyl- 122
5-(*p*-nitrophenyl)-4-(*p*-tolyl)thio- *124*
N-oxides 9, 69, 115, 159, 160
3-oxide 115, 116, 155, 159, 160
2-pentyl- 43
2-(*p*-phenoxyphenyl)- 156
4-(*p*-phenoxyphenyl)- 156
1-phenyl- 8, 120, *126*, 182, **204**, 223
1-phenyl-5-(*p*-tolyl)- 121
1-phenyl-2,4,5-trimethyl- *10*
2-phenyl- 5, 9, *42*, 43, 59, 142, 156, 170, 172, 200
2-phenyl-1,4,5-trimethyl- *9*
4-phenyl- *130*, 145, *158*, **159**
4-phenyl-2-(2'-pyridyl)-5-trifluoromethyl- *152*
4-phenyl-2-(2'-thienyl)-5-(*p*-tolyl)- *152*
4-phenyl-5-(2'-thienyl)-2-(*m*-tolyl)- *152*
4-phenyl-5-(2'-thienyl)-2-(*p*-tolyl)- *152*
4-phenyl-2-(*p*-tolyl)-5-trifluoromethyl- *152*
4-phenyl-5-(*p*-tosyl)- 123
5-phenyl- 99, 100, 206
5-phenyl-4-(*p*-tolyl)thio- *124*

4-phenylamino- 46
5-phenylamino- 127
1-phenylsulfonyl- 209
2-phenylthio-1-trityl- 247
4-phenylthio- 15
4-polyhydroxyalkyl- 152, 156
1-n-propyl- **196**
1-propyl-2,4,5-triphenyl- 131
5-propyl-4-phenyl- 179
1-protected-4-carbinols **220**
1-(2'-pyridyl)- 204
1-(3'-pyridyl)- 204
1-(4'-pyridyl)- 204
1-(4'-pyridyl)-4-(*p*-tosyl)- **122**
5-(2'-pyridyl)-4-(*p*-tolyl)thio- *124*
Schiff base 69
1-SEM- **201**
4-styryl- 128
1-substituted 9, 43, 103, 154, 199
2-substituted 43, 63–65,159,169
4-substituted 99, 123, 126, 198, 206, **219**
sulfones 247
4-sulfonic acid 247
2-sulfonylamino- *67*, 67, 68
1,2,4,5-tetraphenyl- 139
1,2,4,5-tetrasubstituted 7, 8, *10*, 18, *19*, *47*, 48, 54, 58, 60, 66, 70, 98, 101, 104, 115, 127, 128, 131, 132, 139, 143, 160, *173*, 181, 217
4-(2'-thienyl)- 97
4-(2'-thienyl)-2-(*p*-tolyl)- 97
1,1'-thiocarbonyldi- 81, 106
2-thiol (thione) 106, *107*, 108, 245
4-thiol 18, *19*, 101, 115, 119, 124, 246
4-thio-1,2,5-trisubstituted 115
5-thiol 18, 120, 123, 180, 246
1-(*p*-toluoyl)- **205**
1-(*p*-tolyl)- 204
5-(*p*-tolyl)-4-(*p*-tolyl)thio- *124*
1-(*p*-tosyl)- 209
4-(*p*-tosyl)- 120–123, 247
4-(*p*-tosyl)-1-hetaryl- 122
4-(*p*-tosyl)-5-thiol 123, 180, 246
1-(*p*-tosylamino)-2,4,5-trimethyl- *10*
1,2,5-trialkyl- 114
2,4,5-trialkyl- 151, 178
1,2,5-triaryl- **138**
1,4,5-triaryl- 148
2,4,5-triaryl- 151, *152*, 154, **162**, 163, 179
2,4,5-tribromo- 196, 216, 217, 230

Imidazole (*continued*)
  2,4,5-tribromo-1-methyl- **212**
  2,4,5-tribromo-1-vinyl- **208**
  2,4,5-trichloro- 212
  2-trichloromethyl- 240
  2-trideuteriomethyl- *47*
  2-triethylsilyl- 219, 229
  2-trifluoromethyl- *7*, 152
  2-trifluoromethyl-4-methyl- **224**
  4-trifluoromethyl- 16, *17*, *47*, 48, 224
  4-trifluoromethyl-5-methyl- **224**
  5-trifluoromethyl- 16, 17
  2,4,5-triiodo- 214, **233**, 234
  2,4,5-tri-(*p*-methoxyphenyl)- 163
  1,2,5-trimethyl- 181
  1,4,5-trimethyl- 217
  2,4,5-trimethyl 115, 155
  1-trimethylsilyl- 205, 207, **208**
  2-trimethylstannyl- 228
  5-trimethylstannyl- 228
  2,4,5-trinitro- 214, 234
  1,2,5-triphenyl- 125, 138
  1,4,5-triphenyl- 179
  2,4,5-triphenyl- 131, *152*, 153, 162, 168, **179**
  1-triphenylmethyl- 200
  4-triphenylphosphoniomethyl- 140
  1,2,4-trisubstituted 14, 60, 98, 144, 148, 181
  1,2,5-trisubstituted 4, 5, 108, 114, 125, 138, 139, 148, *173*, 216
  1,2,5-trisubstituted-4-thiols 246
  1,4,5-trisubstituted 16, 52, 58, 66, 95, 98, 113, 119–121, 127, 130, 148, 160, *173*, 217
  2,4,5-trisubstituted 14, 15, 43, 46, *47*, 68, 69, 71, 96, 104, 114, 116, 131,133,140,145,153,158,162, 168, 170, 173, 178
  1-vinyl- 116, 183, 208, 209
  2-vinyl- 116, 168
  2-vinyl-4,5-diphenyl- 168

2-imidazoline 63, 64, 138
  2-alkyl- 168
  aromatization 27, **139**, 168
  4-carbomethoxy-1-methyl- **64**
  1-methyl-4-carboxylic acid **64**

3-imidazoline
  2-ethoxy-1-hydroxy- 3-oxide 70
  1-hydroxy- 3-oxide 69

4-imidazolin-2-one 7, 53, 66, 103, 104, 110, 135, 147
  1-alkoxy 53
  4-amino- 147
  1-carbamoyl- 147
  1-cyano- 147
  4,5-dicyano- 67, 68
  4,5-dimethyl- 53, 246
  1,4-dimethyl-5-phenyl- 110

2-imidazolin-4-one 68, 146, 172
  2-anilino-1-phenylimidazolin-4-one **146**
  1-aryl-2-arylamino- 146
  4,5-dialkyl- 172
  1,2-dimethyl-5-phenyl- 172
  1-phenyl-2-(*p*-tolyl)amino- 146

2-imidazolin-5-one **96**

3-imidazolin-2-one
  1-hydroxy-5-methyl-4-phenyl- 70

imidazolin-2-thione 103, 104, 106, *107*, 108, 147, 180
  1-acetylmethyl-5-methyl- *107*
  4-acyl- *107*
  5-acylamino- 180
  1-allyl-5-*t*-butyl- *107*
  1-amino- 180
  5-amino-1-methyl- 180
  1-benzyl-5-*t*-butyl- *107*
  1-benzyl-4-carbomethoxy- *107*
  4-benzyl- *107*
  5-benzyl-4-carbomethoxy- *107*
  5-*t*-butyl-1-methyl- *107*
  5-*t*-butyl-1-methylthioethyl- *107*
  4-carbomethoxy- *107*
  4-carbomethoxy-5-(*o*-chlorophenyl)- *107*
  4-carbomethoxy-5-(*p*-fluorophenyl)- *107*
  4-carbomethoxy-1-methyl- *107*
  4-carbomethoxy-5-phenyl- *107*
  4-carbomethoxy-4-(*p*-tolyl)- *107*
  5-carbomethoxy-1-phenyl- *107*
  4-(*o*-chlorophenyl)- *107*
  1-cyclohexyl-5-isopropyl- *107*
  4-cyclohexyl- *107*
  4,5-dicarbethoxy-1-butyl- *107*
  4,5-dicarbethoxy-1-ethyl- *107*
  4,5-dicarbethoxy-1-isopropyl- *107*
  4,5-dicarbethoxy-1-propyl- *107*

INDEX

1,4-dimethyl- *107*
1,5-dimethyl- *107*
4,5-dimethyl- 53
4-(*o*-fluorophenyl)- *107*
1-heptyl- *107*
4-isobutyl- *107*
4-isopropyl- *107*
5-isopropyl-1-methyl- *107*
4-(*p*-methoxyphenyl)- *107*
5-methyl-1-propyl- *107*
1-pentyl- *107*
4-phenethyl- *107*
4-propyl- *107*
4-(2'-pyridyl)- *107*
4-(3'-pyridyl)- *107*
4-(2'-thienyl)- *107*
4-(*p*-tolyl)- *107*
1-vinyl- 183

imidazolium salts 46, 65, 66, 137, 154, 178, 201, 206
4-amino- 46, 161
1-benzyl-3-methyl-iodide 202
3-benzyl-1,4,5-triphenyl-perchlorate **179**
1-butyl-3-isopropyl-iodide 202
1-butyl-3-methyl-iodide 202
1-butyl-3-propyl-iodide 202
2-carbaldehyde 66
4-carbethoxy- *107*
4-carbomethoxy- *107*
5-carbomethoxy- *107*
1,3-diethyl-iodide 202
1,3-dimethyl-iodide **202**
1,3-disubstituted-2-carbaldehyde 243
1-ethyl-3-methyl-bromide 202
1-ethyl-3-methyl-chloride 202
1-ethyl-3-methyl-iodide 202
1-ethyl-3-methyl perchlorate **202**
1-ethyl-3-methyl-tetraphenylborate **202**
1-methyl-3-phenyl-iodide 202
1-methyl-3-vinyl-iodide 202

imidazopyridines 5
imidazothiazine 59
1-isocyano-2-phenyl-1-(*p*-tosyl)ethene **99**
N-(4-isoxazolyl)formamide **175**
isoxazoles 6, 172
isoxazole-imidazole interconversion **176**
Maquenne synthesis 160
Marckwald synthesis 103, 235
methyl (*E*)- and (*Z*)-3-bromo-2-isocyanocinnamate **99**
N-methylbenzamide acetone **97**
N-methyl-*N*-[2-[[(methylamino)-phenylmethylene]amino]-1-cyclohexen-1-yl]benzamide **8**
N-(5-methyl-4-isoxazolyl)acetamide **174**
N-(5-methyl-4-isoxazolyl)acetamide **177**
o-nitroanilines, acylated 20
oxamides, symmetrical 3
oxamides, unsymmetrical 3
oxazoles, conversion into imidazoles 135, 156, 158, 167, 178
α-oximinoketones 9
N, α-phenylnitrone **130**
phase-transfer catalysis 197, 204
Phillips method 71

photolysis
  of 1-acylimidazoles 6
  of DAMN 14
  of pyrazoles 168
  of 1-vinyltetrazole 170, **171**
Schiff base 9, *10*, 14, 47, 89
thioureas 40, 41, 45, 53, 81, 134, 146
TOSMIC (toluene-*p*-sulfonylmethyl isocyanide) 98, 119–126, 246
α-tosylbenzyl isocyanide **121**
4-tosyloxazolines **125**
2-tributylstannyltetrazole **171**
Ullmann arylation 203, 204
ureas 40, 41, 53, 81, 134, 146, 172
1-vinyltetrazole **171**
Wallach synthesis 3, 5